Capacity and Transport in Contrast Composite Structures

Asymptotic Analysis and Applications

Capacity and Transport in Contrast Composite Structures

Asymptotic Analysis and Applications

A.A. Kolpakov

Novosibirsk State University, Novosibirsk, Russia
Université de Fribourg, Fribourg Pérolles, Switzerland

A.G. Kolpakov

Università degli Studi di Cassino, Cassino, Italy
*Siberian State University of Telecommunications
and Informatics, Novosibirsk, Russia*

CRC Press
Taylor & Francis Group
Boca Raton   London   New York

CRC Press is an imprint of the
Taylor & Francis Group, an **informa** business

CRC Press
Taylor & Francis Group
6000 Broken Sound Parkway NW, Suite 300
Boca Raton, FL 33487-2742

© 2010 by Taylor and Francis Group, LLC
CRC Press is an imprint of Taylor & Francis Group, an Informa business

No claim to original U.S. Government works *100597 5075*

Printed in the United States of America on acid-free paper
10 9 8 7 6 5 4 3 2 1

International Standard Book Number: 978-1-4398-0175-8 (Hardback)

Library of Congress Cataloging-in-Publication Data

Kolpakov, A. A.
    Capacity and transport in contrast composite structures : asymptotic analysis and applications / authors, A.A. Kolpakov, A.G. Kolpakov.
        p. cm.
        "A CRC title."
        Includes bibliographical references and index.
        ISBN 978-1-4398-0175-8 (hardcover : alk. paper)
        1. Composite construction--Mathematics. 2. Structural analysis (Engineering)--Mathematics. 3. Structural frames--Mathematical models. 4. Capacity theory (Mathematics) 5. Asymptotic expansions. I. Kolpakov, A. G. II. Title.

TA664.K65 2010
624.1'8--dc22
                                          2009037001

Visit the Taylor & Francis Web site at
http://www.taylorandfrancis.com

and the CRC Press Web site at
http://www.crcpress.com

CONTENTS

PREFACE

This book is devoted to the analysis of the capacity of systems of closely placed bodies and the transport properties of high-contrast composite structures. This title covers many similar problems well known in natural science, material science and engineering.

The term "transport problem" implies problems of thermoconductivity, diffusion, electrostatics and many other similar problems, which can be described with a scalar linear elliptic equation or a nonlinear equation of elliptic type. For a linear inhomogeneous medium, the transport problem consists of balance equation

$$\text{div}\mathbf{q} = f(\mathbf{x}),$$

constitutive equation

$$q_i = c_{ij}\frac{\partial \varphi}{\partial x_j},$$

which is often written in the form $q_i = -c_{ij}\dfrac{\partial \varphi}{\partial x_j}$, and boundary conditions.

Here φ is the potential, $\nabla\varphi = \left(\dfrac{\partial \varphi}{\partial x_1}, ..., \dfrac{\partial \varphi}{\partial x_n}\right)$ is the driving force, $\mathbf{q} = (q_1, ..., q_n)$ is the flux, c_{ij} is a tensor describing local (microscopic) transport property of the medium (tensor of dielectric constants, tensor of thermoconductivity constants, etc.), n is the dimension of the problem (in the book n takes values 2 or 3).

The equations above can be transformed into one elliptic equation

$$\frac{\partial}{\partial x_i}\left(c_{ij}\frac{\partial \varphi}{\partial x_j}\right) = f(\mathbf{x}),$$

which must be supplied with an appropriate boundary condition.

Table 1 lists several transport problems that are mathematically equivalent. Due to this equivalence we can treat these problems within a common theoretical framework.

In some cases, it is necessary to take into account the nonlinearity of local properties of component(s) of composite. In practice and in nature, we meet various types of nonlinearities. In thermoconductivity, usually, coefficients of thermoconductivity depend on the temperature: $c_{ij} = c_{ij}(\varphi)$ (φ means the temperature). In electrostatics, usually, dielectric constants depend on the electric field: $c_{ij} = c_{ij}(\nabla\varphi)$, ($\varphi$ means the potential of electric field).

Table 1. *List of Phenomena (* asymmetric deformation or torsion).*

Phenomenon	Potential	Driving force	Flux	Local tensor
Heat conduction	Temperature	Temperature gradient	Heat flux	Thermal conductivity
Electrical conduction	Electric potential	Electric field	Current density	Electrical conductivity
Diffusion	Density	Density gradient	Diffusion current density	Diffusivity
Electrostatics	Electric potential	Electric field	Electric displacement	Dielectric permittivity
Magnetostatics	Magnetic potential	Magnetic field	Magnetic induction	Magnetic permittivity
Elasticity theory*	Displacement	Strain	Stress	Elastic moduli
Flow in porous media	Pressure	Weighted fluid velocity	Pressure gradient	Fluid permittivity

The term "composite material" means that the local transport properties (described by the tensor c_{ij}) depend on spatial variable $\mathbf{x}$. Thus, for linear composite materials $c_{ij} = c_{ij}(\mathbf{x})$. For nonlinear composite materials $c_{ij} = c_{ij}(\mathbf{x}, \varphi)$ or $c_{ij} = c_{ij}(\mathbf{x}, \nabla\varphi)$. It would not be correct to call an arbitrary inhomogeneous material a composite material. The term composite material assumes an existence of some structure in material. The structures can be very different: from regular to random, from particles-filled to laminated. Often, the term composite material assumes a property to be solid (to represent a unity). At the same time, systems of bodies / particles in air and liquids (powders, aerosol, suspensions, slurries) should not be separated from the composite material (the mentioned systems consist of at least two components, one of which is bodies / particles and the other component the surrounding medium). This is a reason why we use the term "composite structure" in this book, which designates both composite material and system of bodies / particles.

The systems of bodies and particle-filled composite materials can be treated in the framework of a unique approach. The mathematical models for bodies and particle-filled composites are the same; they are differential equations with discontinuous coefficients (see the equation above). The difference between problems for systems of bodies and composite materials is related to the type of the boundary-value problem: inner boundary-value problems correspond to composite materials and outer boundary-value problems correspond to systems of bodies.

A composite structure has some characteristic dimensions. One dimension is the size of the structure as a whole (so-called macroscopic dimension). We assume the macroscopic dimensional has the order of unity. Another dimension is the size of the structural elements of a composite (so-called microscopic dimension). We denote this dimension $\delta \ll 1$. Note that in many publications devoted to the homogenization theory, the macroscopic dimension is denoted by the symbol ε. Since we present our theory in terms of electrostatics (see below), the symbol ε is reserved in this book for dielectric constant. The number of sizes (often referred to as scales) is not restricted by two. Multi-scale structures are well-known (see, e.g., [30, 283]).

The term "high-contrast" means that transport properties of components of composite material are strongly different. The extremal (and widely used in physics and engineering, see, e.g., [340, 354]) case of high-contrast structures is a system of perfectly conducting bodies / particles.

The book is written in the terms of electrostatics, i.e., we call the solution of the transport problem potential, but not temperature or density, although all the results are valid for thermoconductivity and diffusion problems (as well as for all the problems listed in Table 1). A reason for using the electrostatic terminology is that the transport property of densely packed systems is determined by capacity of the pairs of neighbor bodies (it will be demonstrated below). It explains why capacity stands before the transport properties in the title of the book. It also explains why we discuss most problems keeping in mind the electrostatic problem.

The book presents mathematical treatment to phenomena intensively discussed in literature on natural sciences and engineering. For some problems (for example, the problem of effective properties of nonlinear dielectric) the intensive discussion was started in the last decade. Some problems were known and discussed for more than a century (for example, the problem of the capacity of a system of densely placed bodies). The current progress in the analysis of the mentioned problems was stimulated by progress in the mathematical methods (progress in the theory of partial differential equations, development of the homogenization method, etc.), in computer techniques and finite element computer programs. It is why a considerable part of the book is devoted to mathematics calculations and the presentation of results of numerical computations.

Many problems analyzed in the book were initiated by real world problems. For example, the theory of asymptotic behavior of capacity of a system of closely placed bodies was initiated by a project supported by a consortium of industrial companies (the names of the companies in 1999 were Polyclad and Hadco). The theory of nonlinear high-contrast dielectrics–ferroelectrics composites was initiated by a project supported by the U.S. Department of Energy. The initial stages of the mentioned projects are described in [39, 40, 41, 191].

This book is written on the basis of the authors' results published in Russian and international journals in the 1990s to 2000s. Most Russian scientific journals are translated to English from cover to cover by international publishers. English versions of all the authors' Russian papers included in the list of references can be found on the Internet at *http://www.springer.de* (Springer-Verlag) and *http://*

www.elsevier.com (Elsevier Science Publishers).

The book is structured as follows:

Chapter 1 presents a brief exposition of some asymptotic methods used for analysis of composite structures (composite materials and systems of bodies / particles) with brief historical comments.

Chapter 2 presents results of numerical analysis, which demonstrate specific properties of distributions of local fields in high-contrast composite structures and systems of closely placed bodies. In particular, the existence of "energy necks" in a system of densely packed bodies and closeness of potentials of the bodies determined from solution of the original continuum problem and the "potentials of nodes" determined from the corresponding network model are demonstrated.

Chapter 3 presents asymptotic analysis of the capacity of a system of closely placed bodies. In this chapter, we establish a relationship between the transport problem and the problem of asymptotic behavior of the capacity of a system of closely placed bodies. We do it on the basis of our generalization and mathematic interpretation of the "Tamm shielding effect" for a system of closely placed bodies (for two bodies, the phenomenon was described by the Soviet physicist, Nobel Prize Laureate I.E. Tamm in his book [353] published in 1927). Analysis of the problem leads us to the conclusion that the unique universal property of a system of closely placed bodies is the impossibility of localization of energy outside the channels between the neighbor bodies. As far as Tamm shielding, we found that it is a conditional effect. We demonstrate that the necessary and sufficient condition for existence of Tamm shielding (and, as a result, arising of "energy channels" between neighbor bodies, energy decomposition, network approximation, etc.) is the infinite increasing capacity of a pair of neighbor bodies when the distance between them tends to zero. This is a pure geometrical condition (it depends on the geometry of the bodies only). We note that this condition is not valid for the arbitrary geometry of bodies. As a result, network approximation (network modeling) is not possible for any system of closely placed bodies. Then the capacity (and transport property) of a system of closely placed bodies is controlled not only by material contrast and interparticle distances. The geometry of bodies is an additional necessary control parameter.

In Chapter 4, we put the question: "Do the total flux, energy and capacity (which are characteristics of integral nature) exhaust characteristics of the original continuum model which can be approximated with the corresponding network model?" We demonstrate that the potentials of the bodies can be added to this list (under the condition that the Tamm shielding effect takes place for the bodies under consideration!).

Chapter 5 presents a description of expansion of the method developed in Chapters 3 and 4 for systems of bodies to highly filled contrast composites. In this chapter, we also present some examples of numerical analysis of transport properties of high-contrast highly filled disordered composite material with the network model. The authors think that it would be difficult, if possible, to obtain similar results with a continuum model even using a large computer.

Chapter 6 deals with the mathematical and numerical analysis of special homogenization problems for a nonlinear composite with high-contrast components. The specificity of the problem considered is related not with any restrictions on the original problem (it is just a problem of general form) but with analysis of a special characteristic named the homogenized tunability of composite material. This characteristic is well-known in the electronics industry. From the mathematical point of view, this is (roughly speaking) the measure of nonlinearity of the problem under consideration. This chapter demonstrates that the behavior of effective characteristics of nonlinear composites can differ from the behavior of effective characteristics of linear composites qualitatively. For example, effective (homogenized) tunability can increase significantly when one dilutes nonlinear material with linear inclusions. No analog of this effect exists in linear homogenization theory. The data on the homogenized permittivity presented in this chapter may be of interest for the general theory of composite materials, because they clearly demonstrate that homogenized characteristics can show no correlation with the volume fraction of components of the composite.

Chapter 7 deals with the problem of loss of high-contrast composites.

Chapter 8 is devoted to transport and elastic properties of thin layers, which cover or join solid bodies. This theme is related to the problems considered in Chapters 3 and 4. In particular, the trial functions developed for analysis of thin joints were predecessors of the trial functions used in Chapters 3 and 4.

The authors thank Dr. S.I. Rakin (STU, Novosibirsk) for assistance in research. The authors thank Prof. I.V. Andrianov (RWTH–Aachen), Prof. L. Berlyand (Pennsylvania State University), Prof. V.V. Mityushev (Uniwersytet Pedagogiczny w Krakowie), Prof. A. Gaudiello (Università degli Studi di Cassino), Prof. V.V. Zikov (Vladimir State Humanitarian University) for providing references, useful comments and discussions. The research was supported through Marie Curie actions FP7, project PIIF2-GA-2008-219690.

The authors hope that the book will be used by both applied mathematicians interested in new mathematical methods and engineers interested in prospective materials and design methods. The authors would be happy if the book stimulates the interest of engineering students in mathematics as well as the interest of mathematical students in the problems arising in modern engineering and natural science.

Alexander A. Kolpakov
Alexander G. Kolpakov
Novosibirsk, Russia
Cassino, Italy
2009

Chapter 1

IDEAS AND METHODS OF ASYMPTOTIC ANALYSIS AS APPLIED TO TRANSPORT IN COMPOSITE STRUCTURES

When we consider a medium formed of a large number of small components, a system of closely placed bodies or a medium formed of components with strongly different (contrast) properties, we usually find small or large parameters naturally related to the structures under consideration. Sometimes we found not one but two or even more small or large parameters. For a composite body formed of large number of small components, the natural small parameter is a characteristic dimension of the components (usually, as compared with the dimension of the body). If, in addition, composite material is formed of contrast components, there appears one more parameter — ratio of material characteristics of the components.

If characteristics (either material or geometrical) depend on small or large parameters, the corresponding mathematical models account for these dependences. The mathematical models containing small or large parameters often can be analyzed by using asymptotic methods. The asymptotic methods strongly depend on the specific type of parameter and specific problem. We can divide (very roughly) the asymptotic methods arising in applied sciences into two groups:

1) problems in which geometry depends on a parameter,

2) problems in which material characteristics depend on a parameter.

Examples of the first group problems are asymptotic methods developed for analysis of problems in thin or small diameter domains [70, 75, 182, 282, 336, 360], in singularly perturbed domains [228, 264], in thin layers [229, 294, 317, 337, 338], in junctions of structural elements [44, 90, 130]. Examples of the second group of the methods are classical theory of small perturbation of coefficients of differential equations and integral functionals [153, 164, 334] and the homogenization theory [30, 21, 157]. If material characteristics are periodic with period depending on small

1

parameter, we arrive at the classical theory of homogenization [21, 91, 157]. If material characteristics can be described by random fast oscillation functions, we arrive at the random homogenization [157, 194, 195, 286, 393]. If the variation of material characteristics, in addition, is large, we arrive at so-called "stiff" problems [25, 58, 60, 65, 73, 98, 149, 211, 218, 219, 284, 289] and problems of transmitting through strongly inhomogeneous structures [103, 129].

We present below a brief overview of asymptotic methods, which can be useful for the reader.

1.1. Effective properties of composite materials and the homogenization theory

The problem of computation of overall properties of composite materials has a long history and it has attracted attention of some of outstanding scientists. Historically, analysis of overall properties of composite materials was started with a model of material filled with particles. For example, Poisson [295] constructed a theory of induced magnetism in which the body was assumed to be composed of non-conducting material filled with conducting spheres. Faraday [117] proposed a model for dielectric materials that consists of metallic globules separated by insulating materials. Significant contributions to solution of the problem of computation of overall properties of composite materials were done by Maxwell [227] and Rayleigh [348]. Other well-known 19th century contributors to the field were Clausius [92], Mossotti [261] and Lorenz [215].

In the 20th century many prominent scientists paid attention to the computation of overall properties of mixtures [64, 93, 128, 150, 214], suspensions [111, 112, 202, 310, 375] and systems of bodies and particles [50, 51]. The significant achievement was the theory of bound for effective characteristics of composite materials. The foundations of this theory were laid in the works by Reuss, Voight and Hill [150, 305, 367].

In the 1970s to 1980s, the so-called homogenization method was elaborated and applied to the analysis of composite materials. The foundations of the homogenization theory were laid in the pioneering papers by Spagnolo and Marino [224, 343, 344] published in 1960s, followed by numerous works published in 1970s–1980s. Mention the papers [20, 21, 30, 32, 108, 194, 221, 280, 317, 325, 397] (list is not complete, for additional bibliography information see [30, 21, 157]). The applied directions of the homogenization method are presented in [4, 5, 13, 27, 28, 29, 52, 56, 69, 78, 91, 97, 132, 134, 142, 159, 205, 278, 283, 285, 287, 314, 360, 382]. Applications of the homogenization method provided many important results of both theoretical and engineering significance. Mention theoretical prediction [6, 178] and manufacturing [201] of materials with negative Poisson's ratio and application of the homogenization method to design of composites possessing required overall properties [27, 28, 29].

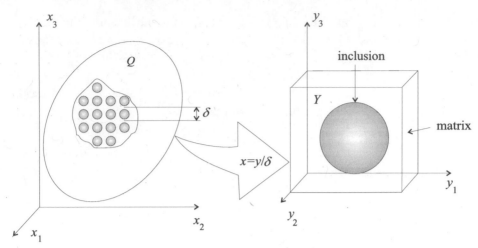

Figure 1.1. *A body of periodic structure and its periodicity cell Y in fast variables.*

The homogenization method for composites of a periodic structure uses various mathematical techniques. The basic techniques are presented in [5, 30, 157, 317]. In the present, various multiscale techniques are developed (see, e.g., [243, 283, 288]) and widely used in applied sciences (see, e.g., [14, 104, 198, 207, 213]).

1.1.1. Homogenization procedure for linear composite materials

In this section, we present basic ideas of the asymptotic expansions method. Consider an inhomogeneous body with a regular distribution of transport properties of those components, see Fig. 1.1.

The transport problem (thermoconductivity, diffusion, etc.) for that body has the form

$$L_\delta T^\delta = \frac{\partial}{\partial x_i}\left(c_{ij}^\delta(\mathbf{x})\frac{\partial T^\delta}{\partial x_j}\right) = f(\mathbf{x}) \text{ in } Q, \tag{1.1}$$

$$T^\delta(\mathbf{x}) = 0 \text{ on } \partial Q. \tag{1.2}$$

Here Q designates the region occupied by the composite material, ∂Q designates its boundary. Here δ is a parameter, which will be associated with the characteristic dimension of inhomogeneity of a composite, see Fig. 1.1. Thus, we consider a problem with parameter (or, in other words, a sequence of problems).

The following standard conditions are applied to the coefficients $c_{ij}^\delta(\mathbf{x})$: for all $\mathbf{x} \in Q$,

$$c_1|\mathbf{z}|^2 \le c_{ij}^\delta(\mathbf{x})z_i z_j \le c_2|\mathbf{z}|^2$$

for any $\mathbf{z} \in R^n$ $(n = 2, 3)$.

Here $0 < c_1, c_2 < \infty$ do not depend on δ. The uniform boundary condition (1.2) does not lead to the loss of generality of our consideration because the homogenized constants do not depend on the type of boundary conditions [30, 157]. We consider here uniform boundary conditions for simplicity.

Let us note that problem (1.1) and (1.2) permits the following formulation: find $T^\delta(\mathbf{x})$ from the solution of the minimization problem,

$$J_\delta(T) + \langle f, T \rangle \to \min, \ T(\mathbf{x}) \in H_0^1(Q), \tag{1.3}$$

where

$$J_\delta(T) = \frac{1}{2} \int_Q c_{ij}^\delta(\mathbf{x}) \frac{\partial T}{\partial x_i}(\mathbf{x}) \frac{\partial T}{\partial x_j}(\mathbf{x}) d\mathbf{x}$$

is a quadratic functional.

In this book, $H_0^1(Q)$ means closure according to the norm

$$\|f\|_{H^1(Q)} = \sqrt{\sum_{i=1}^3 \int_Q \left| \frac{\partial f}{\partial x_i}(\mathbf{x}) \right|^2 d\mathbf{x} + \int_Q f(\mathbf{x})^2 d\mathbf{x}}$$

of a set $C^\infty(Q)$ of finite functions, which are infinitely smooth in Q and vanish in a neighbor of ∂Q (other equivalent definitions of $H_0^1(Q)$ can be found in Appendix A).

The pointed branches in (1.3) signify dual coupling of the elements from $H_0^1(Q)$ and $H^{-1}(Q)$ in the standard duality of these spaces (for details see, e.g., [212]). The dual coupling coincides for sufficiently smooth functions with an inner product in $H_0^1(Q)$ (see [113, 212] for details). The equivalence of problems (1.1) and (1.2), and (1.3) is a well-known fact, see, e.g., [113, 212]. Problem (1.1) and (1.2) can be written in the following (so-called weak) form [212] :

$$-\int_Q q_i^\delta(\mathbf{x}) \frac{\partial \varphi}{\partial x_i}(\mathbf{x}) d\mathbf{x} = \int_Q f(\mathbf{x}) \varphi(\mathbf{x}) d\mathbf{x} \tag{1.4}$$

for any $\varphi(\mathbf{x}) \in H_0^1(Q)$.

In (1.4)

$$q_i^\delta(\mathbf{x}) = c_{ij}^\delta(\mathbf{x}) \frac{\partial T^\delta}{\partial x_j}(\mathbf{x}). \tag{1.5}$$

If a body is formed of many small components, the characteristic dimension δ of components is small: $\delta \ll 1$. Mathematically, this fact is formalized in the form $\delta \to 0$.

As is well known from engineering practice, materials and structures formed of many small components (concrete, wool, suspensions, aerosols) can be regarded as homogeneous ones. Note that most engineering handbooks (except special handbooks on composite structures, see, e.g., [74]) usually present technical constants

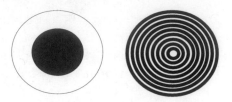

Figure 1.2. *A ring with a circular cartridge (left) and wood-like structure (right).*

of inhomogeneous materials (thermoconductivity coefficients, viscosity, elastic constants) as characteristics of homogeneous material, see, e.g., [197]. It means that one replaces an original inhomogeneous material for a (fictitious) homogeneous material. It is clear that such kind of substitution is possible only in asymptotic sense, when a material is a unity of small components with the characteristics dimension $\delta << 1$ ($\delta \to 0$). Usually, engineer-experimentator does not think about asymptotic, but he cares for taking a specimen for experiment relatively large as compared with the dimension of microstructure [172, 379]. It is clear that choice of specimen is equivalent to acceptance of asymptotic nature of overall characteristics of composite structures. We present two two-component circular structures as example. The volume fractions of the constitutive materials are equal in both structures. The circular structure displayed in Fig. 1.2 (left) cannot be approximated by a homogeneous structure, while the layered circular structure displayed in Fig. 1.2 (right) can be approximated by a homogeneous structure when the characteristic thickness of the layers is small. The layered circular structure displayed in Fig. 1.2 (right) is similar to the structure of wood. The engineering characteristics of woods are given in most hand-books in the form of characteristics of homogeneous materials. These characteristics describe properties of woods adequately and are successfully used in practice.

A homogeneous body (specifically, one which we want to put in correspondence with a composite) is described by the problem

$$LT^{(0)} = \frac{\partial}{\partial x_i}\left(\widehat{c}_{ij}\frac{\partial T^{(0)}}{\partial x_i}\right) = f(\mathbf{x}) \text{ in } Q, \tag{1.6}$$

$$T^{(0)}(\mathbf{x}) = 0 \text{ on } \partial Q, \tag{1.7}$$

or by the minimization problem: find $T^{(0)}(\mathbf{x})$ from the solution of the problem

$$J(T) + \langle f, T \rangle \to \min, \ T(\mathbf{x}) \in H_0^1(Q), \tag{1.8}$$

where

$$J(T) = \frac{1}{2}\int_Q \widehat{c}_{ij}\frac{\partial T}{\partial x_i}(\mathbf{x})\frac{\partial T}{\partial x_j}(\mathbf{x})d\mathbf{x}.$$

Here, $\widehat{c}_{ij}$ are homogenized transport constants describing a homogeneous material (we use the hat symbol " $\widehat{}$ " to mark homogenized characteristic corresponding to local characteristic under consideration). It is clear that the homogenized constants depend on the local transport constants of the composite.

The coupling of operators L_δ and L and functionals J_δ and J is well known. The functionals J_δ and J are the potentials of the corresponding operators, and operators L_δ and L are derivatives, in the sense of Gâteaux [113], of the corresponding functionals.

Asymptotic expansion based approach to the analysis of media with a periodic structure

Periodic materials are less prevalent in nature, where we meet disordered structures of combination of periodicity with various random derivations from it. On the other hand, we meet numerous periodic structures among artificial (man-made) structures.

Let us consider the case when an inhomogeneous body has a periodic structure in coordinates $\mathbf{x}$, with a period (called periodicity cell, unit cell, or basic cell) δY, see Fig. 1.1. For a periodic structure, the factor δ is the dimension of the periodicity cell, see Fig. 1.1. The material characteristics of the indicated medium are described by periodic functions in spatial variable of the following type [30]

$$c_{ij}^{\delta}(\mathbf{x}) = c_{ij}(\mathbf{x}/\delta), \qquad (1.9)$$

where $c_{ij}(\mathbf{y})$ are periodic functions with a periodicity cell Y.

The method of asymptotic expansions is based on the ideas of solving the problem with rapidly oscillating coefficients in the form of the following special series:

$$T^{\delta}(\mathbf{x}) = T^{(0)}(\mathbf{x}) + \sum_{n=1}^{\infty} \delta^n T^{(n)}(\mathbf{x}, \mathbf{y}), \qquad (1.10)$$

where $\mathbf{y} = \mathbf{x}/\delta$ is a "fast" variable and $\mathbf{x}$ is a "slow" variable, i.e., a two-scale expansion is considered. Functions $T^{(n)}(\mathbf{x}, \mathbf{y})$ in (1.10) are assumed to be periodic in variable $\mathbf{y}$ with periodicity cell Y. Function $T^{(0)}(\mathbf{x})$ is a function only of "slow" variable $\mathbf{x}$. By substituting $\mathbf{x}/\delta$ for $\mathbf{y}$, the functions become periodic in $\mathbf{x}$ with periodicity cell δY.

We will seek the solution of problem (1.1) and (1.2) in the form of an asymptotic expansion (1.10). While differentiating, we will separate the variables according to the formula

$$\frac{\partial f(\mathbf{x}, \mathbf{x}/\delta)}{\partial x_i} = f_{,ix}(\mathbf{x}, \mathbf{y}) + \delta^{-1} f_{,iy}(\mathbf{x}, \mathbf{y}), \qquad (1.11)$$

$$\mathbf{y} = \mathbf{x}/\delta. \qquad (1.12)$$

Here and afterward, subscript $,ix$ means $\dfrac{\partial}{\partial x_i}$ and the subscript $,iy$ means $\dfrac{\partial}{\partial y_i}$.

The operator L_δ on the left-hand side of equation (1.1), allowing for the differentiating rule (1.11), can be written as

$$L_\delta = \delta^{-2} L_0 + \delta^{-1} L_1 + L_2, \tag{1.13}$$

where

$$L_0 = \frac{\partial}{\partial y_i} \left(c_{ij}(\mathbf{y}) \frac{\partial}{\partial y_j} \right),$$

$$L_1 = \frac{\partial}{\partial x_i} \left(c_{ij}(\mathbf{y}) \frac{\partial}{\partial y_j} \right) + \frac{\partial}{\partial y_i} \left(c_{ij}(\mathbf{y}) \frac{\partial}{\partial x_j} \right),$$

$$L_2 = c_{ij}(\mathbf{y}) \frac{\partial}{\partial x_i} \frac{\partial}{\partial x_j}.$$

We change in (1.1) operator L_δ for its representation (1.13) and $T^\delta(\mathbf{x})$ for the series (1.10) and then equate the terms of the same order in δ. As a result, we obtain an infinite sequence of problems. The first three of them have the following form:

$$L_0 T^{(0)} = 0, \tag{1.14}$$

$$L_0 T^{(1)} + L_1 T^{(0)} = 0, \tag{1.15}$$

$$L_0 T^{(2)} + L_1 T^{(1)} + L_2 T^{(0)} = f(\mathbf{x}). \tag{1.16}$$

Equation (1.14) is satisfied identically because function

$$T^{(0)} = T^{(0)}(\mathbf{x}) \tag{1.17}$$

does not depend on the variable $\mathbf{y}$, see (1.10).

By virtue of (1.17), (1.15) takes the form

$$\left(c_{ij}(\mathbf{y}) T^{(1)}_{,jy} \right)_{,iy} + (c_{ik}(\mathbf{y}))_{,iy}\, T^{(0)}_{,kx}(\mathbf{x}) = 0. \tag{1.18}$$

By separating the variables $\mathbf{x}$ and $\mathbf{y}$, the solution of (1.18) can be set up as

$$T^{(1)}(\mathbf{x}, \mathbf{y}) = N^k(\mathbf{y}) T^{(0)}_{,kx}(\mathbf{x}) + V(\mathbf{x}), \tag{1.19}$$

where $N^k(\mathbf{y})$ represents a solution of the problem

$$\begin{cases} \left(c_{ij}(\mathbf{y}) N^k_{,jy} + c_{ik}(\mathbf{y}) \right)_{,iy} = 0 \text{ in } Y, \\ N^k(\mathbf{y}) \text{ is periodic in } \mathbf{y} \text{ with periodicity cell } Y, \end{cases} \tag{1.20}$$

and $V(\mathbf{x})$ is an arbitrary function of the argument $\mathbf{x}$.

We call problem (1.20) a cellular problem. It is also called a basic cell or a unit cell problem [30, 157, 318]. Let us consider equations (1.16), in which the function is unknown and $\mathbf{x}$ is a parameter. This problem has a periodic solution if the following equality is fulfilled [21]:

$$\langle L_1 T^{(1)} + L_2 T^{(0)} \rangle = f(\mathbf{x}), \tag{1.21}$$

where

$$\langle \bullet \rangle = \frac{1}{|Y|} \int_Y \bullet\, d\mathbf{x} \tag{1.22}$$

indicates the average value over periodicity cell Y. The average value symbol $\langle\rangle$ should not be confused with the dual coupling symbol $\langle,\rangle$.

From the homogenized equation (1.21), on account of (1.19), we obtain the homogenized (called also averaged or macroscopic) equation (1.6) for $T^{(0)}(\mathbf{x})$ with the boundary conditions (1.7). Simultaneously, we obtain from (1.19) and (1.21) the following formula for computation of the homogenized coefficients $\widehat{c}_{ij}$:

$$\widehat{c}_{ij} = \langle c_{ij}(\mathbf{y}) + c_{ik}(\mathbf{y}) N^j_{,ky}(\mathbf{y}) \rangle. \tag{1.23}$$

It is known (see, e.g., [30]) that

$$T^\delta(\mathbf{x}) \to T^{(0)}(\mathbf{x}) \text{ weakly in } H^1(Q), \tag{1.24}$$

$$T^\delta(\mathbf{x}) \to T^{(0)}(\mathbf{x}) \text{ in } L_2(Q)$$

as $\delta \to 0$, where $T^\delta(\mathbf{x})$ is the solution of the original problem (1.1) and (1.2), and $T^{(0)}(\mathbf{x})$ is the solution of the homogenized problem (1.6) and (1.7). We note that the second limit in (1.24) is the consequence of the first limit and Sobolev embedding theorem [342].

It is also known [30] that

$$T^\delta(\mathbf{x}) - \left(T^{(0)}(\mathbf{x}) + \delta N^j(\mathbf{x}/\delta) \frac{\partial T^{(0)}}{\partial x_j}(\mathbf{x}) \right) \to 0 \text{ in } H^1(Q), \text{ as } \delta \to 0. \tag{1.25}$$

From relations (1.5), (1.11) and (1.25), for the local flux

$$q_i^\delta(\mathbf{x}) = c_{ij}(\mathbf{x}/\delta) \frac{\partial T^\delta}{\partial x_j}(\mathbf{x})$$

the following approximation can be derived:

$$q_i^\delta(\mathbf{x}) - \left(c_{ij}(\mathbf{x}/\delta) + c_{ik}(\mathbf{x}/\delta) N^j_{,ky}(\mathbf{x}/\delta) \right) \frac{\partial T^{(0)}}{\partial x_j}(\mathbf{x}) \to 0 \text{ in } L_2(Q) \text{ as } \delta \to 0. \tag{1.26}$$

It is known (see, e.g., [30, 278]) that

$$\langle \mathbf{q}^\delta(\mathbf{x}) \rangle \to \mathbf{q}^0(\mathbf{x}) \text{ as } \delta \to 0, \tag{1.27}$$

where $\langle \mathbf{q}^\delta \rangle$ is the average value of the local flux $\mathbf{q}^\delta$ and $\mathbf{q}^0$ is the flux determined from the solution of the homogenized problem,

$$q_i^0(\mathbf{x}) = \widehat{c}_{ij} \frac{\partial T^{(0)}}{\partial x_j}(\mathbf{x}). \qquad (1.28)$$

The flux determined by equality (1.28) is called the homogenized flux. One can derive formula (1.27) by averaging (1.26) over periodicity cell with regard to the definition of the homogenized constants (1.23). It follows from (1.24) that in the limit (as $\delta \to 0$), potential (electric potential, temperature, etc.) in a nonhomogeneous material will behave like potential in a homogeneous material with effective transport coefficients given by (1.23). But, it is seen from (1.26) and (1.28), the local flux can differ (and usually it differs) from the homogenized flux. As seen from (1.27), the averaged values of the local flux coincides with the homogenized flux. Note that the actual (substantively existing) flux in a composite is the local flux. The average value of local flux characterizes the overall (macroscopic) transport property of composite material and it can be used to compute the homogenized characteristics of composite.

1.1.2. Homogenization procedure for nonlinear composite materials

The foundations of the homogenization theory for nonlinear operators and functionals were laid in the works [45, 221, 223]. Now, there exist some homogenization procedures for nonlinear composite materials based on $G(\Gamma)$-convergence and multiscale technique [21, 96].

G-limit based approach to analysis of media with a periodic structure

The G-limit approach is a sophisticated mathematical method used in the analysis of homogenization problems and proof of convergence theorems [45, 72, 100, 221]. It was not widely used in applied sciences and engineering previously. Recently, the situation has changed, see, e.g. [300].

We need certain mathematical notations, which were introduced in [221]. Let us denote by V the Banach reflexive space and V^* is a space topologically conjugated with V [113]. Following [221], let us denote by $C_0(V)$ a set of convex functionals in V, which are assumed to take values in $(-\infty, +\infty]$, not identically equal to $+\infty$ and lower semicontinuous (see Appendix A). Let us also denote

$$C(\alpha, v_0, M) = \{f(\mathbf{x}) \in C_0(V) : f(v) \le \alpha(v) \text{ for all } v(\mathbf{x}) \in V, f(v_0) \le M < \infty\},$$

where the functional $\alpha(\mathbf{x}) \in C_0(V)$ is such that $\alpha(v) - \langle v^*, v \rangle$ reaches a minimum in V for any $v^* \in V^*$.

The functional f^*, defined in $v^* \in V^*$ through the equality

$$f^*(v^*) = \sup_{v \in V} (\langle v^*, v \rangle - f(v))$$

is called a conjugate to f.

Definition 1. *The sequence of functionals* $\{f_\delta\} \subset C(\alpha, v_0, M)$ *G-converges to the functional* f, *if*

$$\lim_{\delta \to 0} f_\delta^*(v^*) = f^*(v^*) \text{ as } \delta \to 0 \text{ for any } v^* \in V^*.$$

Definition 2. *The sequence of operators* $L_\delta : V \to V^*$ *G-converges to the operator* $L : V \to V^*$, *if operators* L_δ *and* L *are invertible and for any* $v^* \in V^*$

$$\lim_{\delta \to 0} L_\delta^{-1} v^* = L^{-1} v^* \text{ weakly in } V \text{ as } \delta \to 0.$$

The equivalence of these two definitions of G-convergence was proved in [45].

The abstract G-limit (G-convergence) approach briefly described above can be applied to both linear and nonlinear problems. Mention densely related Γ-convergence method [57, 100].

In the terms of physics, Definition 1 implies the convergence of the energy of the original composite body to the energy of the homogenized body when dimension of inhomogeneities δ becomes small (that is formalized in the form $\delta \to 0$). Definition 2 implies the convergence of the solution (electric potential, temperature, etc., see Table 1) corresponding to the original composite body to the solution corresponding to the homogenized body when $\delta \to 0$. These convergences take place for arbitrary "mass source" v^* in the composite. Note that Definition 2 says nothing about strong convergence of derivatives of the solution of the original problem.

Homogenization procedure for nonlinear composite materials of periodic structure

We consider the following problem: find $T^\delta(\mathbf{x})$ from the solution of the minimization problem,

$$J_\delta(T) + \langle f, T \rangle \to \min, \ T(\mathbf{x}) \in H_0^1(Q), \tag{1.29}$$

where

$$J_\delta(T) = \int_Q G^\delta(\mathbf{x}, \nabla T(\mathbf{x})) \, d\mathbf{x}.$$

A homogeneous body is described by the minimization problem: find $T^{(0)}(\mathbf{x})$ from the solution of the minimization problem,

$$J(T) + \langle f, T \rangle \to \min, \ T(\mathbf{x}) \in H_0^1(Q), \tag{1.30}$$

where

$$J(T) = \int_Q \widehat{G}(\mathbf{x}, \nabla T(\mathbf{x})) \, d\mathbf{x}.$$

The corresponding boundary-value problems are obtained by computation of Gâteaux derivatives (or computation of variations) [113] of the functionals in the left-hand sides of (1.29) and (1.30) and have the form

$$\frac{\partial}{\partial x_i}\left(q_i^{\delta}\left(\mathbf{x}, \frac{\partial T^{\delta}}{\partial x_1}(\mathbf{x}), ..., \frac{\partial T^{\delta}}{\partial x_n}(\mathbf{x})\right)\right) = f(\mathbf{x}) \text{ in } Q, \ T^{\delta}(\mathbf{x}) = 0 \text{ on } \partial Q,$$

and

$$\frac{\partial}{\partial x_i}\left(q_i^{0}\left(\mathbf{x}, \frac{\partial T^{(0)}}{\partial x_1}(\mathbf{x}), ..., \frac{\partial T^{(0)}}{\partial x_n}(\mathbf{x})\right)\right) = f(\mathbf{x}) \text{ in } Q, \ T^{(0)}(\mathbf{x}) = 0 \text{ on } \partial Q,$$

where fluxes

$$q_i^{\delta}(\mathbf{x}, \mathbf{z}) = \frac{\partial G^{\delta}}{\partial z_i}(\mathbf{x}, \mathbf{z}),$$

$$q_i^{0}(\mathbf{x}, \mathbf{z}) = \frac{\partial \widehat{G}}{\partial z_i}(\mathbf{x}, \mathbf{z}).$$

Here $\mathbf{z}$ corresponds to driving force ∇T.

The nonlinear homogenization problem is most investigated in the periodic case when the function $G^{\delta}(\mathbf{x}, \mathbf{z})$ is periodic in spatial variable $\mathbf{x}$ and has the form

$$G^{\delta}(\mathbf{x}, \mathbf{z}) = G^{(0)}(\mathbf{x}/\delta, \mathbf{z}),$$

$G^{(0)}(\mathbf{y}, \mathbf{z})$ is periodic in $\mathbf{y}$ with periodicity cell Y. The sufficient conditions for nonlinear homogenization can be found in [222, 223]. We note that the theory of nonlinear homogenization is not as detailed as the linear homogenization theory (it is naturally, because the nonlinear problems usually are more difficult than linear ones). An exception is the homogenization theory for nonlinear ordinary differential equations, where a relatively complete homogenization theory was developed (in material science, this case corresponds to laminated materials as shown in [159], see also Appendix B).

For composite of periodic structure the function $G(\mathbf{z})$ is determined as follows (we assume that all functions under consideration exist, for details see, e.g., [99, 157]):

$$\widehat{G}(\mathbf{z}) = \min_{N \in V_Y} \int_Y G^{(0)}\left(\mathbf{y}, \nabla(N(\mathbf{y}) + \mathbf{z}\mathbf{y})\right) d\mathbf{y}, \qquad (1.31)$$

where

$$V_Y = \{f(\mathbf{y}) \in H^1(Y) : f(\mathbf{y}) \text{ is periodic in } \mathbf{y} \text{ with periodicity cell } Y\}.$$

In the case under consideration the function $G(\mathbf{z})$ depends on the variable $\mathbf{z}$ only.

The problem (1.31) (under some additional conditions, see, e.g., [100, 222, 223]) is equivalent to a periodic problem for nonlinear partial differential equation corresponding to the functional in the right-hand part of (1.31).

Formula (1.31) expresses the energy form corresponding to the homogenized body through the local energy of the original body. In many cases other relationships of the homogenized and local characteristics can be useful. Mention the relationship of the homogenized and local fluxes, which has the form

$$\int_\Gamma \mathbf{q}^\delta(\mathbf{y})\mathbf{n}dy \to \int_\Gamma \mathbf{q}^0(\mathbf{y})\mathbf{n}dy \text{ as } \delta \to 0, \tag{1.32}$$

for any side Γ of the periodicity cell Y ($\mathbf{n}$ means the normal vector to Γ). Formula (1.32) implies that the fluxes through the volume Y in the original inhomogeneous body and the homogenized body are equal to one another.

The above mentioned relationships between the homogenized and local characteristics are in agreement and predict the same characteristics of the homogenized structure.

1.2. Transport properties of periodic arrays of densely packed bodies

For a history of the problem of mathematical analysis of transport properties of periodic arrays of bodies, turn to Maxwell's book [227], where electric fields in a periodic array of bodies is considered. Maxwell's analysis opens a stage in analysis of the problem, which can be characterized as analysis of transport properties of periodic array of spheres, cylinders and disks. An outstanding contribution in the mathematical analysis of the problem was made by Rayleigh [348] who first studied the problem in mathematically rigorous way, as well as Mossotti [261], Clausius [92], Garnett [128], and Lorenz [215].

Now, one can distinguish two basic methods used to analyze the problem discussed. One is the homogenization method described in Section 1.1.1. We note that the original version of the homogenization method does not assume high contrast of components of composite (nontrivial modifications are required to adopt the homogenization method for contrast composites, see, e.g., [279, 281]) and (it is the main restriction) the homogenization method assumes proportional scaling of all components of composite (see formula (1.12) and Fig. 1.1 illustrating the scaling in the homogenization method). When the inclusions (bodies, particles, etc.) are almost touching, standard homogenization procedures lose the convergence property. It is why other asymptotic methods were developed for analysis of systems of almost touching bodies and similar structures. We present the main ideas, which lay the foundation of the methods developed for analysis of periodic arrays of almost touching bodies and highly filled contrast composite materials.

1.2.1. Periodic media with piecewise characteristics and periodic arrays of bodies

The inhomogeneous media can demonstrate different overall properties in dependence on the local geometry and topology of components of the media. The most well-known example of dependence of the overall properties on the geometry and topology of components of composite material is the percolation phenomenon [138, 170, 293]. Another example is the topology design theory [29].

Periodic and disordered structures are the main types of structures we deal with in practice. A periodic structure is a deterministic structure, which can be obtained by periodic repetition of a typical element (called periodicity or basic cell [30, 159]). The disordered structures (also called topologically disordered structures [143]) are not deterministic. The level of disorder varies in the range from small disorder (small random perturbation of a deterministic, for example periodic, structure) to complete disorder (random structures) [31, 46, 89, 194, 275, 359]. Most of materials, both natural and artificially produced, are partially or completely disordered. Limited number of solid composite materials (we call solid the essentially three-dimensional bodies) of periodic structures are produced using high technologies, see, for example, [88, 230]. At the same time many artificial structures (frames, structural elements of airplanes, ships, etc.) have deterministic structures, usually periodic or quasiperiodic [61, 89, 159, 182, 217, 265, 272, 273, 303].

In periodic systems we can naturally select a typical element — periodicity cell, see Fig. 1.1, which determines the property of the system in whole. It means that the local properties of the periodic system or periodic material (solution of the problem for typical element, usually called local problem) completely determine the overall properties of the periodic system in whole.

In this book, we consider media with piecewise constant characteristics, which correspond to systems of bodies and particle filled composite materials. Material properties of such composites are described by discontinuous (thus, non-differentiable) functions. A typical form of a function describing material properties of particle filled composite material is

$$a(\mathbf{x}) = \begin{cases} a_I \text{ in inclusions,} \\ a_m \text{ in matrix.} \end{cases} \qquad (1.33)$$

The typical graph of the function (1.33) for one inclusion is presented in Fig. 1.3.

Function of the form (1.33) also describes the body. If we consider electrostatic problem for a system of bodies, the distribution of the dielectric characteristics is described just by the function (1.33), where a_I means the dielectric constant of the material of bodies / particles and a_m means the dielectric constant of a substance surrounding the bodies.

In the last decades so called graded composite materials were reported (see, e.g., [139, 155, 255, 333]). In graded composites, there exists a relatively thick intercomponent layer between the basic components of composite. The function $a(\mathbf{x})$, which describes the local material properties of graded composite, changes continuously and has the form

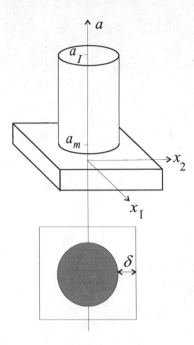

Figure 1.3. *The typical graphs and distribution over the periodicity cell of the functions $a(\mathbf{x})$ for two-dimensional composite of periodic structure described by the function (1.33).*

$$a(\mathbf{x}) = \begin{cases} a_I \text{ in inclusions,} \\ a_0(\mathbf{x}) \text{ in the interface ``particle--matrix'' region,} \\ a_m \text{ in matrix.} \end{cases} \tag{1.34}$$

The typical graph of the functions (1.34) for one inclusion is presented in Fig. 1.4.

If $a_I \ll a_m$ (the properties of the particles are vastly different from the properties of the matrix) then the composite is called a high-contrast composite. In many cases, the highly conducting bodies / particles are replaced by perfectly conducting bodies / particles (with infinite conductivity or, equivalently, zero resistivity). In this case, there remains one parameter, which describes the microstructure of composite. This is called the interparticle distance parameter δ, see Figs 1.3 and 1.4.

1.2.2. Problem of computation of effective properties of a periodic system of bodies

After Maxwell and Rayleigh, the problem of transport properties of periodic arrays of bodies attracted attention of many investigators (it is impossible to present a complete list of references here because it would be very long, some references can be found in [315]). The problem of transport properties of arrays of closely placed

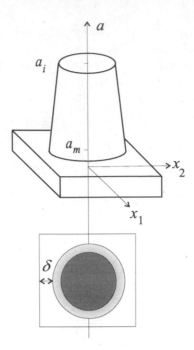

Figure 1.4. *The graphs and distribution over the periodicity cell of the functions $a(\mathbf{x})$ for two-dimensional "graded" composite.*

bodies was analyzed in [36, 114, 166, 231, 232, 235, 236, 253, 296, 349, 399] (the list is not complete). A major contribution into the problem was made by McPhedran and his co-authors, see, e.g., [233, 234, 238, 269, 291, 389]. We emphasize two specific features of problems considered in the above-mentioned papers:

(i) the bodies form a periodic array, which is obtained by periodic translation of one body;

(ii) the bodies have simple geometry (spheres, cylinders, etc).

In the frameworks of directions, many works were devoted to transport properties of array of spheres [10, 106, 166, 232, 236, 319, 320] and circular cylinders (disks) [80, 81, 166, 233, 234, 235]. Transport properties of array of elliptic cylinders [237, 269, 389] and cylinders having square cross-section [11] were considered, see also [383, 384, 398]. Comparison of transport properties of array of periodic spheres and an array of periodic cubes was presented in [8]. In [270] transport properties of square array of coated cylinders were analyzed. Recently arrays of rhombic fibers [9] were considered. In [234] a problem for closely placed, highly conducting cylinders was considered using a technique of complex analysis. Later the methods of complex analysis were effectively applied to analysis of systems of closely placed disks by Mityushev and co-authors [220, 248, 249, 250, 251, 253, 254], see also [292, 311, 312]. In [23] conduction through a granular material was investigated using ensemble averaging and approximate solutions for closely packed spheres.

Most problems were analyzed under the condition of perfect contact between matrix and inclusions, which assumes no jump conditions with respect to potential and normal flux on the surface of conjugation of the matrix and inclusion. The arrays of bodies with not perfect contact were considered in [26, 147, 163, 206, 226, 369].

Method of power series as applied to the computation of transport properties of periodic arrays of bodies

We present the basic ideas of the method of power series widely used for computation of overall properties of periodic structures. We consider a simple-cubic array of identical spheres of radii R embedded into a homogeneous matrix. Denote the periodicity cell of the array by Y. Following to [315], we associate the problem considered with the problem of electrostatic of composite material. An overall external electric field $\mathbf{E}$ is assumed to be applied parallel to Ox_1-axis:

$$\mathbf{E} = (E_0, 0, 0).$$

The potential distribution in the composite satisfies Laplace equation

$$\Delta\varphi = 0 \tag{1.35}$$

both inside and outside the spheres.

Consider a sphere with the center at the point 0. The general power series for the potential in spherical coordinates (r, θ, φ) is the following (see, e.g., [315]): inside the sphere

$$\varphi_I(r, \theta, \varphi) = A_0 + \sum_{l=1}^{\infty} \sum_{m=-l}^{l} (A_{lm}r^l + A_{lm}r^{-l-1})Y_{lm}(\theta, \varphi), \tag{1.36}$$

outside the sphere

$$\varphi_m(r, \theta, \varphi) = C_0 + \sum_{l=1}^{\infty} \sum_{m=-l}^{l} C_{lm}r^l Y_{lm}(\theta, \varphi). \tag{1.37}$$

In (1.36) and (1.37), $Y_{lm}(\theta, \varphi)$ is the spherical harmonics of order (l, m), i.e., solution of Laplace equation in spherical coordinates, which can be represented in the terms of Legendre functions $P_l^m(\cos\theta)$, see [315]

$$Y_{lm}(\theta, \varphi) = \frac{(l - |m|)!}{(l + |m|)!} P_l^{|m|}(\cos\theta)e^{im\varphi}.$$

The conjugation conditions at the interface surface $|\mathbf{x}| = R$ between the sphere and matrix are

$$\varphi_I(\mathbf{x}) = \varphi_m(\mathbf{x}), \tag{1.38}$$

$$a_I \frac{\partial\varphi_I}{\partial\mathbf{n}}(\mathbf{x}) = a_m \frac{\partial\varphi_m}{\partial\mathbf{n}}(\mathbf{x}),$$

where a_I and a_m mean the permittivity of the materials of sphere and matrix, correspondingly. In addition, the function $\varphi_m(\mathbf{x})$ is periodic with periodicity cell Y.

From these conditions (the conjugation conditions and the periodicity condition, as well as the symmetry of solution and the absence of singularities in solution), one can determine the coefficients C_{lm}, A_{lm} and B_{lm}. We present the scheme of solution of the problem following to [315].

From the second condition in (1.38), it follows that

$$A_{lm} = \frac{B_{lm}\left[\dfrac{a_s}{a_m} + \dfrac{l+1}{l}\right]}{R^{2l+1}\left(1 - \dfrac{a_s}{a_m}\right)}. \tag{1.39}$$

Due to symmetry of the array only odd values l in (1.37) and (1.36) and only values of m that are multiples of 4 must be allowed.

The condition of the absence of singularity leads to the equality

$$A_0 + \sum_{l=1}^{\infty}\sum_{m=0}^{2l-1} A_{2l-1,m} r^{2l-1} P_{2l-1}^m(\cos\theta)\cos(m\varphi) - E_0 x_1 =$$

$$= \sum_{l=1}^{\infty}\sum_{m=0}^{2l-1}\sum_{i=0}^{\infty} B_{2l-1,m}\rho_i^{-2l} P_{2l-1}^m(\cos\theta_i)\cos(m\varphi_i). \tag{1.40}$$

In (1.40) the coordinates $(\rho_i, \theta_i, \varphi_i)$ are measured to the center of the i-th sphere. In his treatment of this problem, Rayleigh truncated terms higher than those involving r^3 from the Legendre polynomials, see [315].

The equations with respect to the unknown coefficients $B_{2l-1,m}$ are obtained by equating the partial derivatives of all orders with respect to variable x_1 in (1.40) (see for details [315]). They have the following form

$$\sum_{l=n+1}^{\infty}\sum_{m=0}^{2l-2n-2}\binom{m+2l-1}{2n+1} P_{2l-2n-1}^m(\cos\theta_0)\cos(m\varphi_0)A_{2l-1,m} +$$

$$+ \sum_{l=1}^{\infty}\sum_{m=0}^{2l-1}\sum_{i=1}^{\infty}\binom{2l+2n-m}{2n+1} \rho_i^{-2l-2n-1} P_{2l+2n}^m(\cos\theta_i)\cos(m\varphi_i)B_{2l-1,m} = E_0\delta_{n0},$$

where

$$\binom{n}{r} = \frac{n!}{r!(n-r)!},$$

$(\rho_0, \theta_0, \varphi_0)$ corresponds to a point at the boundary of the periodicity cell Y, and $\delta_{00} = 1$, $\delta_{0n} = 0$ $(n \geq 1)$.

Detailed analysis of the problem can be found in [236]. Analysis of similar problems can be found in [23, 232, 233, 239, 291, 399]. We do not discuss the details of the solution and only note that it is evident that the method of power series works well for small concentrations of spherical or circular inclusions. In this case distribution of potential is similar to distribution of potential in the problem about unique inclusion. Solution for unique inclusion with corrector containing few additional harmonics provides us with accurate solution. If diameter $2R$ of the sphere is close to the size of the periodicity cell Y, the solution of the problem discussed is strongly different from solution for unique inclusion, it is necessary to account for all terms in series and solve the corresponding problems. This fact can be explained easily using some results, which will be discussed in detail below.

1.2.3. Keller analysis of conductivity of medium containing a periodic dense array of perfectly conducting spheres or cylinders

As was noted above, when the bodies are placed relatively far from one another, the approach based on the harmonic series works well and it is necessary to save only a few harmonics in the series to obtain accurate formulas. When the bodies are placed densely, it is necessary to save a large number of harmonics in the series. The results of the paper [174] (see also Chapter 2) explain this fact. We consider a periodic system of disks as an example. When the disks are placed relatively far apart, the distribution of energy around each disk is a smooth function of polar angle close to distribution of energy around a single disk, see Fig. 1.5 (left).

When the disks are placed densely, the energy as a function of polar angle looks like a singular function, see Fig. 1.5 (right). It is known that a function like that shown in Fig. 1.5 (left) usually can be approximated well with a small number of harmonics and it is necessary to save a large number of harmonics in series to approximate a function like shown in Fig. 1.5 (right).

It would be natural to manipulate with functions like those shown in Fig. 1.5 (right) without using technique of power series. This was done in [166]. In the previously mentioned paper Keller reported that *"The previous results of Maxwell, of Rayleigh and Meredith and Tobias are not valid near the singularity"* (this is for almost touching bodies) and derived formulas for transport properties of periodic arrays of circular disks and spheres different from the Maxwell's formula. Also it was reported that new asymptotic formulas derived in [166] agree well with the numerical results [165].

Keller's analysis was based on a hypothesis (formally incorrect, see below, nevertheless very fruitful) about the form of flux between two closely placed disks (spheres). We employ Keller method to derive an approximate formula for the flux between two disks (the i-th and the j-th) of radii R placed at the distance δ_{ij}, see Fig. 1.6. Although the original Keller analysis was given for two spheres, we present corresponding computations for two disks. We do it in order to demonstrate the dimensional sensitivity of the problem (i.e., existence of some differences in properties

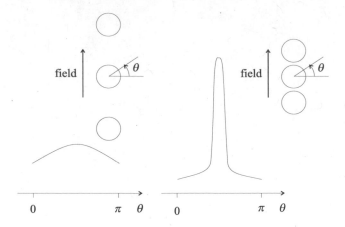

Figure 1.5. *Typical distribution of density of the local energy around disk as a function of polar angle θ: left – dilute composite, right – composite with densely packed disks.*

of solutions of two- and three-dimensional problems).

We approximate the disks by the tangential parabolas

$$y = \frac{\delta_{ij}}{2} + \rho\frac{x^2}{2},$$

and

$$y = -\left(\frac{\delta_{ij}}{2} + \rho\frac{x^2}{2}\right),$$

where

$$\rho = \frac{1}{R} \tag{1.41}$$

is curvature of the disks.

The distance $H(x)$ between the parabolas is

$$H(x) = \delta_{ij} + \rho x^2. \tag{1.42}$$

We assume that the matrix is uniform and dielectric constant of the material of the matrix is equal to ε. Following [166], we assume that the potential in the region between the disks has the form

$$\varphi(\mathbf{x}) = \frac{(t_i - t_j)y}{H(x)}.$$

Then the local flux in the region between the disks has the form (here $\mathbf{x} = (x, y)$)

$$\mathbf{v}(\mathbf{x}) = \varepsilon\nabla\varphi(\mathbf{x}) = \varepsilon\left(0, \frac{t_i - t_j}{H(x)}\right). \tag{1.43}$$

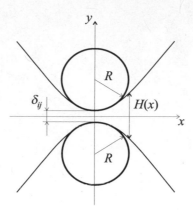

Figure 1.6. *Two neighbor disks of equal radii R.*

That is the flux is proportional to the difference of potentials of the disks and inversely proportional to the distance between the disks.

Then the total (integral) flux between the disks can be calculated as follows (this is the flux through Ox-axis):

$$J_{ij} = \varepsilon(t_i - t_j) \int_{-\infty}^{\infty} \frac{dx}{\delta + \rho x^2} = \varepsilon \frac{t_i - t_j}{(\rho\delta_{ij})^{1/2}} \arctan\left[\left(\frac{\rho}{\delta_{ij}}\right)^{1/2} x\right]\Bigg|_{-\infty}^{\infty}. \quad (1.44)$$

We write in (1.44) the limits of integration $-\infty$ and ∞. It is not a good choice, because formula (1.43) determines the flux between disks only. Following [166] we have to change the limits of integration $-\infty$ and ∞ for $-R$ and R, where R is the radius of the disks. We underline that in the case under consideration there is no difference if we integrate from $-\infty$ to ∞ or from $-R$ to R because the integral (1.44) converges at infinity and its behavior with respect to δ_{ij} does not depend on the choice of the limit of integration. Note that for spheres (the case analyzed in [166], see also below) the corresponding integral diverges at infinity.

If the sum of the curvatures ρ is not small, and the distance δ_{ij} between the disks is small, then the value of arctangent in the right-hand side (1.44) is equal to π. Taking into account (1.41), we obtain

$$J_{ij} = \varepsilon\pi\sqrt{\frac{R}{\delta_{ij}}}(t_i - t_j). \quad (1.45)$$

This formula can be derived from the exact formula for capacity of two disks as the leading term in asymptotic in δ_{ij}. This exact solution (capacity of two cylinders) can be found in [340].

Keller's approach [166] accounted, in fact, the phenomenon of localization of fields in high-contrast highly filled composite (see Chapter 2 for detailed discussion of the localization phenomenon). This approach takes into account the field in the channel between the neighbor spheres/disks and neglects the field outside the channel. This idea strongly distinguishes Keller's approach from Rayleigh's approach based on the accounting of the fields corresponding to all bodies of the system.

In 1927, in his book [353], I.E. Tamm made note on the capacity of two closely placed bodies. Tamm's idea (see Chapter 3 for details) as well as Keller's approach assumes that the effect of localization of fields is universal property of high-contrast highly filled composite materials. This assumption was accepted until the papers [176, 183, 185] demonstrated that the localization of energy (which is the necessary and sufficient condition of network approximation) is not the universal but conditional property of perfectly conducting closely placed bodies / particles. In the papers mentioned it was demonstrated that the localization phenomenon is related to the asymptotic property of capacity of pair of neighbor bodies. In other words, the existence or not existence of the localization phenomenon is determined by the geometry of the bodies. Thus, in the analysis of problems under consideration, the geometry of the bodies must be taken into account in addition to high-contrast of components and high filled matrix with inclusions. These results naturally relate the problem of transport properties of a system of closely placed bodies / particles to the notion of capacity and Tamm shielding effect.

Some imperfections of the simplified method

We indicate some principal imperfections of the simplified approach presented above. The field $\mathbf{v}(\mathbf{x})$ (1.43) is divergence free:

$$\operatorname{div}\mathbf{v}(\mathbf{x}) = 0,$$

unfortunately, it is not a potential field. The necessary condition of potentiality [308]

$$\frac{\partial v_x}{\partial y} = \frac{\partial v_y}{\partial x}$$

is not satisfied for the field (1.43), because

$$\frac{\partial 0}{\partial y} \neq \frac{\partial \left(\dfrac{t_i - t_j}{H(x)} \right)}{\partial x}$$

if $H(x) \neq const.$

Analyzing flux between two closely placed spheres, Keller [166] used the following approximation for potential between the spheres function (here $\mathbf{x} = (x, y, z)$, $t_i = 1$, $t_i = -1$, and notations corresponds to Fig. 1.7):

$$\varphi(\mathbf{x}) = \frac{z}{\delta + \dfrac{x^2 + y^2}{R}}. \tag{1.46}$$

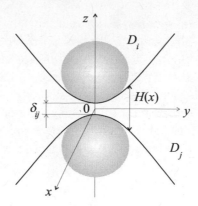

Figure 1.7. *Two neighbor spheres.*

Unfortunately, the function (1.46) does not satisfy Laplace equation because for (1.46)

$$\frac{\partial^2}{\partial^2 x}\left(\frac{z}{\delta + \dfrac{x^2 + y^2}{R}}\right) = \frac{2z}{R} \cdot \frac{-\delta + \dfrac{3x^2 - y^2}{R}}{\left(\delta + \dfrac{x^2 + y^2}{R}\right)^3},$$

$$\frac{\partial^2}{\partial^2 y}\left(\frac{z}{\delta + \dfrac{x^2 + y^2}{R}}\right) = \frac{2z}{R} \cdot \frac{-\delta + \dfrac{-x^2 + 3y^2}{R}}{\left(\delta + \dfrac{x^2 + y^2}{R}\right)^3},$$

$$\frac{\partial^2}{\partial^2 z}\left(\frac{z}{\delta + \dfrac{x^2 + y^2}{R}}\right) = 0,$$

and finally, we have for the function (1.46)

$$\Delta\varphi(\mathbf{x}) = \frac{4z}{R} \cdot \frac{-\delta + \dfrac{x^2 + y^2}{R}}{\left(\delta + \dfrac{x^2 + y^2}{R}\right)^3} \neq 0,$$

where $\mathbf{x} = (x, y, z)$.

Two-dimensional analog of the Keller approximation for potential between two closely placed disks, which has the form

$$\varphi(\mathbf{x}) = \frac{x}{\delta + \dfrac{y^2}{R}},$$

also does not satisfy Laplace equation (above we set $\mathbf{x} = (x, y)$).

The rigor solution of the problem can be obtained in the framework of the asymptotic approach, where the functions similar to (1.43) and (1.46) are used not as exact solutions but as trial functions for variational estimates. Such approach was realized first in so-called Kozlov's model of high-contrast composite [196].

Now, we consider two spheres D_i and D_j subjected to the potentials t_i and t_j, see Fig. 1.7. Following [166], we approximate the spheres with the tangential paraboloids

$$z = \pm \left(\frac{\delta}{2} + \rho \frac{r^2}{2} \right)$$

and assume that the local field in the region between the disks in cylindrical coordinates has the form

$$\mathbf{v}(r, z) = \varepsilon \left(0, 0, \frac{t_i - t_j}{\delta + \rho r^2} \right). \tag{1.47}$$

The transport coefficient (capacity in terms of [166]) is computed as the total flux through Oxy-plane:

$$C_{ij}^{(2)} = 2\varepsilon\pi \int_0^\infty \mathbf{v}_x(x, r) r \, dr = 2\varepsilon\pi (t_i - t_j) \int_0^\infty \frac{r \, dr}{\delta + \rho r^2} =$$

$$= \frac{\varepsilon\pi}{\rho} (t_i - t_j) \ln(\delta + \rho r^2) \Big|_\delta^\infty. \tag{1.48}$$

The integral in (1.48) diverges at infinity because

$$\lim_{r \to \infty} \ln(\delta + \rho r^2) = \infty.$$

We note that in two-dimensional case the corresponding integral converges, see Section 1.2.3. It means that the method used is dimension sensitive. The explanation of this fact was given later in the framework of the field localization idea [174], see also Section 2.3.

Integral in the right-hand side of (1.48) was computed in [166]. The leading term of the right-hand part of (1.48) in δ is $-\varepsilon\pi R(t_i - t_j) \ln \delta$ (we remind that $R = 1/\rho$). In [166], it was demonstrated that capacity of two closely placed spheres computed using series [371] has the leading term $-\varepsilon\pi R \ln \delta$ in δ. Thus, the flux between two closely placed spheres is proportional to the capacity of these spheres and the differences on potentials of the spheres. Then the transport property of the spheres is described by the capacity of these spheres.

The capacity relates the difference of potentials $t_i - t_j$ of spheres and electric charge Q:

$$Q = C_{ij}^{(2)} E. \tag{1.49}$$

On the other hand, the capacity relates the energy $\mathcal{E}$ of electric field and the difference of potentials $t_i - t_j$:

$$\mathcal{E} = \frac{C_{ij}^{(2)}(t_i - t_j)^2}{2}. \tag{1.50}$$

It is known from the general theory of electrostatic [340, 354] that both the definitions (1.49) and (1.50) of the capacity are equivalent. The computation presented above corresponds to the first definition (1.49). It is seen that the simplified approach under natural restriction on the limits of integration provides us with correct asymptotic of capacity of closely placed bodies if one uses definition (1.49).

It would be interesting to know the predictions of the simplified approach corresponding to another definition (1.50). For this reason, we compute capacity $C_{ij}^{(2)}$ by using definition (1.50) and the approximation for potential presented above. We compute the energy $\mathcal{E}$ corresponding to the field $\mathbf{v}(r, z)$ (1.47) (three-dimensional case). It is

$$\mathcal{E} = 2\pi\varepsilon \int_0^\infty \frac{1}{2}|\mathbf{v}|^2(r, z)r\, dr =$$

$$= 2\pi\varepsilon(t_i - t_j)^2 \int_0^\infty \frac{r\, dr}{(\delta + \rho r^2)^2} = \frac{(t_i - t_j)^2}{2}\left[\frac{2\pi\varepsilon R}{\delta}\right]. \tag{1.51}$$

The expression in the square brackets in (1.51) demonstrates no correlation with the formula $C_{ij} = -\varepsilon\pi R \ln \delta$ [166] for the capacity of two closely placed spheres.

In two-dimensional case, computation of the energy $\mathcal{E}$ of the flux (1.43) (two-dimensional case) leads to the following result:

$$\mathcal{E} = \varepsilon\frac{(t_i - t_j)^2}{2}\int_{-\infty}^\infty \frac{dx}{(\delta + \rho x^2)^2} = \tag{1.52}$$

$$= \varepsilon\frac{(t_i - t_j)^2}{2}\cdot\left[\frac{x}{2\delta(\delta + \rho x^2)} + \frac{\arctan\left(x\sqrt{\frac{\rho}{\delta}}\right)}{2\delta^{3/2}\rho^{1/2}}\right]\Bigg|_{-\infty}^\infty =$$

$$= \frac{(t_i - t_j)^2}{2}\left[\varepsilon\frac{\pi}{2\delta}\sqrt{\frac{R}{\delta}}\right].$$

The expression in the square brackets in (1.52) demonstrates no correlation with the formula $C_{ij} = \varepsilon\pi\sqrt{\frac{R}{\delta}}$ [166] for the capacity of two closely placed disks.

Thus, we conclude that computation of flux using simplified approach [166] leads to a realistic result and indicates the relationship of transport property of a system of densely placed disks / spheres with the notion of capacity. But computation of energy using simplified approach leads to nonrealistic results, which demonstrates no correlation with the capacity. We note that in the original Keller publication [166], only fluxes were computed for disks and spheres. We note that this incompatibility means that Keller approach based on notion of flux and Tamm approach based on the notion of energy–capacity cannot be reconciled directly. The discrepancy indicated appealed for a rigorous mathematical analysis of the problem. The incompatibility between computation of capacity through flux and energy can be removed by accurate construction of trial functions [41, 176, 184, 185] (see also Chapter 3).

Deriving formula (1.45) from the exact formula for capacity of two parallel cylinders

Taking into account the problem in the computation of capacity using the simplified method, it would be useful to have rigor formulas for computation of capacity of closely placed bodies. Such formula can be derived elementary enough for two parallel circular cylinders (disks) of the same radii. It can be found in the course of electrostatics that the capacity $C_{ij}^{(2)}$ of two parallel circular cylinders (indexed by i and j) of the radii R per unit length of the cylinders is (see, e.g., [339])

$$C_{ij}^{(2)} = \frac{\pi \varepsilon}{\operatorname{arccosh}\left(\dfrac{2R + \delta_{ij}}{2R}\right)}. \tag{1.53}$$

Here δ_{ij} is the distance between the cylinders.

We can rewrite (1.53) as

$$\frac{2R + \delta_{ij}}{2R} = \cosh\left(\frac{\pi \varepsilon}{C_{ij}^{(2)}}\right) = \frac{e^{\frac{\pi \varepsilon}{C_{ij}^{(2)}}} + e^{-\frac{\pi \varepsilon}{C_{ij}^{(2)}}}}{2}. \tag{1.54}$$

Expanding the right-hand side of (1.54) into Tailor's series, we have

$$1 + \frac{\delta_{ij}}{2R} \approx 1 + \frac{\left(\dfrac{\pi \varepsilon}{C_{ij}^{(2)}}\right)^2}{2}. \tag{1.55}$$

After gathering the terms in (1.55), we obtain that

$$C_{ij}^{(2)} = \pi \varepsilon \sqrt{\frac{R}{\delta_{ij}}}.$$

1.2.4. Kozlov's model of high-contrast media with continuous distribution of characteristics. Berriman–Borcea–Papanicolaou network model

Discrete networks have been used as analogs of continuum problems in various areas of physics and engineering for a long time. Many outstanding mathematicians and physicists paid attention to the problems, as we now understand, directly related to the network modeling (see, e.g., [1, 7, 33, 95, 192, 268, 324]). Well-known and widely-used resistor network models (see, e.g., [315, 388]) and spring network models (analogs of the well-known Cauchy–Born model [50, 51]) are used to describe great variety of phenomena from molecular dynamics to composite materials.

While network models are widely used in material sciences, the mathematical aspect was not developed for a long time. Mathematical analysis of high-contrast composite materials with network model was started with a model of material with coefficients describing by differentiable function (at least twice the differentiation of material characteristics with respect to the spatial variables is necessary for the theory discussed in this section). Substantial progress in the analysis of high-contrast composites with continuously distributed properties and understanding the mathematical nature of the approximation problem was achieved in the pioneering works [47, 48, 49]. These works employed Kozlov's model of high-contrast media [196].

In Kozlov's model [196] of high-contrast composite material, it is assumed that the material's properties of the composite can be described by a function of the form

$$a(\mathbf{x}) = c(\mathbf{x})e^{S(\mathbf{x})/\delta^2}, \tag{1.56}$$

where $c(\mathbf{x}) > 0$, and $S(\mathbf{x})$ is non-negative differentiable function. The last condition means that Kozlov's model describes media with continuously distributed characteristics only.

Fig. 1.8 illustrates the behavior of function (1.56) as δ tends to zero in one-dimensional case. In the points where $S(\mathbf{x}) = 0$, $a(\mathbf{x}) = c(\mathbf{x})$ for any δ, while

$$a(\mathbf{x}) \to \infty \text{ as } \delta \to 0$$

in other points. Thus, function (1.56) can be used as a model of high-contrast material as $\delta \to 0$.

Note that similar behavior is demonstrated by numerous functions of the form

$$a(\mathbf{x}) = c(\mathbf{x})F\left(\frac{S(\mathbf{x})}{\delta}\right), \tag{1.57}$$

with a function $F(s)$ such that

$$F(s) \to \infty \text{ as } s \to \infty.$$

In Fig. 1.9 plots of the functions

$$a(x) = e^{25\sin(x)}$$

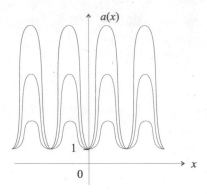

Figure 1.8. *The evolution of Kozlov function $a(x) = c(x)e^{S(x)/\delta^2}$ as δ tends to zero in one-dimensional case.*

and

$$a(x) = (\sin(x) + 1)^{36}$$

are shown for $x \in [0, \pi]$. When periodically repeated, these plots correspond to the graphs of the functions $e^{25|\sin(x)|}$ and $(|\sin(x)| + 1)^{36}$.

The first function is Kozlov's exponential function (in the case under consideration $c(x) = 1$, $S(\mathbf{x}) = |\sin(x)| + 1$, and $\delta^2 = 0.2$). The second function is a trigonometric polynomial on the interval $[0, \pi]$. The plots look very similar and any of them can be used to model high-contrast composite material. The choice of adequate form of the function $a(\mathbf{x})$ in Kozlov's model is an open problem for now. From the physical point of view, composite materials with similar distribution of local material characteristics have to demonstrate similar effective characteristics.

Now, we present briefly the Berriman–Borcea–Papanicolaou approach to analysis of continuum model of high-contrast composites with network models. Detailed presentation of Berriman–Borcea–Papanicolaou theory of network approximation

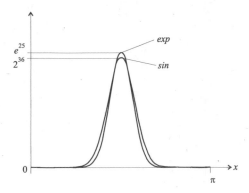

Figure 1.9. *The plots of the functions $e^{25\sin(x)}$ (marked by exp) and $(\sin(x) + 1)^{36}$ (marked by sin).*

for continuum transport problem can be found in [47, 48, 49].

As $\delta \to 0$, (1.56) describes a high-contrast composite material. In [47] formula (1.56) is used to describe the local resistivity of a periodic medium. It was assumed that the function $S(\mathbf{x})$ has a specific form: periodic, differentiable, and has two maxima and two minima inside the periodicity cell, see Fig. 1.10. These assumptions guarantee the existence of a *saddle point* $\mathbf{x}_s$ between the points of maxima and minima.

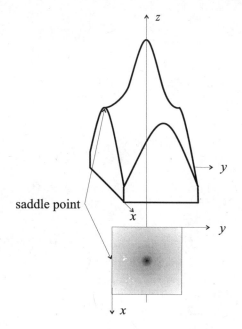

Figure 1.10. *The typical graphs and distribution over the periodicity cell of the functions $a(\mathbf{x})$ for two-dimensional composite described by Kozlov's function.*

A two-dimensional boundary-value problem corresponding to Kozlov's model was analyzed in [47, 48, 49]. In the mentioned papers, the problem of justification of network model was clearly formulated first and a network corresponding to Kozlov's model was constructed. The problem considered in [48] has the form

$$\mathrm{div}\big(c(\mathbf{x})e^{-S(\mathbf{x})/\delta}\nabla\varphi\big) = 0 \text{ in } Q,$$

$$c(\mathbf{x})e^{-S(\mathbf{x})/\delta}\frac{\partial\varphi}{\partial\mathbf{n}}(\mathbf{x}) = I(\mathbf{x}) \text{ on } \partial Q, \qquad (1.58)$$

where $I(\mathbf{x})$ satisfies the condition

$$\int_{\partial Q} I(\mathbf{x})d\mathbf{x} = 0.$$

In (1.58), $\mathbf{n}$ denotes the outward unit normal vector to ∂Q.

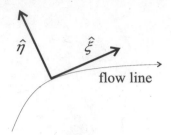

Figure 1.11. *Local coordinate system along the flow line.*

To construct trial functions, the authors [48] introduced flow lines going through the saddle points of the function $S(\mathbf{x})$ and local coordinates ξ and η along and perpendicular to the flow lines, see Fig. 1.11.

Then

$$S(\mathbf{x}) = S(\xi, \eta) = S(\xi, 0) + \frac{k(\xi)}{2}\eta^2 + \dots,$$

where $k(\xi)$ is curvature in the direction perpendicular to the flow line.

The flow lines divide the domain into "cells," see Fig. 1.12, which forms the network.

The trial function in primal problem is constructed as $\varphi(\mathbf{x}) = const$ inside the cells and

$$\varphi(\mathbf{x}) = -\frac{f}{2}\mathrm{erf}\left(\frac{\eta}{\sqrt{\dfrac{2\delta}{k(\xi)}}}\right) + const, \tag{1.59}$$

near the flow lines. In (1.59) f means the change of $\varphi(\mathbf{x})$ when $\mathbf{x}$ goes from one cell to another one through the flow line.

The trial function in a dual problem is constructed as

$$\mathbf{j}(\mathbf{x}) = \frac{f}{2\sqrt{\pi\delta}}e^{\frac{-k(\xi)\eta^2}{2\delta}}\left[\sqrt{k(\xi)}\widehat{\xi} - \frac{\eta}{2\sqrt{k(\xi)}}\frac{dk(\xi)}{d\xi}(1 + O(\delta))\widehat{\eta}\right],$$

where $\widehat{\xi}$ and $\widehat{\eta}$ are the unit tangential and normal vectors, respectively (see for details [48]).

We use the standard notation:

$$o(1) \to 0 \text{ as } \delta \to 0, O(1) \le const < \infty \text{ as } \delta \to 0, \tag{1.60}$$

$$\frac{o(\delta)}{\delta} \to 0 \text{ as } \delta \to 0, \frac{O(\delta)}{\delta} \le const < \infty \text{ as } \delta \to 0.$$

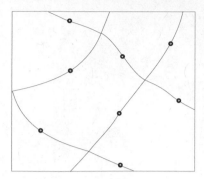

Figure 1.12. *The cells formed by flow lines (solid points indicate the positions of the saddle points).*

Using these trial functions, the authors of [47, 48, 49] found that the effective conductivity R between the adjacent cells of the composite material is expressed in the form

$$R = c(\mathbf{x}_s)\sqrt{\frac{k^+}{k^-}}, \qquad (1.61)$$

where k^+ and k^- are the principal curvatures at the saddle point $\mathbf{x}_s$ of the function $c(\mathbf{x})$.

Writing the balance equation with the conductivities (1.61), the authors of [47, 48, 49] arrive at the finite-dimensional (network) model for the problem (1.58).

It follows from (1.61) that the curvatures k^+ and k^- determine the conductivity of the high-contrast material with continuously distributed properties.

Berriman–Borcea–Papanicolaou approach cannot be applied to media with point-wise characteristics or systems of bodies. It is evident due to the formula (1.61), which relates overall transport properties of Kozlov's model to saddle point for the function (1.33). Evidently, there exists no saddle point for the functions (1.33) and (1.34). Another point of view is that saddle point for the function (1.33) and (1.34) is degenerated and $k^+ = k^- = 0$. As a result, formula (1.61) does not work.

The technical restriction of Berriman–Borcea–Papanicolaou approach is that the trial function was constructed in the neighbor of saddle point and cannot be continued into the whole region. Because Kozlov's model belongs to the class of problems, where the energy localization effect takes place, it brought no problems to network construction in [47, 48, 49]. Note that the energy localization is not a universal property of high-contrast composite materials [176, 185]. The existence (or non-existence) of the effect of energy localization depends not on the property of being densely packed, but on the capacity (this means the shape) of the bodies. The notion of shape of bodies is not compatible with Kozlov's model, which involves only in-point characteristics – curvatures in saddle point(s). It is evident that one can construct various (and very different) functions with the same set of saddle points and the same curvatures in these points. The approach based on the notion of capacity involves the shape of bodies.

1.3. Disordered media with piecewise characteristics and random collections of bodies

In a disordered system, we cannot select a typical element generating the system. For disordered systems, a so-called representative element is introduced [89, 131, 365]. The representative element is determined as the minimal piece of the system or material, which saves the properties of the material in whole. The problem of the representative element (how large is the piece of the system that saves the properties of the system in whole) is not solved completely for now. There are experimental, numerical and theoretical results on the subject, predicting that the representative element must have dimension along one direction not smaller than ten characteristic dimensions of the constitutive elements. Dimension in 20 to 30 characteristic dimensions of the constitutive elements along one direction is accepted as assuredly sufficient for representative element. It means that the representative element must include more than 100 constitutive elements in two-dimensional case and more than 1000 constitutive elements in three-dimensional case.

1.3.1. Disordered and random system of bodies

A disordered system of bodies is naturally related with a graph. In percolation theory [138, 170] often graphs of regular geometry are used for modeling of materials or structures, see Fig. 1.13 (left). The real structures correspond to graphs of general geometry Fig. 1.13 (right).

One can try to calculate effective properties of random high contrast composites using numerical methods by directly solving the partial differential equations with rapidly oscillating coefficients (e.g., using finite elements or integral equations, see [20, 81, 368], or Fourier transforms [115, 241]). Then two questions arise. First, how

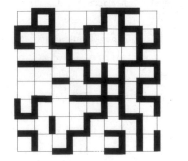 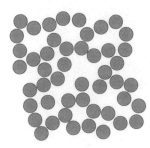

Figure 1.13. *Disordered structures: random square array model (widely used in percolation theory) (left) and a system of randomly distributed disks (right).*

large the dimension of the problem is (since the problem is highly inhomogeneous, the corresponding grid has to have a large number of grid points). Second, how to take into account the high contrast in the properties of the phases. In particular, it is necessary to address the issue of concentrated fluxes between closely spaced bodies, which requires the use of a very fine mesh in the domains of concentration (see also [208] for an analytical study of the concentration phenomenon). It leads to a further increase in the computational complexity of the problem. Finally, it is necessary to repeat this computation many times in order to obtain an average over a large number of random configurations (i.e., to collect statistical information).

1.3.2. Homogenization for materials of random structure

The problem of computation of transport properties of disordered materials and random arrays of bodies attracted attention of numerous investigators. Various methods for computation of transport properties of inhomogeneous materials of random structure were developed, see, e.g., [80, 86, 127, 167, 168, 274, 292, 356, 378].

Solution of the problem of computation of transport properties of disordered materials is related to the theory of homogenization for random structures [54, 55, 157, 258, 286, 313, 393]. In spite of progress in theory of random homogenization, the problem is not solved completely for now, in particular, there is no effective method (except direct numerical computations [20, 80, 241]) for computation of overall properties of random composite with large (0.2 to 0.5) contents of particles. For example, the approach developed by Kozlov [194, 397], assumed construction of a special random field similar to the field $N^k(\mathbf{y})$ introduced by (1.20). It was proved that the field exists but no effective method for construction of the field was proposed. Various contributions to the theory of random homogenization were done in [46, 258, 286, 356, 359, 378]. Note that standard numerical methods, such as finite element method or finite difference method, are not effective for analysis of disordered materials if the number of structural elements is large. This problem is discussed in detail in Chapter 2.

It is accepted that the problem of random homogenization, in spite of efforts of many researchers, is not so fully investigated as the problem of homogenization for periodic structures.

Qualitative analysis of composite materials filled with randomly placed disks was done in [86, 292]. For composites of random structure filled with densely packed disks, computations of overall transport properties were done in [41, 184]. The theory presented in [41, 184] can be considered as a "random" analog of theories developed for composites with a periodic arrays of densely packed disks, see, e.g., [166, 233, 234]. Note that it would be necessary to develop a new approach for analysis of random systems of densely packed disks. The procedures developed for periodic arrays of bodies (like the one presented in Section 1.2.2) essentially use the condition of periodicity. As a result, these procedures cannot be modified for the analysis of random systems of bodies. The general approach, which makes it possible to analyze random systems of various densely packed bodies / particles, will be presented in Chapter 3.

1.3.3. Network approximation of the effective properties of a high-contrast random dispersed composite

Above we described the main steps in the analysis of the Maxwell–Keller problem (we call this the problem of computation of transport properties of a system of regular array of *dense packed* bodies) before the paper [41]. It seemed that the main challenge in the network modeling is constructing asymptotic method for disordered structures, which was based on the usual for material science parameters. One of the necessary conditions is that the method can be applied to particle filled composites – systems with piecewise distribution of material properties. The paper [41] presented a complete analysis of the formulated problem for a disordered system of closely placed circular disks. Solution of the problem was given in the terms of "relative interparticle distance." The paper started with a mathematical analysis of the *general* Maxwell–Keller problem (we call this the problem for *nonperiodic* system of densely packed bodies). In [38], the approach developed in [41] was expanded to medium filled with spheres. The methods of the network approximation developed in [41] were modified for vectorial problems. Note that vectorial problems (elasticity problem, problems in hydrodynamics, etc.) for periodic contrast structures traditionally attracted attention of the researcher [102, 119, 121, 136, 218, 276, 277, 289, 332, 362, 381, 387] (the list is not complete). In [35], the method of network approximation was applied to the computation of effective viscosity of concentrated suspensions. We do not discuss the mentioned problems in detail because it would take us away from the subject of this book. A separate book is planned that will cover the themes mentioned above.

1.4. Capacity of a system of bodies

Since the notion of capacity is widely used in this book, we present some classical definitions related to this topic. We consider a system of N perfectly conducting bodies $\{D_i, \ i = 1, ..., N\}$ and denote the part of R^n not occupied by bodies (the "perforated" R^n) by

$$Q = R^n \backslash \bigcup_{i=1}^{N} D_i. \tag{1.62}$$

The electrostatic problem for this system has the following form ($n = 2, 3$ is dimension of the problem):

$$\Delta\varphi = 0 \text{ in } Q; \tag{1.63}$$

$$\varphi(\mathbf{x}) = t_i \text{ on } \partial D_i, \ i = 1, \ldots, N; \tag{1.64}$$

$$|\varphi(\mathbf{x})| \le const < \infty \text{ as } |\mathbf{x}| \to \infty \ (n = 2), \tag{1.65}$$

$$\varphi(\mathbf{x}) \to 0 \text{ as } |\mathbf{x}| \to \infty \ (n = 3). \tag{1.66}$$

The electrical charge $Q^{(i)}$ of the i-th body belonging to the system under consideration is

$$Q^{(i)} = -\sum_{i=1}^{N} \int_{\partial D_i} \varepsilon \frac{\partial \varphi}{\partial \mathbf{n}}(\mathbf{x}) d\mathbf{x}, \tag{1.67}$$

where ε means the permittivity of the medium outside the bodies D_i $(i = 1, \ldots, N)$.
We denote $\varphi^{(i)}(\mathbf{x})$ solution of (1.63)–(1.66) for the case

$$t_i = 1, \ t_j = 0 \text{ for } i \ne j.$$

Then solution (1.63)–(1.66) can be written as

$$\varphi(\mathbf{x}) = \sum_{i=1}^{N} t_i \varphi^{(i)}(\mathbf{x}). \tag{1.68}$$

Substituting (1.68) into (1.67), we obtain

$$Q^{(i)} = -\sum_{j=1}^{N} t_j \int_{\partial D_i} \varepsilon \frac{\partial \varphi^{(j)}}{\partial \mathbf{n}}(\mathbf{x}) d\mathbf{x}. \tag{1.69}$$

Introducing matrix $\tilde{C}_{ij}$ by the equality

$$\tilde{C}_{ij} = -\int_{\partial D_i} \varepsilon \frac{\partial \varphi^{(j)}}{\partial \mathbf{n}}(\mathbf{x}) d\mathbf{x},$$

we can write the equality (1.69) as

$$Q^{(i)} = \sum_{j=1}^{N} \tilde{C}_{ij} t_j. \tag{1.70}$$

The matrix $\tilde{C}_{ij}$ is symmetric and positively defined matrix.
Formula (1.70) can be written in the form

$$Q^{(i)} = C_{ij} t_i + \sum_{i \ne j} C_{ij}(t_i - t_j). \tag{1.71}$$

Comparison of (1.70) and (1.71) leads to the following relations

$$C_{ii} = \sum_{j=1}^{N} \tilde{C}_{ij},$$

$$C_{ij} = -\tilde{C}_{ij} \text{ for } i \neq j.$$

Using the matrix $\tilde{C}_{ij}$, one can express the energy

$$\mathcal{E} = \frac{\varepsilon}{2} \int_{Q} |\nabla\varphi(\mathbf{x})|^2 d\mathbf{x}$$

of electric field in the form

$$\mathcal{E} = \frac{\varepsilon}{2} \sum_{i=1}^{N} \sum_{j=1}^{N} \tilde{C}_{ij} t_i t_j.$$

Boundary value problem (1.63)–(1.66) is equivalent to the minimization problem

$$I(\varphi) = \frac{1}{2} \int_{Q} |\nabla\varphi(\mathbf{x})|^2 d\mathbf{x} \to \min, \qquad (1.72)$$

considered on functional space

$$V = \Big\{ \varphi(\mathbf{x}) = t_i \text{ on } D_i, i = 1, \ldots, N;$$

$$|\varphi(\mathbf{x})| \leq const < \infty \text{ as } |\mathbf{x}| \to \infty \ (n = 2), \varphi(\mathbf{x}) \to 0 \text{ as } |\mathbf{x}| \to \infty \ (n = 3) \Big\}.$$

The dual formulation of the problem has the form of maximization problem

$$J(\mathbf{v}) = -\frac{1}{2} \int_{Q} \mathbf{v}^2(\mathbf{x}) d\mathbf{x} + \sum_{i=1}^{N} t_i \int_{\partial D_i} \mathbf{v}(\mathbf{x}) \mathbf{n} d\mathbf{x} \to \max, \qquad (1.73)$$

considered on functional space

$$W = \{\mathbf{v}(\mathbf{x}) = (v_1(\mathbf{x}), \ldots, v_n(\mathbf{x})) \in L_2(Q) : \operatorname{div}\mathbf{v} = 0, \mathbf{v}(\mathbf{x}) \to 0 \text{ as } |\mathbf{x}| \to \infty\}.$$

The extremal principles (1.72) and (1.73) were used to develop various methods of approximate solution of the electrostatic problems (including finite element method) and obtain bounds on capacity of bodies of complex geometry and systems of bodies, see, e.g., [298, 304, 361].

We note that for the both cases $n = 2$ and $n = 3$ the following estimate for electric field holds:

$$|\nabla\varphi(\mathbf{x})| \leq \frac{C}{|\mathbf{x}|^2},$$

where $C < \infty$ (see, e.g., [366]).

Asymptotic of capacity of a system of bodies

The bodies can be placed in the space in various ways. The most evident classification of a system of bodies is based on the distance between the bodies. This classification includes two extremal cases:

1) the distances between all bodies are large as compared with the characteristics dimension of the bodies,

2) the distances between the neighbor bodies are small as compared with the characteristics dimension of the bodies.

In the theory of composite materials these cases correspond to dilute (case 1) and densely packed (case 2) composites [246, 315]. Both the problems are asymptotic problems.

The first problem was investigated in detail. Its solution is given, in particular, by the formulas known as Maxwell–Garnett, Clausius–Mossotti and Lorentz–Lorenz formulas. Many refined formulas were derived after the named investigators, see, e.g., [32, 264]. Solution of the problem can be approximated by using a combination of the electrostatic problem (1.63)–(1.66) for a single body, see, e.g., [203].

Our attention will be concentrated on the problems of the second type. The special case of this problem is the asymptotic problem of capacity of a system of closely placed (densely packed) bodies. This is formulated as follows: *investigate asymptotic behavior of a system of bodies under condition that the distances between the neighbor bodies tend to zero.* This will be the problem of our special interest.

Observing extensive literature related to capacity of systems of bodies, the authors have found little number of publications directly related to the problem of asymptotic of capacity of a system of closely placed bodies. Mention papers [141, 364] related to the capacity of two closely placed spheres. The mentioned above publications on the transport properties of regular arrays of closely placed bodies, in fact, densely related to the problem of asymptotic of capacity of a system of closely placed bodies. Nevertheless, direct indication on the relationship of the transport properties of regular arrays of closely placed spheres and asymptotic of capacity of two closely placed spheres was made in [166], only.

Chapter 2

NUMERICAL ANALYSIS OF LOCAL FIELDS IN A SYSTEM OF CLOSELY PLACED BODIES

In this chapter, we present results of numerical analysis of electric fields in and around closely placed bodies. Although numerical analysis provides us with the information about a special system of bodies / particles, the results of numerical analysis may be used as heuristic arguments for developing a general approach to the problem. In other words, numerical analysis can play a role of natural (physical) experiment. Considering closely placed bodies or highly-filled composite materials, we have difficulties with experimental observations of fields because it is necessary to observe and measure fields in small gaps between closely placed bodies / particles. This is a problem of technical nature. But there is another problem of principal value. With modern experimental technique we can observe many types of fields but not all of them. For example, there exist methods for measuring temperature or strains in experimental specimen (although these are not trivial experiments [74]). But, to the best knowledge of the authors, there are no effective methods for observation of distribution of local energy. The numerical analysis allows us to observe numerous physical characteristics which can not be observed in a physical experiment. In the problems related to closely placed bodies or highly-filled composite materials, the problem of the "energy channel" plays very important role. The idea of the energy channel emerged in a paper [166] and it was formulated in [41] as the fundamental property of a system of closely placed systems of disks. Nevertheless, the energy channels were not observed in experiment (the authors assume that the energy channels cannot be observed in experiment for the reasons mentioned above). The first image of the energy channel between disks was obtained using numerical method in [174]. The images of numerical energy channel presented in [174] were in full agreement with the theoretical predictions.

The effect of energy concentration between the closely placed inclusions in high-contrast composite material attracts attention of numerous researchers. The papers

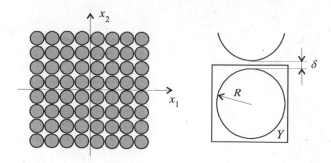

Figure 2.1. *A periodic system of disks (left) and periodicity cell Y (right).*

[49, 119, 166, 234, 183] (the list is not complete) tackle this effect in the analytical context. The paper [183] demonstrates that it is necessary to distinguish the phenomenon of concentration of fields and the phenomenon of localization of fields. In addition, the paper [183] demonstrates that the high-contrast and close placing of bodies are necessary but not sufficient conditions for the energy localization and approximation of the continuum problem with network models. That is why the verification of the effect of energy localization is a problem of both scientific and practical values.

The authors assume that Dykhne's experiment [257] can be considered as indirect proof of existence of the energy localization effect. Argument in the favor of such interpretation of the result of Dykhne's experiment will be presented in Section 3.6.1.

Analyzing specific problem formulated above, the authors use standard finite element method and the general purpose finite element computer programs. The high accuracy solutions presented below were obtained due to the use of a modern computer, small number and simple geometry of bodies and the special mesh refinement.

2.1. Numerical analysis of two-dimensional periodic problem

As was noted in Section 1.2, the periodic arrays of disks (cylinders) and spheres were used to compute transport properties of systems of bodies first. The periodic arrays show themselves as suitable models for analysis of properties of local fields at microscopic level. We start our presentation also with an analysis of periodic problem for a system of closely placed disks. We consider a periodic system of disks $\{D_i, i = 1, \ldots\}$, which do not overlap or touch each other, see Fig. 2.1, and perforated domain

$$Q = R^2 \backslash \bigcup_{i \in N} D_i,$$

which is referred to matrix.

We consider electrostatic equation with respect to the potential $\varphi(\mathbf{x})$ (in the case of planar problem $\mathbf{x} = (x_1, x_2)$, in the spatial problems $\mathbf{x} = (x_1, x_2, x_3)$)

$$\text{div}(\varepsilon(\mathbf{x}) \nabla \varphi) = 0 \text{ in } R^2. \tag{2.1}$$

The dielectric constant $\varepsilon(\mathbf{x})$ is taken in the form

$$\varepsilon(\mathbf{x}) = \begin{cases} \varepsilon_I \text{ in inclusions } D_i, \ i \in N. \\ \varepsilon_m \text{ in matrix } Q. \end{cases}$$

If the problem (2.1) is considered in the classical sence then the conjugation conditions similar to (1.38) should be futher stated. If it is considered in the sence of distributions (see Appendix A), the conjugation conditions are assumed.

The problem involves two parameters: the contrast c of the components of the composite material, which we define as the ratio

$$c = \frac{\varepsilon_I}{\varepsilon_m}, \tag{2.2}$$

and the relative distance δ^* between the disks, which we define as the ratio

$$\delta^* = \frac{\delta}{2R}, \tag{2.3}$$

where δ is the absolute distance between the disks and R is the radius of the disks, see Fig. 2.1.

For a composite filled with densely packed disks possessing high dielectric constant, the contrast is large:

$$c \gg 1, \tag{2.4}$$

and the relative distance is small:

$$\delta^* \ll 1. \tag{2.5}$$

We assume that an overall electric field is applied to the composite and it has the form

$$\mathbf{E} = (0, 1)$$

(i.e., we consider planar problem and $\mathbf{E}$ is a unit field parallel to Ox_2-axis).

The periodicity of the system of bodies makes it possible to construct solution of the original problem (2.1) in R^2 if it would solve the following problem, which determines potential $\phi(\mathbf{x})$ on the periodicity cell $Y = [-1,1]^2$ displayed in Fig. 2.1:

$$\text{div}(\varepsilon(\mathbf{x})\nabla\phi) = 0 \text{ in } Y = [-1,1]^2, \tag{2.6}$$

$$\phi(x_1, \pm 1) = \pm 1,$$

$$\frac{\partial\phi}{\partial\mathbf{n}}(\pm 1, x_2) = 0.$$

We use standard notations

$$\text{div} = \sum_{i=1}^{n} \frac{\partial}{\partial x_i},$$

$$\nabla = \sum_{i=1}^{n} \frac{\partial}{\partial x_i}\mathbf{e}_i,$$

$$\frac{\partial}{\partial\mathbf{n}} = \sum_{i=1}^{n} n_i\frac{\partial}{\partial x_i},$$

where $\mathbf{n}$ is the outward unit normal vector to a corresponding surface.

The relationship between the functions $\varphi(\mathbf{x})$ and $\phi(\mathbf{x})$ is evident. The function $\phi(\mathbf{x})$ is a restriction of the functions $\varphi(\mathbf{x})$ on the periodicity cell Y. In its turn, the potential $\varphi(\mathbf{x})$ in the periodicity cell

$$Y_{ij} = Y + 2i\mathbf{e_1} + 2j\mathbf{e_2},$$

obtained by translation of Y to the vector $2i\mathbf{e_1} + 2j\mathbf{e_2}$ (i, j are integers), is given by the formula

$$\varphi(\mathbf{x}) = \phi(\mathbf{x}) + 2j\mathbf{e_2}.$$

The boundary value problem (2.6) is a special type of the cellular problem (1.20). The problem (2.6) (as well as other problems in this section) was solved numerically using ANSYS finite element software [385]. The linear plane three degrees of freedom finite elements were used (we do not discuss general finite element procedures in detail here, details can be found in [22, 123, 125, 154, 385]). Selected solutions were compared with the solutions presented in [191], which were obtained in another way. The solutions were in full agreement with one another.

The problem (2.6) was solved for the contrast c in the range from 2 to 100000 and the relative distance between the disks δ in the range from 0.05 to 0.005. In the numerical computations, the dielectric constant of the matrix was taken to be of the unit value. In Figs. 2.2 and 2.3 the distribution of the double local energy

$$2E(\mathbf{x}) = \varepsilon(\mathbf{x})|\nabla\phi|^2$$

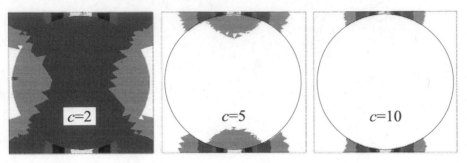

Figure 2.2. *The density of energy distribution over the periodicity cell. Formation of the energy channel when contrast c increases.*

(the density of the local energy multiplied by two) is shown for the relative distance between the disks δ=0.05 and various values of the contrast c (the values of c are indicated in figures).

It is seen from Fig. 2.2 that when the value of contrast is small ($c = 2$), the energy is "spread" over the whole periodicity cell. When the contrast increases then, we observe two processes, see Figs 2.2 and 2.3:

1. the energy "leaves" the inclusion;

2. outside the inclusion, the energy is localized in specific regions, which we call *energy channels*.

It is seen from Fig. 2.2 that formation of the energy channel as a geometrical object starts at the value of the contrast about 10. For the contrast greater than 10, the energy channel is clearly observable and has stable geometry but the numerical value of the local energy in the channel depends on the value of the contrast.

As the contrast increases, the distribution of the energy in the channel becomes stable (tends to a limit distribution, which depends only on δ (see Fig. 2.3 and Table 2.1). In Table 2.1 we present values indicated in the legends for pictures presented in Fig. 2.3 for contrasts $c = 1000$ and $c = 10000$. Figure corresponding to $c = 100000$

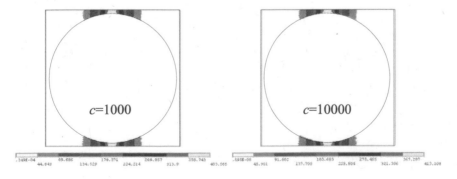

Figure 2.3. *The density of energy distribution over the periodicity cell. Stabilization of the energy channel.*

Table 2.1. *ANSYS Scales Corresponding to Fig. 2.3.*

c	1	2	3	4	5	6	7	8	9	10
1000	0	44.8	89.7	134.5	179.4	224.2	269.2	313.0	358.7	403.6
10000	0	45.9	91.8	137.7	183.6	229.5	275.4	321.3	387.2	413.1
100000	0	45.9	91.8	137.7	183.6	229.5	275.4	321.3	387.2	413.1

is not displayed (it coincides with Fig. 2.3 (right) completely). In accordance with our computations, the distribution of the density of the energy in the channel can be accepted as stable for the contrast $c \geq 1000$. In other words, if contrast is greater 1000, one can accept the inclusions as "perfectly conducting."

Finally, we present pictures displaying distribution of the energy in periodicity cell for square and hexagonal arrays of disks subjected to overall electric field of the form $\mathbf{E} = (1, 1)$ for contrast $c = 1000$.

For both the cases, we observe the completely developed energy necks, connecting the displayed disk with its neighbors, see Fig. 2.4.

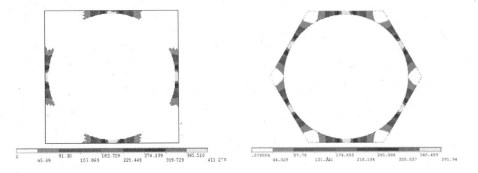

Figure 2.4. *The energy channels in a composite with square (left) and hexagonal (right) packed circular inclusions, $\delta^* = 0.05$.*

2.2. Numerical analysis of three-dimensional periodic problem

In the previous section, we presented results devoted to numerical analysis of transport problem in high-contrast composite. We also mention papers [80, 137] devoted to the similar problems. Most numerical results were obtained for two-dimensional problems. There is a reason to believe that most of the results obtained for two-dimensional problems are valid for three-dimensional case. But no experimental or numerical verification of this hypothesis was done, to the best knowledge of the authors.

Now, we present a result of numerical computation, demonstrating energy channels in three-dimensional problem. We consider simple cubic array of spheres, see

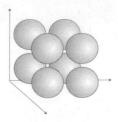

Figure 2.5. *A fragment of simple cubic array of spheres.*

Fig. 2.5, which is three dimensional analog of planar array displayed in Fig. 2.1.

The numerically solved problem is the boundary-value problem for Laplace equation for $1/4$ of periodicity cell. The periodicity cell is cube $Y = [-1,1]^3$ with spherical inclusion

$$D_0 = \{|\mathbf{x}| \le R\} < 1$$

in its center. The boundary conditions correspond to application of unit overall electric field in all three directions, i.e., $\mathbf{E} = (1,1,1)$.

Then, potential t_{ijk} of the sphere

$$D_{ijk} = D_0 + 2i\mathbf{e_1} + 2j\mathbf{e_2} + 2k\mathbf{e_3}$$

obtained by translation of the sphere D_0 to the vector $2i\mathbf{e_1} + 2j\mathbf{e_2} + 2k\mathbf{e_3}$ (i, j, k are integers) is

$$t_{ijk} = 2(i + j + k).$$

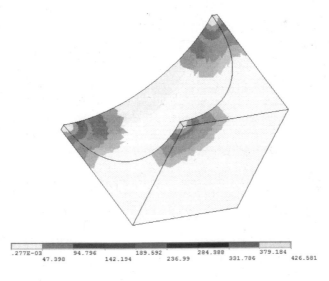

.277E-03 94.796 189.592 284.388 379.184
47.398 142.194 236.99 331.786 426.581

Figure 2.6. *Distribution of double local energy* $2E = \varepsilon(\mathbf{x})|\nabla(\mathbf{x})\phi|^2$ *in three-dimensional periodicity cell for simple cubic array of spheres.*

The relative interparticle distance in our numerical computation is equal to 0.05 (the radius of inclusion $R = 0.95$ and contrast $c = 1000$).

The result of our numerical computation is presented in Fig. 2.6. The energy channels are clearly seen in the figure.

Prospective of traditional numerical methods as applied to analysis of high-contrast highly filled composites

To carry out accurate solution of the problem (2.6) for one disk, we used 5000 finite elements and adopted finite element mesh for the geometry of the problem (about 2/3 of the finite elements were concentrated in the thin region where energy channels arise). So, we spend large computational resources solving a problem for one disc. Analyzing a system consisting of 100 disks, we would arrive at a problem involving about million finite elements. This problem can be solved with a computer, but it is already near the limit of capability for modern computers. A similar three-dimensional problem cannot be solved with a modern computer. There were attempts to develop a method for solving contrast problems using Fourier transforms [241, 242] and series technique [254]. The authors doubt that such methods can be effective for analysis of high-contrast densely packed system of bodies / particles. This doubt is based on the fact that the mentioned methods do not take into account the energy localization phenomenon. The energy localization phenomenon is discussed in detail in Section 2.3. In few words, the energy in high-contrast densely packed system of bodies / particles is localized in small domains between the bodies / particles. The methods mentioned above are not adopted for a processing of specific small domains, where energy density is very large. The methods mentioned process the whole domain. As a result, as applied to the problem under consideration, these methods spend computational resources to process domains which do not play a key role (and possibly play no role) in the transport process and (probably) do not allocate sufficient resources to process small domains with high accuracy, which determine the transport process. This conclusion was supported by information obtained from Prof. V.V. Mityushev [252] about convergence of series for closely placed disks. The authors were informed [252] that series technique, which works well for systems without "near touching" disks, demonstrates poor convergence in the presence of near touching disks. Note that series technique is most stable with respect to near touching situations because it can (at least theoretically) manipulate even with δ–functions.

2.3. The energy concentration and energy localization phenomena

Observing the results of our numerical computations, see Figs. 2.3, 2.4, and 2.6, we found that the local energy takes large values (as large as 400 for the example considered above, see Table 2.1) in the energy channels. At the same time, it takes

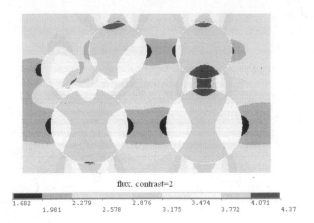

flux, contrast=2

| 1.682 | | 2.279 | | 2.876 | | 3.474 | | 4.071 | |
| | 1.981 | | 2.578 | | 3.175 | | 3.772 | | 4.37 |

Figure 2.7. *Numerically computed flux* $|\nabla\varphi(\mathbf{x})|$ *in a non-contrast composite (the contrast $c = 2$). Concentration of field can be observed.*

small values (as small as 10^{-3}) outside the energy channels. It leads us to the conclusion that it is necessary to distinguish the effect of concentration and the effect of localization of fields when we analyze a system of closely placed bodies.

The well-known effect of concentration of field in composite material is the result of inhomogeneity of the material. Talking about the "concentration" phenomenon we assume that the field takes relatively large value in one region but not small outside this region. Fig. 2.7 illustrates the effect of the energy concentration in particle filled composite subjected to overall field $\mathbf{E} = (0, 1)$. In the example, contrast $c = 2$ and relative distances vary from 0.03 to 0.3.

The elasticity theory (see, e.g., [193, 216]) provides us with another example of strain–stress field concentrations. The strain–stress field in the elasticity theory can be infinite in the tip of a crack, but the elastic energy is distributed over the whole region. In other words, the energies stored in various regions of the same characteristic dimension (even in the region containing the tip of a crack) have similar values. So, the defining property of the effect of concentration of a field is that the absolute value of the field is large in a specific region, but integral values of the field in the region and outside the region have similar values.

In our computations, for small δ^* and not very large c, we observe the picture (see picture "$c = 2$" in Fig. 2.2) displaying the relatively uniformly distributed energy field. The increasing of c transforms the distribution of the energy field, see Fig. 2.2. For large c, we observe the qualitative transformation of the energy distribution as compared with the energy distribution for not large c (compare pictures "$c = 2$" in Fig. 2.2 and "$c = 1000$", "$c = 10000$" in Fig. 2.3). In pictures "$c = 1000$" and "$c = 10000$", we see that the almost all energy is collected in specific regions (in the energy channels) and negligible value of the energy remains outside these specific regions.

Thus, the case we are dealing with differs from the phenomenon usually referred

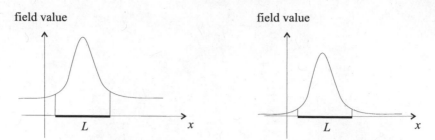

Figure 2.8. *The typical distributions corresponding to concentration and localization phenomena: left – field concentrates in the region L; right – field localizes in the region L.*

to as concentration. In order to categorize this case, we introduce the notion of localization (more exactly, asymptotic localization) of field.

Definition 3. *The gathering of a field (not absolutely all the field but most part of the field) in some specific regions is referred to as the effect of localization of field.*

Definition 4. *If an effect of localization of field takes place under condition $\delta \to 0$, it is referred to as the effect of asymptotic localization of field.*

The difference between the concentration and localization phenomena is illustrated by Fig. 2.8. For the function displayed in Fig. 2.8 (left) the area of the curvilinear trapezoid with the base L and the area of subgraph outside this trapezoid have similar values. In our terms, this is the case of concentration in the region L. For the function displayed in Fig. 2.8 (right) the area of subgraph outside the curvilinear trapezoid with the base L is sufficiently small as compared with the area of the trapezoid. In our terms, this is the case of localization in the region L.

In the numerical examples presented above, the regions of the energy localization have the form of energy channels: a domain situated between the neighboring closely placed disks and spheres.

In the definition above, although we tolerate for the most part, the field that gathers in the region of localization, the field outside the region of localization does not vanish. Also, the stronger type of localization happens when most of the energy gathers in the region of localization and, in addition, the field leaves other regions.

Definition 5. *The effect of asymptotic localization for which field tends to zero as $\delta \to 0$ outside the region of localization is referred to as the effect of strong asymptotic localization of field. The localization, which is not strong is referred to as the effect of relative asymptotic localization.*

In the case under consideration, we meet a combination of strong and relative localization for the local energy. It is seen from Fig. 2.9 that the local energy not only increases in the channels but it leaves the regions outside the channels. But it does not leave these regions absolutely (i.e., it does not tend to zero in

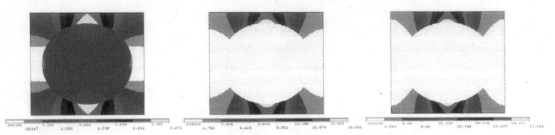

Figure 2.9. *The evolution of the local energy distribution when contrast c increases (from left to right $c = 2$, $c = 10$, and $c = 100$).*

the region outside the channels in the matrix), although its value decreases when contrast c increases. ANSYS color scale divides the interval of all displayed values in ten subintervals [385]. So, the minimum color in the figures corresponds to values less than 10% of the maximum value. We see that this zone of lower 10% increases. At the same time, the local energy leaves the high-contrast inclusion absolutely. In order to observe this effect visually, we present in Fig. 2.10 the values of cut-off function $\min\{\varepsilon(\mathbf{x})|\nabla\phi(\mathbf{x})|^2, C\}$. A similar situation takes place when the interparticle distance δ decreases, see Fig. 2.11.

Table 2.2. *The Maximal Local Energy Value (Fig. 2.9).*

contrast c	2	10	100
the maximal local energy	3.073	15.252	27.266

We summarize our observations as follows: as $\delta \to 0$,

1. the local energy absolutely leaves the inclusions possessing high transport property (dielectric constant, conductivity, etc.),

2. in the matrix relative asymptotic localization of the energy is observed in the gaps between the neighbor inclusions (we call such gaps the energy channels).

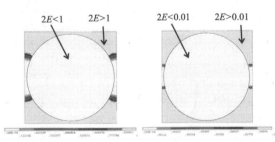

Figure 2.10. *The values of the cut-off function for $C = 1$ (left) and $C = 0.01$ (right). In the zone, where doubled energy $2E = \varepsilon(\mathbf{x})|\nabla\phi(\mathbf{x})|^2 < C$, the local energy vanishes in the inclusion and does not vanish outside inclusion. The color changes from white to black when value increases.*

Estimate of the width of the energy channel

The energy channels for disks are displayed in Fig. 2.3 and Fig. 2.4. For spheres, the energy channels are displayed in Fig. 2.6. These figures as well as the other ones presented in [174, 177] lead to the conclusion that for disks, circular cylinders and spheres of radius R the width of the energy channel does not exceed $R/2$.

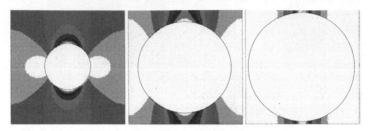

Figure 2.11. *The evolution of the local energy distribution when δ decreases (from left to right $\delta^* = 0.2$, $\delta^* = 0.8$, and $\delta^* = 0.95$).*

Table 2.3. *The Maximal Local Energy Value (Fig. 2.11).*

contrast δ	0.2	0.80	0.95
the maximal local energy in the channel	5.209	27.266	299.611

2.4. Which physical field demonstrates localization most strongly?

In accordance with the results presented in the previous sections, the localization effect arises in a system of closely placed bodies. Considering the electrostatic problem (as well as the other transport problems) we meet various fields: electric potential, electric field, dielectric displacement and local energy. Not all of these fields demonstrate the localization property. As far as the fields demonstrating the localization property, they demonstrate it more or less strongly. To clarify this problem, we solved the problem (2.6) for the periodic system of disks displayed in Fig. 2.1 and computed electric field, dielectric displacement and local energy. The distributions of the fields are displayed in Fig. 2.12.

The following conclusions can be made on the basis of analysis of Fig. 2.12.

1) The dielectric displacement $|\varepsilon(\mathbf{x})\nabla\phi(\mathbf{x})|$ does not demonstrate the localization property. Note, that although the normal component of dielectric displacement $\varepsilon(\mathbf{x})\dfrac{\partial\phi(\mathbf{x})}{\partial\mathbf{n}}$ has to be continuous, the absolute value of the dielectric displacement $|\varepsilon(\mathbf{x})\nabla\phi(\mathbf{x})|$ could not share this property.

2) The electric field $\nabla\phi(\mathbf{x})$ demonstrates the localization property, see also Fig. 2.2.

3) The field demonstrating the localization property most strongly is the local energy $\frac{1}{2}\varepsilon(\mathbf{x})|\nabla\phi(\mathbf{x})|^2$.

2.5. Numerical analysis of potential of bodies in a system of closely placed bodies with finite element method and network model

The numerical analysis of the periodic problem presented in the previous section verifies the existence of the energy channels and estimates the values of contrast and relative distance between disks when the energy concentration phenomenon in high-contrast dense packed composite transforms into the energy localization effect.

The phenomenon of the energy localization laid the foundation of network models for high-contrast dense packed composite materials [41, 49, 183]. We call the network model a finite-dimensional model, which approximate (in any sense) the original continuum model. Network models are well-known and widely used in natural science for a long time (see, e.g., [50, 315]). Nevertheless, the problems of rigorous computation of parameters of network models, justifying the approximation and even the criterion of existence of network approximation have not been examined carefully. These problems are treated in this book.

2.5.1. Analysis of potential of bodies belonging to an alive net

In this section, we consider the problem of approximation of the potentials of bodies determined from the continuum electrostatic problem

$$\operatorname{div}(\varepsilon(\mathbf{x})\nabla\varphi) = 0 \text{ in } Q = [-a, a]^2, \tag{2.7}$$

$$\varphi(x_1, \pm a) = \pm 1,$$

$$\frac{\partial\varphi}{\partial\mathbf{n}}(\pm a, x_2) = 0$$

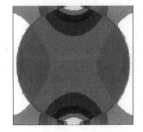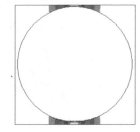

Figure 2.12. *From left to right: Distributions of electric field $|\nabla\phi(\mathbf{x})|$, dielectric displacement $\varepsilon(\mathbf{x})|\nabla\phi(\mathbf{x})|$, and the double local energy $2E = \varepsilon(\mathbf{x})|\nabla\phi(\mathbf{x})|^2$ over the periodicity cell Y for periodic system of disks.*

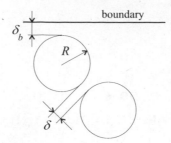

Figure 2.13. *Definition of the geometrical characteristics.*

with the potentials determined from the network model. We consider this problem for alive net (i.e., for a net containing bodies placed near the boundaries $x_2 = \pm a$ subjected to a nonzero difference of potentials).

The approximation of the potentials of the bodies determined from the continuum problem (2.7) with the potentials determined from corresponding network problem (if it exists) will be an additional confirmation of existence of the energy localization effect (it is evident that the energy concentration, without the energy localization, is not sufficient for the network approximation). In addition, it is interesting to receive information about distribution of potential in high-conducting bodies. In the previous publications on the subject, the constant values of potentials in the bodies were assumed. The potentials, no doubt, are constant in perfectly conducting bodies. In contrast (but not perfectly conducting) bodies, the potentials are not equal to constants exactly. The numerical analysis makes it possible to estimate the variation of potential inside specific bodies.

It is known [41, 315] that for domain filled with circular disks the network model exists and the total energy of the network model approximates the total energy of the original problem. The network model corresponding to the problem (2.7) consists of the Kirchhoff-type equations

$$\sum_{j \in N_i} C_{ij}^{(2)} (t_i - t_j) = 0, \ i = 1, ..., N, \tag{2.8}$$

for the interior nodes corresponding to the disks placed inside the domain Q, see Figs 2.14, 2.15, and the conditions

$$t_i = -1, \quad t_i = 1 \tag{2.9}$$

for the nodes belonging to the top and the bottom of the domain Q, where potential is applied, N is the total number of the disks, N_i means the indices of the disks adjacent to the i-th disk (the disks, which are the neighbors of the i-th disk).

In the network model (2.8) and (2.9) t_i implies the potential of the i-th node. The coefficient $C_{ij}^{(2)}$ is taken to be equal to the capacity of the neighboring pair (i-th and j-th) of disks in R^2 (the reason for this choice is explained in Chapter

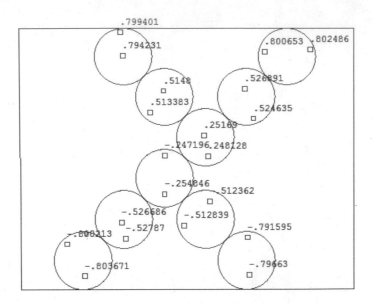

Figure 2.14. *The system of disks with values of potentials determined from the so-lution of continuum problem. The number inside the disks indicates the value of the numerically computed potential $\varphi(\mathbf{x})$ in the place marked by small square (the picture is the copy of computer screen).*

3). For the pair disk–disk it is equal to $\pi\sqrt{\dfrac{R}{\delta}}$ [166, 340]. For the disk near the boundary (the top or bottom boundary of the domain $[-a, a]^2$) it is equal to the capacity of the pair disk–plane $\pi\sqrt{\dfrac{2R}{\delta_b}}$. Here R denotes the radius of the disks (in the numerical computations in this section $R = 1$); δ is the distance between disks, δ_b is the distance between near-boundary disk and the boundary, see Fig. 2.13. The structure of composite material is displayed in Fig. 2.14.

The corresponding network is displayed in Fig. 2.15. The accuracy of approximation of the problem (2.7) with the problem (2.8), (2.9) is described through the differences of potentials of the disks and the nodes.

We denote by $\{t_i^{cont}(\mathbf{x}), \ i = 1, ..., N\}$ the potentials of the i-th disk determined from solution of the continuum problem (2.7) and by $\{t_i^{net}, \ i = 1, ..., N\}$ solution of the network problem (2.8), (2.9). It follows in no way that

$$t_i^{cont}(\mathbf{x}) = t_i^{net},$$

because $t_i^{cont}(\mathbf{x})$ and t_i^{net} are determined from solution of different problems (really, very different problem (2.7) and problem (2.8) and (2.9)).

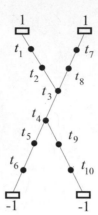

Figure 2.15. *The network model corresponding to the continuum structure displayed in Fig. 2.14.*

We introduce the relative accuracy of the network approximation as

$$\varepsilon_r = \max_{i=1,\ldots,N} \left\| \frac{t_i^{cont}(\mathbf{x}) - t_i^{net}}{t_i^{cont}} \right\|_{C(D_i)} \tag{2.10}$$

and the absolute accuracy of the network approximation as

$$\varepsilon_a = \max_{i=1,\ldots,N} \left\| t_i^{cont}(\mathbf{x}) - t_i^{net} \right\|_{C(D_i)}. \tag{2.11}$$

The problems (2.7) and (2.8) and (2.9) were solved for the contrast in the range from 10^3 to 10^6 and the relative distance between the disks in the range from 0.05 to 0.001.

Notion of the potential of disk

In Fig. 2.14, numerically computed values of the potential are shown for various points inside the disks. The number near a small square means the potential in the center of the corresponding square. The potentials were computed for the contrast $c = 10000$ and the relative distance between the disks is about 0.01. It is seen that the potential inside a specific disk is close to a constant value. It makes possible to "legalize" the notion of the "potential of a disk" t_i^{cont} as the characteristic value of $t_i^{cont}(\mathbf{x})$ for $\mathbf{x} \in D_i$ (i.e. $t_i^{cont}(\mathbf{x}) \approx t_i^{cont}$ in D_i). We note that the constancy of the potential inside a specific disk is the basic hypothesis of network modeling approaches [41, 49, 183, 315]. Introducing the notion of the potential of a disk, we arrive at the question about the relationship of potentials of the nodes $\{t_i^{net}, i = 1, \ldots, N\}$ and potentials of the disks $\{t_i^{cont}, i = 1, \ldots, N\}$.

In Tables 2.4 to 2.6 values of t_i^{cont} and t_i^{net} computed for various values of relative distances δ and δ_b and contrast c are presented. From Tables 2.4 to 2.6, it is seen that the accuracy becomes smaller when the distances δ^* and δ_b^* decrease. Thus,

Table 2.4. *The Relative Distances:* $\delta^*=0.01$, $\delta_b^*=0.01$. *Accuracies:* $\varepsilon_r = 0.08$, $\varepsilon_a = 0.03$.

i	1	2	3	4	5	6	7	8	9	10
t_i^{net}	0.81	0.54	0.27	-0.27	-0.54	-0.81	-0.81	-0.54	-0.54	-0.81
t_i^{cont}, $c=10^3$	0.80	0.51	0.25	-0.25	-0.53	-0.80	-0.80	-0.52	-0.51	-0.79
t_i^{cont}, $c=10^6$	0.79	0.51	0.25	-0.25	-0.52	-0.80	-0.80	-0.52	-0.51	-0.80

Table 2.5. *The Relative Distances:* $\delta^*=0.0054$, $\delta_b^*=0.005$. *Accuracies:* $\varepsilon_r = 0.074$, $\varepsilon_a = 0.02$.

i	1	2	3	4	5	6	7	8	9	10
t_i^{net}	0.80	0.52	0.25	-0.26	-0.53	-0.81	-0.81	-0.53	-0.52	-0.80
t_i^{cont}, $c=10^3$	0.80	0.54	0.27	-0.27	-0.54	-0.80	-0.80	-0.54	-0.54	-0.80
t_i^{cont}, $c=10^6$	0.80	0.52	0.26	-0.26	-0.53	-0.81	-0.81	-0.53	-0.52	-0.80

one can make the conclusion that solution of the network problem (2.8) and (2.9) approximates solution of the continuum problem (2.7) in the disks.

Note that the quantities (2.10) and (2.11) tend to zero slow enough. The absolute accuracy is about 0.05 for the relative distances of about 0.03; about 0.03 for the relative distances of about 0.01, about 0.02 for the relative distances about 0.005, and about 0.01 for the relative distances of about 0.001.

The relative distance between disks in the range between 0.03 and 0.01 can be realized in practice. It corresponds to very dense packed disks. The relative distances in the range between 0.005 and 0.001 are not realized in practice (at least the author has no information about real composite materials with so closely placed disks).

Table 2.6. *The Relative Distances:* $\delta^*=0.00097$, $\delta_b^*=0.001$. *Accuracies:* $\varepsilon_r = 0.037$, $\varepsilon_a = 0.01$.

i	1	2	3	4	5	6	7	8	9	10
t_i^{net}	0.80	0.53	0.26	-0.26	-0.53	-0.81	-0.80	-0.53	-0.53	-0.80
t_i^{cont}, $c=10^4$	0.81	0.54	0.27	-0.27	-0.54	-0.81	-0.81	-0.54	-0.54	-0.81

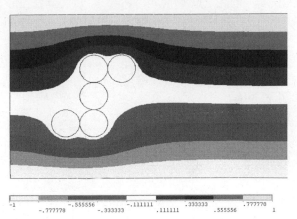

Figure 2.16. *Distribution of the potential* $\varphi(\mathbf{x})$ *for an insulated chain of disks.*

2.5.2. Analysis of potential of bodies belonging to an insulated net

In this section, we present results of numerical computations for a system of disks containing an insulated net (i.e., a net which does not contain bodies placed near the boundaries $x_2 = \pm a$ subjected to a nonzero difference of potentials or near disks belonging in an alive net, see Fig. 2.16).

Solution of the equations (2.8) is $t_i^{net} = const$ for the disks belonging in the insulated net. We present the solution of the boundary-value problem (2.7) for an insulated chain formed of five disks. The neighbor disks in the chain are placed closely, at the same time the chain is placed at a relatively long distance from the top and bottom boundaries, which are subjected to voltage.

In Fig. 2.16 one can see the distribution of the potential in the domain containing an insulated chain. Potentials $+1$ to -1 are applied to the sides $x_2 = a$ or $x_2 = -a$ of the domain. It is seen that potentials remain the same value (with the accuracy of the scale bar) for all the disks in the insulated chain.

2.5.3. Conjecture of potential approximation for non-regular array of bodies

Our computations demonstrate that potentials in contrast disks have small (about 0.1% for the case displayed in Fig. 2.14) variation inside particular disks. The assumption of constant potential in contrast bodies (disks, spheres, etc.) is accepted for a long time. We only confirm the validity of that hypothesis with our numerical computations. The new problem is the relationship of the potentials of the disks determined from the solution of the continuum problem and corresponding network problem. Earlier, the equality $t_i^{cont} = t_i^{net}$ was accepted as an obvious property of a system of bodies with no critical discussion.

We underline that this problem arises as a nontrivial problem in the case of non-regular distribution of disks. This problem also arises as a nontrivial problem in the case of non-uniform geometry of the bodies (see, e.g., Fig. 2.17) and it does arise for periodic system of disks, spheres and similar symmetric bodies. When a periodic array of symmetric bodies is considered, one can determine the potential of the bodies without solution of any problem (continuum or network) but using arguments based on the symmetry and periodicity of the problem only.

Relying on the results of the numerical analysis presented in this section, we can conclude that solution of the problem (2.8) and (2.9) approximates solution of the problem (2.7) in the disks for values of the relative distances 0.03 or less and for values of contrast 1000 or greater. For the values of the relative distances (0.01 − 0.03) realized in practice the absolute accuracy of approximation is about 0.03 and the relative accuracy of approximation is about 0.08.

These results lead us to the hypothesis that potentials determined from the original continuum problem can be approximated by potentials of the nodes of network model. This conjecture was formulated in general form in [174] and proved for a disordered system of perfectly conducting circular disks in [186]. Later it was found [176] that the network approximation is densely related to the notion of capacity and the existence of network approximation is governed by the condition (3.1) presented in Section 3.1.1 in Chapter 3. Using the technique based on the idea of Tamm shielding effect, we will prove the theorem about the potential approximation for bodies of general geometry in Chapter 4.

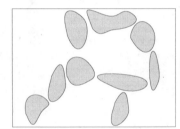

Figure 2.17. *A system of bodies of non-uniform geometry.*

2.6. Energy channels in nonperiodic systems of disks

The basic assumption of the network modeling approach [41, 176] is the effect of the energy localization (previously not clearly separated from the phenomenon of energy concentration) in the gaps between the neighbor inclusions in high-contrast dense packed composites. In order to observe this phenomenon in a disordered system of bodies, the distribution of the density of energy was numerically computed and displayed on the computer screen.

For technical reasons this multicolor picture is not suitable for reproduction. We give a verbal description of the colored picture. On the computer screen, it was clearly seen that almost all the domain outside the gap between inclusions is filled with the color corresponding to near zero value and only small domains between the disks are filled with the colors corresponding to large (from 10 to 1000) values.

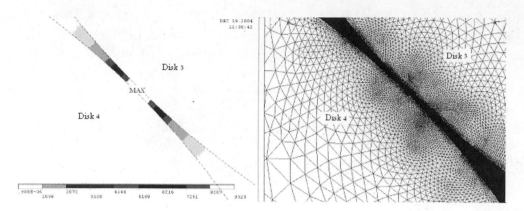

Figure 2.18. *Distribution of the density of the energy in the channel between two disks (left) and finite elements mesh in the channel between the disks (right). The numbers of the disks are indicated in Fig. 2.15.*

Note that the overall density of the energy has the order of unity. The black and white picture is displayed in Fig. 2.18.

Fig. 2.18 (left) displays the distribution of the energy in the channel between disks 4 and 5 (see numbers of the disks in Fig. 2.15). The energy channel is clearly seen in the figure.

In Fig. 2.18 (right) the finite element mesh in the disks and between the disks is shown. The dark color between the disks indicates that the finite elements between the disks are very small (the dark color indicates finite element boundaries). The number of the finite elements between the disks is estimated as 4000. Such fine mesh was taken to make an accurate computation.

Chapter 3

ASYMPTOTIC BEHAVIOR OF CAPACITY OF A SYSTEM OF CLOSELY PLACED BODIES. TAMM SHIELDING. NETWORK APPROXIMATION

This chapter is devoted to the mathematical analysis of the problem formulated and numerically analyzed in the previous chapter. We consider a collection of disordered perfectly conducting bodies and investigate asymptotic behavior of capacity of this system under the condition where characteristic distances δ between the neighbor bodies are small.

3.1. Problem of capacity of a system of bodies

The problem of capacity of many bodies arose simultaneously with the notion of capacity and it is presented in many textbooks on electrostatics, see, e.g., [203, 340, 353, 354]. Solution of the problem strongly depends on the shapes, positions and number of the bodies. Solution of the problem for two infinite planes can be found in the secondary school textbooks, solutions for spheres and cylinders (disks) can be found in university textbooks. Analysis of capacity of periodic system of disks or spheres or capacity of a system of bodies of complex geometry is the subject matter of research papers. In recent decades, a significant progress in the analysis of electrostatic problems and computation of capacity of systems of bodies was related to the progress in numerical methods (see, e.g., [247, 316, 322, 363]) and the increasing power of computers. Nevertheless, for disordered three-dimensional systems of many bodies, the electrostatics problem cannot be solved with standard computational software even by using a powerful modern computer.

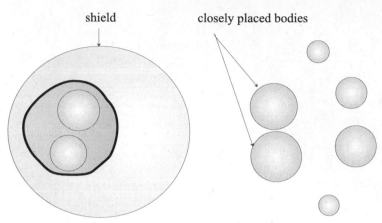

Figure 3.1. *Shielding: (left) the classical shielding, (right) the shielding as described by Tamm.*

In this section, we present a brief exposition of the basic analysis of asymptotic behavior of capacity of a system of closely placed bodies developed in [176, 184, 185].

3.1.1. Tamm shielding effect

Analyzing the problem of capacity of many closely placed bodies, the authors found that it is closely related to the effect described by the Soviet physicist, Nobel Prize Laureate I.E. Tamm in his book [353], see [354] for English translation of the book. The first edition of the book was published in 1927. We call this effect Tamm shielding effect. We found that Tamm shielding effect is the keystone phenomenon for transport processes in structures with piecewise material characteristics [41], i.e., for systems of bodies, particle filled composites, powders, etc.

In the book [353], Tamm discussed two sorts of shielding. The first is the classical shielding when one perfectly conducting body – the shield, surrounds other bodies, see Fig. 3.1 (left). The second (this is referred as Tamm shielding effect in this book) is described as the independence (more exactly, weak dependence, see Section 3.6.2 for the original Tamm formulation) of capacity of two closely placed bodies on other bodies situated not close to them, see Fig. 3.1 (right). In other words, Tamm shielding decomposes the original problem (i.e. "separate" an interaction of neighbor closely placed bodies from other interactions).

At first sight, it seems that on the basis of the notion of Tamm shielding effect it is possible to develop a complete asymptotic theory of capacity of a system of many bodies. This conclusion is not completely correct. Investigating the problem, the authors found that Tamm shielding effect, as it was originally formulated by I.E. Tamm [353, 354], does not take place for every system of closely placed bodies [183, 185]. The authors found the necessary and sufficient condition of Tamm shielding effect, which has the following form. We denote $C_{ij}^{(2)}$ the capacity of a pair of bodies

(i-th and j-th) in R^n ($n = 2, 3$). Tamm shielding effect takes place if condition

$$C_{ij}^{(2)} \to \infty \text{ as } \delta \to 0 \qquad (3.1)$$

is satisfied.

If condition (3.1) takes place for neighbor bodies belonging to a system of bodies, then some characteristics of the original continuum problem can be approximated by the corresponding characteristics determined from finite dimensional (network) problem. The formal definitions and theorems are presented below.

3.1.2. Two-scale geometry of the problem

The high density of packing of the bodies leads to appearance of two characteristic dimensions (two scales of geometrical nature) in the problem. One (small) dimension is the size of the domain (called the energy channel in this book) between the neighbor bodies. Another (not small) dimension is the size of domains outside the channels, see Fig. 3.2. Usually, the existence of two scales make the problem more complex as compared with a one-scale problem. But if a method is developed which successfully accounts for the specifics of the problem, a two-scale problem can be made simple as compared with the original problem. Usually, the specific method is an appropriate asymptotic method, which accounts for the role of the scales through construction of solution in a special form. Examples of application of asymptotic methods to analysis of partial differential equations in thin domains can be found in the papers [179, 180, 228, 229, 337, 338]. The papers mentioned were devoted to transport and elastic properties of thin domains, flux in thin channels and similar problems. The problem of interaction of many bodies was not analyzed in the mentioned papers. The results and methods developed in the mentioned papers cannot be simply expanded to the problem considered in this chapter.

3.1.3. The physical phenomena determining the asymptotic behavior of capacity of a system of bodies

As was noted, after the paper [183] it became clear that the existence (as well as nonexistence) of the phenomena traditionally related to the system of closely placed bodies (such as Keller asymptotic, Tamm shielding, network approximation) is controlled by the condition (3.1).

The existence of condition (3.1) means that the energy localization in the channels between neighbor bodies is not a universal phenomenon, but it is a conditional effect, which takes place only if condition (3.1) is satisfied. Condition (3.1) depends on the shape of bodies and, in addition, it is dimension sensitive. Spheres and disks satisfy the condition (3.1). It explains why the methods based on direct construction of solution were so effective for spherical and circular inclusions.

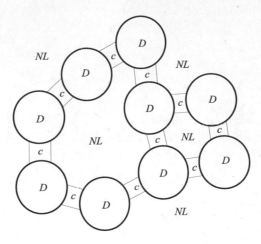

Figure 3.2. *The basic objects of asymptotic analysis of the capacity of many bodies problem: D – bodies, c – channels between the neighbor bodies, NL – domains outside the bodies and the necks.*

What is the *universal* phenomenon in the problem of asymptotic capacity of many bodies (if it exists)? To answer this question, we remember that the electrostatic problem under consideration is the boundary-value problem for Laplace equation. In a fixed domain, Schauder-type estimate takes place [321] for solution of the boundary-value problem for Laplace equation. Below, we will demonstrate that Schauder estimate (in the form of Ladizenskay–Ural'tseva theorem [200] modified for a system of moving domains) takes place for regions outside the channels between the neighbor bodies. We call these zones NL (No Localization) zones. The existence of NL zones, in which localization of energy is impossible in any case, is the unique universal property of a system of bodies.

Condition (3.1) controls both global and local properties of a system of bodies. Being satisfied, it leads to infinite increasing of the total energy of the system of bodies. Simultaneously, it leads to *localization* of the energy.

Under condition (3.1), the energy is localized in the channels between the neighbor bodies. But there is a great variety of channels between two neighbor bodies. We solved this problem of the variety of the channels by establishing the fact that the energies accumulated by various channels between two neighbor bodies do not depend asymptotically on the specific channel. We demonstrated that energies accumulated by various channels between two neighbor bodies are asymptotically equivalent to the energy of these two bodies in whole R^n.

The phenomenon of the energy localization arises in the system of highly conducting closely placed bodies if and only if additional condition (3.1) is satisfied. We give brief explanation on the term "additional" above. The investigations in the field of arrays of densely packed bodies, which were carried out before [183], may give the impression that control parameter of the problem is δ (this parameter is

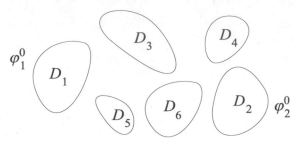

Figure 3.3. *A system of bodies.*

called interparticle distance after [41]). This conclusion would be incorrect. The interparticle distance does control transport in a system of closely placed bodies only under the condition (3.1). The interparticle distance controls the transport in a system of smooth geometrically isotropic bodies like disks or spheres, but not in the case of arbitrary geometry of the bodies. In the general case, it is necessary to take into account individual geometry of bodies. Various characteristics exist related to the geometry of a body. In the case under consideration, the capacity is the required characteristic. Note that our approach solves the well-known problem of dependence of capacity and transport properties of a system of closely placed bodies on the geometry of the bodies automatically.

If the condition (3.1) is realized, the energy of the ensemble of bodies asymptotically decomposes. In the case when only two bodies in the system are closely placed, this decomposition leads to the phenomenon of Tamm shielding in the form described originally by I.E. Tamm [353, 354]. If many bodies in the system are closely placed, the phenomenon of decomposition leads to the network approximation for continuum problem.

In the following sections, we present detailed analysis of the problem of asymptotic behavior of capacity of a system of many closely placed bodies.

3.2. Formulation of the problem and definitions

This section contains formulation of the problem (in the differential form and in the form of extremal problems for integral functionals) and definitions characterizing the topology of networks.

In Chapters 3 and 4, for the sake of brevity, we assume the local transport coefficient (permittivity, conductivity, etc.) $\varepsilon = 1$.

3.2.1. Formulation of the problem

We consider a set of nonintersecting and not touching convex bodies (domains) $\{D_i, i = 1, \ldots, N\} \subset R^n$ (n=2, 3 denotes the dimension of the problem under consideration) with piecewise smooth boundaries, see Fig. 3.3.

We denote the part of R^n not occupied by the bodies (the perforated R^n) by

$$Q = R^n \setminus \bigcup_{i=1}^{N} D_i. \qquad (3.2)$$

and consider the following boundary value problem:

$$\Delta \varphi = 0 \text{ in } Q; \qquad (3.3)$$

$$\varphi(\mathbf{x}) = t_i \text{ on } D_i, \ i \in I = \{3, \dots, N\}; \qquad (3.4)$$

$$\int_{\partial D_i} \frac{\partial \varphi}{\partial \mathbf{n}} d\mathbf{x} = 0, i \in I; \qquad (3.5)$$

$$\varphi(\mathbf{x}) = \varphi_1^0 \text{ on } D_1, \ \varphi(\mathbf{x}) = \varphi_2^0 \text{ on } D_2; \qquad (3.6)$$

$$|\varphi(\mathbf{x})| \leq const < \infty \text{ as } |\mathbf{x}| \to \infty \ (n = 2); \qquad (3.7)$$

$$\varphi(\mathbf{x}) \to 0 \text{ as } |\mathbf{x}| \to \infty \ (n = 3). \qquad (3.8)$$

We denote $\mathbf{x} = (x_1, x_2)$ if $n = 2$ and $\mathbf{x} = (x_1, x_2, x_3)$ if $n = 3$;

$$\Delta = \sum_{i=1}^{n} \frac{\partial^2}{\partial^2 x_i}$$

is Laplace operator; and $\mathbf{n}$ is the outward unit normal to the boundary of the domain. We assume (it does not restrict the generality of our consideration) that

$$\varphi_1^0 < \varphi_2^0.$$

The unknowns in the problem (3.3)–(3.8) are function $\varphi(\mathbf{x})$ in domain Q and real numbers $\{t_i, i \in I\}$ (potentials of the bodies $\{D_i, i \in I\}$), I is the set of indices of bodies, which are not subjected to direct application of voltage. The belonging of potentials $\{t_i, i \in I\}$ of the bodies $\{D_i, i \in I\}$ to unknowns is the main difference of the problem under consideration from the classical problem of electrostatics discussed in Section 1.4.

Condition (3.6) implies that potentials (electrical potential, temperature, and so on) are held to be equal to φ_1^0 and φ_2^0 on the bodies D_1 and D_2, respectively.

Mathematically speaking, the condition of perfect conductivity is usually formulated as

$$\varphi = t_i \text{ on } \partial D_i$$

(in the form of boundary condition) but physically it appears as condition (3.4), which means that a field (electrical potential, temperature and so on, see Table 1 in Preface) is constant in the every perfectly conducting body D_i. The possibility of not distinguishing ∂D_i and D_i in (3.4) is in agreement with the physical nature of the problem. The boundary condition (3.5) means that the perfectly conducting bodies do not contain sources or sinks.

The difference of the formulation of the condition (3.7), (3.8) "at infinity" for dimension $n = 2$ and $n = 3$ is a well-known fact from electrostatic theories and the theory of partial differential equations (see, e.g., [94]).

Boundary value problem (3.3)–(3.8) is equivalent to the minimization problem

$$I(\varphi) = \frac{1}{2} \int_Q |\nabla\varphi(\mathbf{x})|^2 dx \rightarrow \min, \ \varphi(\mathbf{x}) \in V, \tag{3.9}$$

where the set V is defined as follows:

$$V = \left\{ \varphi(\mathbf{x}) \in H^1(Q) : \varphi(\mathbf{x}) = \varphi_1^0 \text{ on } D_1, \varphi(\mathbf{x}) = \varphi_2^0 \text{ on } D_2; \right. \tag{3.10}$$

$$\varphi(\mathbf{x}) = t_i \text{ on } D_i, i \in I;$$

$$\left. |\varphi(\mathbf{x})| \leq const < \infty \text{ as } |\mathbf{x}| \rightarrow \infty \ (n = 2), \varphi(\mathbf{x}) \rightarrow 0 \text{ as } |\mathbf{x}| \rightarrow \infty \ (n = 3) \right\}.$$

In (3.10) $\{t_i, i \in I\}$ mean not fixed constants. Note that the condition (3.5) can obtained by integrating by parts in Euler–Lagrange equation for functional (3.9) with the use of condition (3.4). Really, the variation of the functional (3.9) is

$$\delta I(\varphi) = \int_Q \nabla\varphi(\mathbf{x})\nabla\delta\varphi(\mathbf{x})dx,$$

where $\delta\varphi(\mathbf{x})$ means the variation of function.

Euler–Lagrange equation corresponding to (3.9) is

$$\int_Q \nabla\varphi(\mathbf{x})\nabla\delta\varphi(\mathbf{x})dx = 0. \tag{3.11}$$

Integrating by parts in the left-hand side of (3.11), we have (it is known that operator ∇ and variation δ are interchangeable)

$$\int_Q \nabla\varphi(\mathbf{x})\nabla\delta\varphi(\mathbf{x})dx = -\int_Q \Delta\varphi(\mathbf{x})\delta\varphi(\mathbf{x})dx + \sum_{i=1}^{N} \int_{\partial D_i} \frac{\partial\varphi}{\partial\mathbf{n}}(\mathbf{x})\delta\varphi(\mathbf{x})dx. \tag{3.12}$$

Since $\varphi(\mathbf{x}) \in V$, then

$$\delta\varphi(\mathbf{x}) = 0 \text{ on } \partial D_1, \partial D_2,$$

$$\delta\varphi(\mathbf{x}) = \delta t_i \text{ on } \partial D_i \ (i \geq 3), \tag{3.13}$$

where δt_i means variation of real number t_i.

Taking into account (3.13) and (3.12), we can rewrite (3.11) as

$$-\int_Q \Delta\varphi(\mathbf{x})\delta\varphi(\mathbf{x})dx + \sum_{i=3}^{N} \delta t_i \int_{\partial D_i} \frac{\partial\varphi}{\partial\mathbf{n}}(\mathbf{x})dx = 0.$$

Since $\delta\varphi(\mathbf{x})$ is an arbitrary function and δt_i $(i \geq 3)$ are arbitrary numbers, we arrive at the equations (3.3)–(3.5).

Definition 6. *The energy $\mathcal{E}$ of the electric field corresponding to the collection of bodies $\{D_i, i = 1, \ldots, N\}$ is defined as Dirichlet integral*

$$\mathcal{E} = \frac{1}{2}\int_Q |\nabla\varphi^{cont}(\mathbf{x})|^2 dx, \tag{3.14}$$

where $\varphi^{cont}(\mathbf{x})$ is solution of the continuum problem (3.3)–(3.8) and domain Q is defined by (3.2).

The energy $\mathcal{E}$ (3.14) can be also computed as

$$\mathcal{E} = \frac{1}{2}\min_{\varphi\in V}\int_Q |\nabla\varphi(\mathbf{x})|^2 dx. \tag{3.15}$$

The energy is densely related to the notion of capacity of collection of bodies $\{D_i, i = 1, \ldots, N\}$ (see, e.g., [340, 354]). Computation of capacity of a collection of many bodies is a difficult task not resolved for now. At the same time, for a pair of bodies in whole space, the problem of computation of capacity is solved for bodies of various shapes and dimensions [340, 354]. For some pairs of bodies (disks, spheres, ellipsoids), solutions were obtain in explicit form. Numerous solutions were obtained in two-dimensional case by using the method of function of complex variable.

Solvability of the problems (3.3)–(3.8) and (3.9)

The functional $I(\varphi)$ defined by the formula (3.9) is continuous and strictly convex on V. The functional space $H^1(R^n)$ is Hilbert space, then it is reflexive [210]. The set $V \subset H^1(Q)$ (3.10) is convex and closed subset of the reflexive functional space $H^1(R^n)$. It is known (see, e.g., [122]) that under these conditions there exists unique function $\varphi^{cont}(\mathbf{x}) \in V$, such that

$$I(\varphi^{cont}) = \min_{\varphi\in V} I(\varphi).$$

It means that the minimization problem (3.9) has unique solution $\varphi^{cont}(\mathbf{x})$. This is also solution of the boundary-value problem (3.3)–(3.8).

3.2.2. Primal and dual problems and ordinary two-sided estimates

We start our analysis of capacity of a system of closely placed bodies with the writing of dual extremal formulation of the problem (3.3)–(3.8) and the corresponding two-sided estimates (we call them *ordinary* two-sided estimates), which are analogs of well-known Voight [367] and Reuss [305] upper- and lower-sided estimates.

Primal problem and ordinary upper-sided estimate

The primal problem is the minimization problem (3.9) and (3.10). The corresponding ordinary upper-sided estimate is

$$\mathcal{E} \leq \frac{1}{2} \int_Q |\nabla \varphi(\mathbf{x})|^2 d\mathbf{x} \tag{3.16}$$

for any $\varphi(\mathbf{x}) \in V$.

Dual problem and ordinary lower-sided estimate

We define the following functional space

$$W = \Big\{ \mathbf{v}(\mathbf{x}) = (v_1(\mathbf{x}), \dots, v_n(\mathbf{x})) \in L_2(Q) : \mathrm{div}\mathbf{v}(\mathbf{x}) \in L_2(Q), \tag{3.17}$$

$$\mathbf{v}(\mathbf{x}) \leq \frac{const}{|\mathbf{x}|^2} \text{ as } |\mathbf{x}| \to \infty, \ \int_{\partial D_i} \mathbf{v}(\mathbf{x})\mathbf{n}d\mathbf{x} = 0, i \in I \Big\}.$$

The conditions in definition (3.17) correspond to the conditions (3.5), (3.7) and (3.8). For a function from W, their trace on a smooth (piecewise-smooth) surface of dimension $(n-1)$ can be determined (see, e.g., [212, 113]). The trace determined in accordance with [212] allows us to apply Green's formula (the differentiating by parts) to a pair of functions from V and W. Although the conditions "at infinity" with respect to the potential $\varphi(\mathbf{x}) \in V$ are different for $n = 2$ and $n = 3$ (see conditions (3.7) and (3.8)), the corresponding conditions for $\mathbf{v}(\mathbf{x}) \in W$ (associated with $\nabla\varphi(\mathbf{x})$) are the same for $n = 2$ and $n = 3$.

We note that the considered problem has a classical solution (see, e.g., [94]) and traces can be interpreted as values of functions in classical sense. We use Sobolev spaces $H^1(Q)$ and the corresponding trace theorems because they are naturally related to the notion of energy, which is the main object in our consideration.

To obtain dual problem, we use the Legendre transform (see Appendix A)

$$\frac{1}{2}\mathbf{x}^2 = \max_{\mathbf{v} \in R^n} \left(\mathbf{v}\mathbf{x} - \frac{1}{2}\mathbf{v}^2 \right). \tag{3.18}$$

We write (3.15) in the form

$$\mathcal{E} = \frac{1}{2} \min_{\varphi \in V} \int_Q |\nabla \varphi(\mathbf{x})|^2 d\mathbf{x} = \tag{3.19}$$

$$= \frac{1}{2} \min_{\varphi \in V} \max_{\mathbf{v} \in W} \int_Q \left(\mathbf{v}(\mathbf{x}) \nabla \varphi(\mathbf{x}) - \frac{1}{2} \mathbf{v}^2(\mathbf{x}) \right) d\mathbf{x} =$$

$$= \frac{1}{2} \max_{\mathbf{v} \in W} \min_{\varphi \in V} \int_Q \left(\mathbf{v}(\mathbf{x}) \nabla \varphi(\mathbf{x}) - \frac{1}{2} \mathbf{v}^2(\mathbf{x}) \right) d\mathbf{x}.$$

Since the quadratic functional $I(\varphi)$ satisfies the conditions of Proposition 5.2 Chapter III [113], we can interchange max and min in (3.19).

Since the term $\frac{1}{2} \mathbf{v}^2$ in (3.19) does not depend on $\nabla \varphi(\mathbf{x})$, we can write (3.19) in the form

$$\mathcal{E} = \frac{1}{2} \max_{\mathbf{v} \in W} \left(\min_{\varphi \in V} \int_Q \mathbf{v}(\mathbf{x}) \nabla \varphi(\mathbf{x}) - \frac{1}{2} \int_Q \mathbf{v}^2(\mathbf{x}) \right) d\mathbf{x}. \tag{3.20}$$

Integrate by parts in the first integral in (3.20), we obtain

$$\mathcal{E} = \frac{1}{2} \max_{\mathbf{v} \in W} \left[\min_{\varphi \in V} \left(-\int_Q \mathrm{div} \mathbf{v}(\mathbf{x}) \varphi(\mathbf{x}) d\mathbf{x} + \sum_{i=1}^{N} \int_{\partial D_i} \varphi(\mathbf{x}) \mathbf{v}(\mathbf{x}) \mathbf{n} d\mathbf{x} \right) - \tag{3.21} \right.$$

$$\left. - \frac{1}{2} \int_Q \mathbf{v}^2(\mathbf{x}) d\mathbf{x} \right].$$

The integrals

$$\int_{\partial D_i} \varphi(\mathbf{x}) \mathbf{v}(\mathbf{x}) \mathbf{n} d\mathbf{x} = 0$$

for $i \in I$ in accordance with the boundary condition (3.5). For the remaining integrals, corresponding to $i = 1$ and $i = 2$, we know values of the function $\varphi(\mathbf{x})$, see the boundary condition (3.6). These values are φ_1^0 and φ_2^0, correspondingly.

For convenience in writing formulas below, we introduce function $\varphi^0(\mathbf{x})$ as

$$\varphi^0(\mathbf{x}) = \begin{cases} \varphi_1^0 \text{ on } D_1, \\ \varphi_2^0 \text{ on } D_2. \end{cases} \tag{3.22}$$

Using the function (3.22), we can write (3.21) as follows

$$\mathcal{E} = \frac{1}{2} \max_{\mathbf{v} \in W} \left[\min_{\varphi \in V} \left(-\int_Q \mathrm{div} \mathbf{v}(\mathbf{x}) \varphi(\mathbf{x}) d\mathbf{x} \right) + \tag{3.23} \right.$$

$$\left. + \int_{\partial D_1 \bigcup \partial D_2} \varphi^0(\mathbf{x}) \mathbf{v}(\mathbf{x}) \mathbf{n} d\mathbf{x} - \frac{1}{2} \int_Q \mathbf{v}^2(\mathbf{x}) d\mathbf{x} \right].$$

The first term in (3.23) is linear in $\varphi(\mathbf{x})$ and it is equal to $-\infty$ if $\operatorname{div}\mathbf{v}(\mathbf{x}) \neq 0$. To understand this, it is sufficient to take a function $\psi(\mathbf{x}) \in V$, such that

$$\int_Q [\psi(\mathbf{x}) - \varphi^0(\mathbf{x})] \operatorname{div}\mathbf{v}(\mathbf{x})d\mathbf{x} \neq 0$$

and then take

$$\varphi(\mathbf{x}) = t[\psi(\mathbf{x}) - \varphi^0(\mathbf{x})] + \varphi^0(\mathbf{x}),$$

where t is an arbitrary number, as a trial function. Note that $\varphi(\mathbf{x}) \in V$ for any $t \in R$ if $\psi(\mathbf{x}) \in V$. Thus, we can assume that $\psi(\mathbf{x}) - \varphi^0(\mathbf{x})$ is an arbitrary function from $H_0^1(Q)$. For such choice of trial function

$$\int_Q \varphi(\mathbf{x}) \operatorname{div}\mathbf{v}(\mathbf{x})d\mathbf{x} = \tag{3.24}$$

$$= t \int_Q [\psi(\mathbf{x}) - \varphi^0(\mathbf{x})] \operatorname{div}\mathbf{v}(\mathbf{x})d\mathbf{x} + \int_Q \varphi^0(\mathbf{x}) \operatorname{div}\mathbf{v}(\mathbf{x})d\mathbf{x}.$$

Because t is arbitrary, then expression (3.24) can take any values from $-\infty$ to $+\infty$ if

$$\int_Q [\psi(\mathbf{x}) - \varphi^0(\mathbf{x})] \operatorname{div}\mathbf{v}(\mathbf{x})d\mathbf{x} \neq 0$$

and the minimal value is $-\infty$. The only exception is the case when

$$\int_Q [\psi(\mathbf{x}) - \varphi^0(\mathbf{x})] \operatorname{div}\mathbf{v}(\mathbf{x})d\mathbf{x} = 0 \tag{3.25}$$

for every $\psi(\mathbf{x}) - \varphi^0(\mathbf{x}) \in V$.

The set V is dense in $L_2(Q)$ [212]. Then, from the equality (3.25) it follows that

$$\operatorname{div}\mathbf{v}(\mathbf{x}) = 0 \text{ in } Q. \tag{3.26}$$

Thus, the maximum in (3.23) is attained on the functions satisfying the condition (3.26). In formulas $\operatorname{div}\mathbf{v}(\mathbf{x}) = 0$ means (3.26).

As a result, we transform (3.15) into the following equality:

$$\mathcal{E} = \max_{\substack{\mathbf{v} \in W_p, \\ \operatorname{div}\mathbf{v}=0}} \left(-\frac{1}{2} \int_Q \mathbf{v}^2(\mathbf{x})d\mathbf{x} + \int_{\partial D_1 \bigcup \partial D_2} \varphi^0(\mathbf{x})\mathbf{v}(\mathbf{x})\mathbf{n}d\mathbf{x} \right). \tag{3.27}$$

The ordinary lower-sided estimate associated with (3.27) is the following:

$$\mathcal{E} \geq -\frac{1}{2} \int_Q \mathbf{v}^2(\mathbf{x})d\mathbf{x} + \int_{\partial D_1 \bigcup \partial D_2} \varphi^0(\mathbf{x})\mathbf{v}(\mathbf{x})\mathbf{n}d\mathbf{x} \tag{3.28}$$

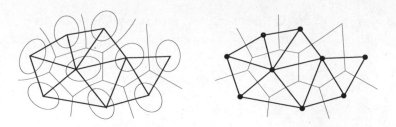

Figure 3.4. *Voronoi tessellations (shown in hair lines) and Delaunay graph in dimension two (shown in heavy lines): for bodies (left) and points (right).*

for any $\mathbf{v}(\mathbf{x}) \in W$ such that

$$\operatorname{div}\mathbf{v}(\mathbf{x}) = 0. \tag{3.29}$$

For the solution $\varphi^{cont}(\mathbf{x})$ of the continuum problem (3.9), (3.10) (or, that is the same, problem (3.3)–(3.8)), equality

$$I(\varphi^{cont}) = J(\mathbf{v}) \tag{3.30}$$

holds.

Here

$$\mathbf{v}(\mathbf{x}) = \nabla\varphi^{cont}(\mathbf{x}), \tag{3.31}$$

and $\mathbf{v}(\mathbf{x})$ is solution of the problem

$$J(\mathbf{v}) \to \max, \ \mathbf{v}(\mathbf{x}) \in W,$$

where

$$J(\mathbf{v}) = -\frac{1}{2} \int_Q \mathbf{v}^2(\mathbf{x})d\mathbf{x} + \int_{\partial D_1 \bigcup \partial D_2} \varphi^0(\mathbf{x})\mathbf{v}(\mathbf{x})\mathbf{n}d\mathbf{x}.$$

Equality (3.30) follows from the definitions of the functionals $I(\varphi)$ and $J(\mathbf{v})$ and Green's formula

$$\int_Q |\nabla\varphi^{cont}(\mathbf{x})|^2 d\mathbf{x} = \int_{\partial D_1 \bigcup \partial D_2} \frac{\partial\varphi^{cont}}{\partial\mathbf{n}}(\mathbf{x})\varphi^0(\mathbf{x})d\mathbf{x},$$

following from the conditions (3.5), (3.7) and (3.8).

3.2.3. The topology of a set of bodies, Voronoi–Delaunay method

The notion of neighboring bodies / particles plays an important role in our consideration. While for periodic arrays of bodies the notion of neighbors is obvious, for disordered (non-periodic) arrays the formal definition of neighbors requires an effort.

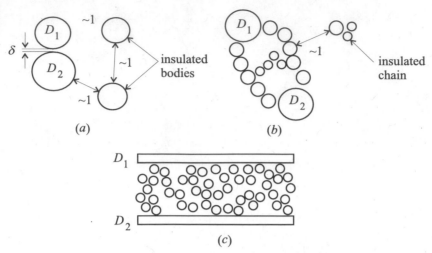

Figure 3.5. *Topologies of collection of bodies: (a) two closely placed bodies – the case described by Tamm [354], (b) nets and chains, (c) a capacitor-like network (a model of capacitor with layer filled with particles).*

The concept of neighboring bodies

The concept of neighboring bodies is formalized by invoking Voronoi–Delaunay method (see, e.g., [15]). Voronoi tessellations (also referred to as Wigner–Seitz cells [315]) corresponding to a particular body (domain) is defined as a set of points that are closer to the given body (domain) than to the remaining ones. The bodies that belong to neighboring Voronoi tessellations are referred to as *neighboring* ones. A body belonging to a Voronoi tessellation can be represented by a point called node. The graph with the edges connecting the neighboring nodes is referred to the Delaunay graph corresponding to the given system of Voronoi tessellations, see Fig. 3.4. The Delaunay graph is a connected one.

It is well-known that for a given set of generating points on a plane (space) the Voronoi tessellations are polygons (polyhedra).

The typical structures of set of bodies

Bodies can form a great variety of geometrical structures in space and on a plane. In Fig. 3.5 some typical structures are shown. We describe some types of structures formed of closely placed bodies (domains). Here and below domain means the domain occupied by the corresponding body.

We assume that the positions of the bodies (domains) $\{D_i, i = 1, \ldots, N\}$ depend on a parameter $\delta : D_i = D_i(\delta)$ and there exist limit positions $\{D_i(0), i = 1, \ldots, N\}$ for the bodies (domains):

$$D_i(\delta) \to D_i(0) \text{ as } \delta \to 0.$$

Denote $\delta_{ij} = \delta_{ij}(\delta)$ the distance between the i−th and j−th bodies (domains).

Definition 7. *Let* $\{D_{i(k)}(\delta),\ k = 1,\ldots,M\}$ *be a subset of the set* $\{D_i(\delta), i = 1,\ldots,N\}$ *such that*

$$\delta_{i(k),i(k+1)}(\delta) \to 0 \text{ for all } k = 1,\ldots,M$$

as $\delta \to 0$. *Such subset is referred to the chain of bodies.*

Definition 8. *A chain of bodies, which has joints or loops, is referred to a network of bodies. A body (a subsystem of bodies), which does not belong in a chain or a network is (are) referred to an insulated body (bodies).*

Fig. 3.6 illustrates this definition.

Definition 9. *If a chain (net) contains bodies* D_1 *and* D_2, *it is referred to as the alive chain (net). If a chain (net) does not contain bodies* D_1 *or* D_2 *it is referred to as the insulated chain (net).*

In other words, the alive chain (net) is subjected to direct application of a difference of potentials (a voltage is applied directly to the elements of the chain (net)). The insulated chain (net) is not subjected to direct application of a difference of potentials. Nevertheless, the potentials of the bodies forming an insulated chain (net) can differ from one another. See Fig. 3.5 (*a*, *b*) for illustration of the definitions.

Evidently, the original set of bodies $\{D_i(\delta),\ i = 1,\ldots,N\}$ decomposes into a system of chains, nets and insulated bodies. We assume that the bodies under consideration form C nets (chains). We denote the k-th network (chain) by $CH(k), k = 1,\ldots,C$. In the considered case there exists only one alive network (chain). We denote it by

$$K = CH(1).$$

We will omit the parameter δ so it does not lead to confusion.

3.3. Heuristic network model

Analysis of numerical solutions of the problems similar to (3.3)–(3.8) presented in Chapter 2 leads to conclusion that for a system of dense packed bodies, the well-known phenomenon of concentration of fields (see, e.g., [216, 241]) asymptotically (as $\delta \to 0$) transforms into the phenomenon of energy localization of fields in the channels between bodies. Fig. 3.7 (left) illustrates the energy localization phenomenon. As has been noted in Section 2.4, the local energy demonstrates the localization property more strongly than other related fields.

Figure 3.6. *Chain (left) and net (right).*

The energy localization effect appears in an alive network (chain) only, see Section 2.5.2. In view of this observation (which will be rigorously justified below for n-dimensional case, $n = 2,3$), we construct a finite-dimensional analogue of the original continuum problem (3.3)–(3.8). An important note here is that the energy localization leads to energy decomposition phenomenon. The main part of energy is collected in the energy channels, which do not intersect one another, see Fig. 3.7 (right). Then the total energy (more exactly, the main part of the total energy) is the sum of energies accumulated in the energy channels.

The phenomena of energy localization and energy decomposition are not universal property of a system of closely placed bodies. As was noted in [183] (see also Section 3.7 below), one has no reason to expect that the energy channel will arise between any closely placed bodies. The necessary and sufficient condition of arising of the energy channel and related phenomena sounds as (3.1).

As a result, distribution of the energy in a system of closely placed bodies is associated with a network (weighted graph)

$$\mathbf{G} = \{\mathbf{x}_i, C_{ij}^{(2)};\ i, j \in K\},$$

where $\{\mathbf{x}_i; i \in K\}$ are nodes and K are indices corresponding to the bodies forming the network; $C_{ij}^{(2)}$ describes the transport properties of the edges of the network. We naturally arrive at the question: *"What is the right value for $C_{ij}^{(2)}$?"*

For a pair of disks the answer is

$$C_{ij}^{(2)} = \pi \sqrt{\frac{R}{\delta_{ij}}}. \tag{3.32}$$

This is Keller's formula derived for periodic array of disks in [166] and justified for an arbitrary disordered system of disks in [41]. We show how to modify this

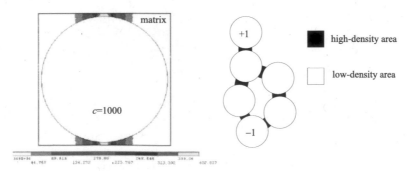

Figure 3.7. *Left: distribution of density of energy in and around a body in a system of closely placed bodies (contrast $c = 1000$, relative interparticle distance $\delta^* = 0.01$). The energy channel is clearly seen. Right: illustration of the energy decomposition phenomenon in a system of closely placed bodies.*

formula for a pair of bodies of arbitrary shape. We recall Tamm assumption [354] that the capacity of closely placed bodies weakly depends on other bodies. We make the next step and assume that closely placed bodies weakly interact with the bodies, which are not placed near these bodies. When we say weakly, we assume that interaction is weak as compared with the interaction of closely placed bodies. In other words, interaction of two closely placed bodies as elements of a system of many bodies is similar to interaction of these two bodies in R^n. The interaction of two closely placed bodies in R^n is described by capacity. Then, we must take $C_{ij}^{(2)}$ equal to the capacity of the pair of the bodies D_i and D_j in R^n.

It is known that the specific transport characteristics of the channels between two disks and two spheres are asymptotically (as $\delta \to 0$) equivalent to capacity of the pair of disks or spheres, see [166] and Section 1.2.3. It gives additional support to our assumption.

The alive network is unique in the case described by the problem (3.3)–(3.8). The bodies $\{D_i, i \in I \backslash K\}$ are insulated. One can formulate a problem similar (3.3)–(3.8) with potentials applied to more than two bodies. In this case some (more then one) alive nets can appear. We consider this case in Section 5.1.3 which is devoted to composite materials.

The heuristic expression for the total energy $\mathcal{E}_d$ accumulated in the channels between the neighbor bodies is the sum of energies accumulated in the capacitors formed by the pairs of the neighbor bodies. It has the form

$$\mathcal{E}_d = \frac{1}{4} \sum_{i=1}^{N} \sum_{j \in N_i} C_{ij}^{(2)} (t_i - t_j)^2. \tag{3.33}$$

We denote by N_i the set of nodes adjacent to the i–th node. When we carry out summation in indices i, j, we count every channel twice. Therefore a factor $1/4$ appears in (3.33) instead of usual $1/2$.

In addition, we neglect the energy accumulated in insulated (i.e., not alive) chains (nets) $CH(k)$ $(k > 1)$ (if it exists). We demonstrate below in Section 3.5.3 that this neglect is possible in an asymptotic sense. Then (3.33) can be rewritten as (remind that $K = CH(1)$ is the alive chain (net))

$$\mathcal{E}_d = \frac{1}{4} \sum_{i \in K} \sum_{j \in N_i} C_{ij}^{(2)} (t_i - t_j)^2. \tag{3.34}$$

Minimizing the energy $\mathcal{E}_d$ (3.34), we arrive at the following Kirchhoff equations for the interior nodes K of the alive chain (network)

$$\sum_{j \in N_i} C_{ij}^{(2)} (t_i - t_j) = 0, \ i \in K. \tag{3.35}$$

The balance equations (3.35) are supplemented with the boundary conditions

$$t_1 = \varphi_1^0, \ t_2 = \varphi_2^0. \tag{3.36}$$

We denote $\{t_i^{net}, i \in K\}$ solution of the problem (3.35) and (3.36). We note that solution of the problem (3.35) and (3.36) determines the potentials of the bodies belonging to the alive network K, only.

Often network models are derived from the condition of balance of fluxes defined as

$$p_{ij} = C_{ij}^{(2)}(t_i - t_j) \qquad (3.37)$$

in the nodes of network model. In this case the resulting equations are the same – (3.35). Taking into account that the energy demonstrates the localization property (thus, the decomposition property) most strongly (see Section 2.4), we suppose derive network models from the minimum of energy principle.

The problem (3.35) and (3.36) often is referred to as discrete network problem corresponding to the continuum problem (3.3)–(3.8).

3.4. Proof of the principle theorems

In this section, we present the proof of three densely related theorems. In common words, the first theorem about NL (No Localization) zones says that large energy in a system of closely placed bodies (under condition that voltage applied to a system is fixed) cannot be accumulated in domain outside the channels. The second theorem about the asymptotic equivalence of capacities says that under condition (3.1), the capacities with respect to various channels (definition of capacity with respect channel is given below, see Definition 12), including the classical capacity in R^n, are equivalent as $\delta \to 0$. The third theorem says that under condition (3.1) the energy computed with the heuristic network problem (3.35) and (3.36) approximates the energy computed with the continuum problem (3.3)–(3.8).

3.4.1. Principles of maximum for potentials of nodes in network model

The following discrete principle of maximum (similar to the principle of maximum for harmonic function [19]) holds for the alive chain (net).

Lemma 1. *The solution $\{t_i^{net}, i \in K\}$ of the problem (3.35) and (3.36) satisfies the inequality*

$$\varphi_1^0 \le t_i^{net} \le \varphi_2^0$$

for any $i \in K$.

Proof. For the interior vertices $\mathbf{x}_i$, $i \in I$, the system (3.35) can be rewritten as follows

$$\sum_{j \in N_i} C_{ij}^{(2)}(t_i^{net} - t_j^{net}) = 0. \qquad (3.38)$$

From (3.38) we have the following analog of the mean value theorem for harmonic functions:

$$t_i^{net} \sum_{j \in N_i} C_{ij}^{(2)} = \sum_{j \in N_i} C_{ij}^{(2)} t_j^{net}, \ i \in K, \tag{3.39}$$

where summation is taken over all adjacent (neighboring) vertices.

We call (3.39) an analog of the mean value theorem because it relates the value t_i^{net} of the potential at a given vertex $\mathbf{x}_i$ to the values of the potential t_j^{net} at the adjacent vertices $\mathbf{x}_j$, $j \in N_i$. From the formula (3.39) it follows that the maximum value cannot be achieved in an interior site. We prove this by contradiction. Suppose that the maximum value M is achieved at some interior vertices $i \in K$. Then

$$t_j^{net} \leq M$$

for all vertices adjacent to the i-th vertex, i.e., for all $j \in N_i$.

Because $C_{ij} \geq 0$, we have for the right-hand side of (3.39) the following non-strict inequality:

$$\sum_{j \in N_i} C_{ij}^{(2)} t_j^{net} \leq M \sum_{j \in N_i} C_{ij} \tag{3.40}$$

in which the equality is possible only if

$$t_j^{net} = M \text{ for all } j \in N_i.$$

Since $t_i^{net} = M$ by assumption, we obtain from (3.40) that $M \leq M$. If at least one t_j^{net}, $j \in N_i$ in (3.40) is smaller than M, (3.40) leads to the impossible inequality $M < M$. Thus,

$$t_j^{net} = M \text{ for all } j \in N_i.$$

Since the graph $\mathbf{G}$ is connected (since it is a Delaunay graph), it follows that $t_j^{net} = M$ for all vertices of the graph. But the graph $\mathbf{G}$ includes the vertices $\mathbf{x}_1$ and $\mathbf{x}_2$ corresponding to bodies D_1 and D_2, where $t_1^{net} = \varphi_1^0$ and $t_2^{net} = \varphi_2^0$, respectively. This leads us to a contradiction, which means that the maximum value can be achieved only in $\mathbf{x}_2$ and this maximum value is φ_2^0.

The same argument works for the minimum value φ_1^0. **Proof is completed.**

Lemma 1 establishes the principle of maximum for network model. It is valid not for all bodies, but only for bodies belonging to an alive net.

We will need an estimate for potentials $\{t_i^{cont}, i \in I\}$ of bodies determined from continuum model (3.3)–(3.8). For these potentials a principle of maximum is also valid.

Lemma 2. *For solution of the original continuum problem (3.3)–(3.8) the following inequalities hold*

$$\varphi_1^0 \leq t_i^{cont} \leq \varphi_2^0 \text{ for all } i \in I, \tag{3.41}$$

$$\varphi_1^0 \leq \varphi(\mathbf{x}) \leq \varphi_2^0 \text{ for any } \mathbf{x} \in Q. \tag{3.42}$$

Proof. It is sufficient to prove point 1 (the item 2 of the lemma evidently follows from the item 1 and the principle of maximum for harmonic function [19]). We assume that $\varphi(\mathbf{x})$ takes the maximal value on a domain D_i and this value $t_i^{cont} > \varphi_2^0$ (we have assumed that $\varphi_2^0 > \varphi_1^0$). Consider the outward unit normal vector $\mathbf{n}$ to ∂D_i. The function $\varphi(\mathbf{x})$ does not increase along $\mathbf{n}$ (t_i^{cont} is the maximal value in accordance with our assumption), then

$$\frac{\partial \varphi}{\partial \mathbf{n}}(\mathbf{x}) \leq 0 \text{ on } \partial D_i.$$

From this and (3.5), we obtain

$$\frac{\partial \varphi}{\partial \mathbf{n}}(\mathbf{x}) = 0 \text{ on } \partial D_i.$$

Now we consider a domain S, such that $D_i \subset S$ and S does not intersect other domains D_j, $j \neq i$, see Fig. 3.8.

The function $\varphi(\mathbf{x})$ satisfies the following equalities:

$$\Delta \varphi = 0 \text{ in } D_i; \tag{3.43}$$

$$\varphi(\mathbf{x}) = t_i^{cont} \text{ on } \partial D_i; \tag{3.44}$$

$$\frac{\partial \varphi}{\partial \mathbf{n}}(\mathbf{x}) = 0 \text{ on } \partial D_i; \tag{3.45}$$

$$\Delta \varphi = 0 \text{ in } S \backslash D_i. \tag{3.46}$$

The equality (3.43) follows from the condition

$$\varphi(\mathbf{x}) = t_i^{cont} \text{ in } D_i.$$

Now we demonstrate that $\varphi(\mathbf{x})$ determined by (3.43)–(3.46) is a harmonic function in domain S in the sense of distribution [212]. We denote by $C_0^\infty(S)$ the set of infinitely differentiable functions with compact support belonging to S [212]. We want to demonstrate that the equality

$$\int_S \varphi(\mathbf{x}) \Delta \xi(\mathbf{x}) d\mathbf{x} = 0 \tag{3.47}$$

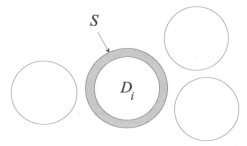

Figure 3.8. *Definition of the domain S.*

is satisfied for any function $\xi(\mathbf{x}) \in C_0^\infty(S)$.

Integrating (3.47) by parts twice over the domain S with account of (3.43)–(3.46), we obtain the following equality

$$\int_S \varphi(\mathbf{x})\Delta\xi(\mathbf{x})dx = \int_{D_i} \Delta\varphi(\mathbf{x})\xi(\mathbf{x})dx + \int_{S\backslash D_i} \Delta\varphi(\mathbf{x})\xi(\mathbf{x})dx + \qquad (3.48)$$

$$+ \int_{\partial D_i} \left[\varphi(\mathbf{x})\frac{\partial\xi}{\partial\mathbf{n}}(\mathbf{x})\right]_i dx - \int_{\partial D_i} \left[\frac{\partial\varphi}{\partial\mathbf{n}}(\mathbf{x})\xi(\mathbf{x})\right]_i d\mathbf{x}$$

Here $[\bullet]_i$ means "the jump on ∂D_i" – the difference of values of a function on the opposite sides of the surface ∂D_i.

The trial function $\xi(\mathbf{x}) \in C_0^\infty(S)$ and its derivatives have zero jump on ∂D_i. By virtue of (3.44), (3.45) and condition

$$\varphi(\mathbf{x}) = t_i^{cont} \text{ in } D_i,$$

the function $\varphi(\mathbf{x})$ and its normal derivative have zero jump on ∂D_i. The volume integrals in the right-hand side of (3.48) are equal to zero by virtue of (3.43) and (3.46). Then the right-hand side of (3.48) is equal to zero and equality (3.47) is valid. Thus, $\varphi(\mathbf{x})$ is a harmonic function in domain S in the sense of distribution. It is known (see, e.g., [94, 366]) that a function harmonic in the sense of distribution is harmonic in the classical sense. Then $\varphi(\mathbf{x})$ is harmonic in the classical sense in the domain S. As a result, we obtain that $\varphi(\mathbf{x})$ satisfies the following two equalities:

$$\Delta\varphi = 0 \text{ in } S, \qquad (3.49)$$

$$\varphi(\mathbf{x}) = t_i^{cont} \text{ in } D_i \subset S.$$

It is known (see, e.g., [94, 366]) that a harmonic function is an analytical function (i.e., it can be represented in the form of a converging power series). Then $\varphi(\mathbf{x})$ is an analytical function. For analytical function $\varphi(\mathbf{x})$, from (3.49) it follows that $\varphi(\mathbf{x}) = t_i^{cont}$ in whole domain S.

Now we use the standard procedure to arrive at a contradiction. We take domain S in such a way that it touches a body (domain) D_j nearest to D_i. If D_i does not coincide with D_2, we obtain that

$$t_i^{cont} = t_j^{cont}$$

and continue this procedure until S touches the body (domain) D_2. When S touches D_2, we obtain

$$t_j^{cont} = \varphi_2^0.$$

On another hand (by the assumption)

$$t_i^{cont} > \varphi_2^0.$$

This contradiction proves that the assumption $t_i^{cont} > \varphi_2^0$ is not correct. Thus, $t_i^{cont} \leq \varphi_2^0$. In a similar manner we can prove the inequality $t_i^{cont} \geq \varphi_1^0$.

The inequality (3.42) is a direct consequence of the inequality (3.41) and the principle of maximum for harmonic functions [19]). **Proof is completed**.

Solution of the finite-dimensional problem (3.35) and (3.36) and its properties described in Lemma 1 and Lemma 2 will be used to obtain refined two-sided estimates below.

3.4.2. Electrostatic channel and trial function

Now, we describe a channel between neighbor bodies and corresponding trial function, which are constructed using the solution of a special electrostatic problem. We consider the electric field in R^n produced by two neighbor bodies D_i and D_j with potentials t_i and t_j respectively. Potential $\varphi^{(ij)}$ of this electric field is solution of the following boundary-value problem:

$$\Delta\varphi^{(ij)} = 0 \text{ in } R^n\backslash(D_i \bigcup D_j); \tag{3.50}$$

$$\varphi^{(ij)}(\mathbf{x}) = t_i \text{ on } D_i, \ \varphi^{(ij)}(\mathbf{x}) = t_j \text{ on } D_j;$$

$$|\varphi^{(ij)}(\mathbf{x})| \leq const < \infty \text{ as } |\mathbf{x}| \to \infty \ (n = 2);$$

$$\varphi^{(ij)}(\mathbf{x}) \to 0 \text{ as } |\mathbf{x}| \to \infty \ (n = 3).$$

The problem (3.50) is referred to electrostatic problem in R^n for a pair of bodies [340, 354]. Solution of this problem usually is used to introduce the notion of capacity of various types of capacitors (see [340, 354] for the physical introduction and [59, 372] for the mathematical introduction of the capacity).

Definition 10. *The function $\varphi^{\pm 1}(\mathbf{x})$ means solution of the problem (3.50) corresponding to potentials $t_i = 1/2$ and $t_j = -1/2$ applied to bodies D_i and D_j.*

The quantity

$$C_{ij}^{(2)} = \int_{R^n\backslash(D_i \bigcup D_j)} |\nabla\varphi^{\pm 1}(\mathbf{x})|^2 d\mathbf{x} \tag{3.51}$$

is called the capacity of the pair of bodies (domains) D_i and D_j in R^n (or capacity of capacitor formed of the bodies D_i and D_j).

The problem (3.50) was investigated numerous researchers and properties of its solutions are known (see, e.g., [340]). In particular, it is known that the problem (3.50) has a unique solution.

Now we construct a special channel S_{ij} between two closely placed bodies D_i and D_j. It is constructed in the terms of lines of force. The surface at which $\varphi^{(ij)}(\mathbf{x}) = const$ is called the equipotential surface. The lines normal to the equipotential

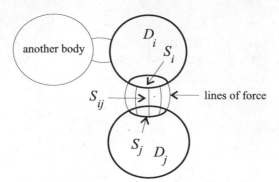

Figure 3.9. *Neighbor bodies and the electrostatic channels between the bodies.*

surfaces (or, the lines in which tangential vector in the point $\mathbf{x}$ is parallel to $\nabla\varphi^{\pm 1}(\mathbf{x})$) are called the lines of force. Outside of $D_i \bigcup D_j$, the equipotential surfaces and lines of force form an orthogonal coordinate system.

For two bodies, the structure of the equipotential surfaces and lines of force is known and has the form shown at Fig. 3.9. The lines of force issuing from the neighborhood S_i of the pole of the body D_i are terminated in the neighborhood S_j of the pole of the body D_j. We call the poles the nearest points of the bodies D_i and D_j. The lines of force passing from S_i to S_j fill a domain S_{ij}. Domain S_{ij} has in width dimension of the domains S_i and S_j. For small δ (tending to zero) and small (but fixed) S_i and S_j, the domain S_{ij} forms an isolated channel between the bodies D_i and D_j. Note once more that we call domains S_i and S_j small not because they tend to zero, but because it is possible to take these domains so small that the corresponding channel S_{ij} does not intersect with other channels.

Definition 11. *A channel S_j between two bodies D_i and D_j with lateral boundaries formed by the linear of force determined from solution of the electrostatics problem (3.50) in R^n is called the electrostatic channel, see Fig. 3.9.*

A trial function constructed using electrostatic channel(s) will be referred to the electrostatic trial function.

3.4.3. Refined lower-bound estimate

Now we refine estimates (3.16) and (3.28) by constructing special trial functions accounting the specificity of the problem under consideration. The idea of the construction of the trial functions is the following. In a domain between neighboring bodies (in the channel), the trial functions are taken equal to "the electrostatics" solution. Outside the channels, the trial functions are equal to zero. The solution of problem (3.35)–(3.36) will be used to satisfy balance equations for trial functions.

The lower-sided estimate has the form (3.28), where the trial function $\mathbf{v}(\mathbf{x})$ must

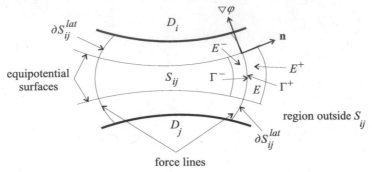

Figure 3.10. *The electrostatic channel S_{ij} between two neighbor bodies.*

satisfy the conditions

$$\mathrm{div}\mathbf{v}(\mathbf{x}) = 0 \text{ in } Q, \tag{3.52}$$

$$\int_{\partial D_i} \mathbf{v}(\mathbf{x})\mathbf{n}dx = 0, i \in I, \tag{3.53}$$

$$|\mathbf{v}(\mathbf{x})| \le \frac{const}{|\mathbf{x}|^2} \text{ as } |\mathbf{x}| \to \infty \; (n = 2, 3). \tag{3.54}$$

To construct a trial function, we consider two neighboring bodies (i−th and j−th) shown in Fig. 3.10. Note that the choice of directions of coordinate axes is of no significance because all quantities in (3.28), (3.52)–(3.54) are invariant with respect to rotation of the coordinate axes.

Estimate of the volume integral in (3.28)

Usually, a body has several neighbors, see Fig. 3.2. We choose the width of the channels, see Fig. 3.9, such that the channel S_{ij} does not intersect other channels and at the same time its width is greater than a positive number S_0, which does not depend on δ. It always can be done for closely placed bodies (domains).

We consider the following function:

$$\mathbf{v}(\mathbf{x}) = \begin{cases} \nabla\varphi^{(ij)}(\mathbf{x}) \text{ for } \mathbf{x} \in S_{ij}, \\ 0 \text{ for } \mathbf{x} \in R^n \backslash (S_{ij} \bigcup D_i \bigcup D_j) \end{cases} \tag{3.55}$$

Lemma 3. *For function (3.55), the equality*

$$\mathrm{div}\mathbf{v} = 0 \; in \; R^n \backslash (D_i \bigcup D_j)$$

holds.

Proof. We take a trial function $\xi(\mathbf{x}) \in C_0^\infty(R^n \backslash (D_i \bigcup D_j))$ ($C_0^\infty(R^n \backslash (D_i \bigcup D_j))$ means the set of infinitely differentiable functions with their supports supp $\xi(\mathbf{x})$

belonging to $R^n \backslash (D_i \bigcup D_j)$). With regard to definition (3.55) of function $\mathbf{v}(\mathbf{x})$, in every domain inside or outside the channel S_{ij}, the equality

$$\operatorname{div} \mathbf{v}(\mathbf{x}) = 0 \qquad (3.56)$$

holds.

Really, inside the channel

$$\operatorname{div} \mathbf{v}(\mathbf{x}) = \operatorname{div}(\nabla \varphi^{(ij)}(\mathbf{x})) = \Delta \varphi^{(ij)}(\mathbf{x}) = 0.$$

Outside the channel $\mathbf{v}(\mathbf{x}) = 0$, then $\operatorname{div} \mathbf{v}(\mathbf{x}) = 0$.

We consider the case when supp $\xi(\mathbf{x}) \bigcap \Gamma \neq 0$, where surface Γ is a part of the lateral surface ∂S_{ij}^{lat} of the channel S_{ij}, see Fig. 3.10. In other words, one part of the set supp $\xi(\mathbf{x})$ belongs to the channel S_{ij} and another part lies outside S_{ij}. Surface Γ is a boundary between two domains S_{ij} and $R^n \backslash (S_{ij} \bigcup D_i \bigcup D_j)$ in which function (3.55) is determined in different ways. We consider the integral $\int_E \mathbf{v}(\mathbf{x}) \nabla \xi(\mathbf{x}) dx$, where $E \subset$ supp ξ is a neighborhood of surface Γ, see Fig. 3.10. We have

$$\int_E v_i(\mathbf{x}) \frac{\partial \xi}{\partial x_i}(\mathbf{x}) dx = - \int_{E^+} \operatorname{div} \mathbf{v}(\mathbf{x}) \xi(\mathbf{x}) dx + \qquad (3.57)$$

$$+ \int_{E^-} \operatorname{div} \mathbf{v}(\mathbf{x}) \xi(\mathbf{x}) dx + \int_{\partial E^+ \bigcup \partial E^-} \mathbf{v}(\mathbf{x}) \xi(\mathbf{x}) \mathbf{n} dx,$$

where E^+ and E^- are domains in which the surface Γ partitions the domain E, see Fig. 3.10, and ∂E^+ and ∂E^- are the boundaries of the domains ∂E^+ and ∂E^-. By virtue of the equalities

$$\operatorname{div} \mathbf{v}(\mathbf{x}) = 0 \text{ in } E^+, \qquad (3.58)$$

$$\operatorname{div} \mathbf{v}(\mathbf{x}) = 0 \text{ in } E^-,$$

the volume integral in (3.57) vanishes. We consider now a function $\xi(\mathbf{x}) \in C_0^\infty(E)$, such that $\xi(\mathbf{x}) = 0$ on the boundaries ∂E^+ and ∂E^- of the domains E^+ and E^-, except the surface $\Gamma \subset \partial S_{ij}$. The surface Γ has two sides Γ^+ and Γ^- turning to the domains E^+ and E^-, respectively, see Fig. 3.10. With regard to (3.58), the right-hand side of (3.57) is equal to

$$\int_{\Gamma^+} \mathbf{v}(\mathbf{x}) \xi(\mathbf{x}) \mathbf{n} dx + \int_{\Gamma^-} \mathbf{v}(\mathbf{x}) \xi(\mathbf{x}) \mathbf{n} dx. \qquad (3.59)$$

On the side Γ^- of the surface Γ function $\mathbf{v}(\mathbf{x}) = 0$, see (3.55). Thus, only one term remains in (3.59). It is

$$\int_{\Gamma^+} \mathbf{v}(\mathbf{x}) \xi(\mathbf{x}) \mathbf{n} dx. \qquad (3.60)$$

Surface Γ is formed by the lines of force. The vector $\mathbf{n}$ normal to Γ is tangential to the equipotential surfaces. Then, the vectors $\mathbf{n}$ and $\mathbf{v}(\mathbf{x}) = \nabla\varphi^{(ij)}(\mathbf{x})$ are perpendicular one to another on Γ^+, and

$$\mathbf{v}(\mathbf{x})\mathbf{n} = \nabla\varphi^{(ij)}(\mathbf{x})\mathbf{n} = 0 \text{ on } \Gamma^+, \tag{3.61}$$

Then, integral (3.60) (and integral (3.57), as a consequence) is equal to zero. **Proof is completed**.

The equality (3.61) means that there is no flux through the lateral boundary of the electrostatic channel. This property explains why the electrostatics channels are suitable for the analysis of the problem considered. Lemma 3 is not correct for arbitrary (not electrostatics) channel S_{ij}.

Next, it is necessary to satisfy condition (3.53) for trial function. By virtue of the construction of the function $\mathbf{v}(\mathbf{x})$ (see formulas (3.55) and (3.61)), we have

$$\int_{S_i \cup S_j} \mathbf{v}(\mathbf{x})\mathbf{n}\,d\mathbf{x} = \int_{S_i \cup S_j} \frac{\partial\varphi^{(ij)}}{\partial\mathbf{n}}(\mathbf{x})\,d\mathbf{x} =$$

$$= -\int_{S_{ij}} \Delta\varphi^{(ij)}(\mathbf{x})\,d\mathbf{x} = 0.$$

Then,

$$\int_{S_i} \frac{\partial\varphi^{(ij)}}{\partial\mathbf{n}}(\mathbf{x})\,d\mathbf{x} = -\int_{S_j} \frac{\partial\varphi^{(ij)}}{\partial\mathbf{n}}(\mathbf{x})\,d\mathbf{x}. \tag{3.62}$$

Multiplying the equality

$$\Delta\varphi^{(ij)} = 0 \text{ in } S_{ij} \tag{3.63}$$

from (3.55) by $\varphi^{(ij)}(\mathbf{x})$ and integrating the result by parts in S_{ij} taking into account the remaining conditions from (3.50), we obtain the following equality:

$$0 = \int_{S_{ij}} \Delta\varphi^{(ij)}(\mathbf{x})\varphi^{(ij)}(\mathbf{x})\,d\mathbf{x} = \tag{3.64}$$

$$= -\int_{S_{ij}} |\nabla\varphi^{(ij)}(\mathbf{x})|^2 d\mathbf{x} + \int_{S_i \cup S_j} \frac{\partial\varphi^{(ij)}}{\partial\mathbf{n}}(\mathbf{x})\varphi^{(ij)}(\mathbf{x})\,d\mathbf{x}.$$

From (3.64), using that

$$\varphi^{(ij)}(\mathbf{x}) = t_i \text{ on } D_i,$$
$$\varphi^{(ij)}(\mathbf{x}) = t_j \text{ on } D_j$$

(see definition of the function $\varphi^{(ij)}(\mathbf{x})$ (3.50)), we have

$$-\int_{S_{ij}} |\nabla\varphi^{(ij)}(\mathbf{x})|^2 d\mathbf{x} = t_i \int_{S_i} \frac{\partial\varphi^{(ij)}}{\partial\mathbf{n}}(\mathbf{x}) d\mathbf{x} + t_j \int_{S_j} \frac{\partial\varphi^{(ij)}}{\partial\mathbf{n}}(\mathbf{x}) d\mathbf{x}. \qquad (3.65)$$

With allowance for (3.62), we obtain from (3.65) the following equality:

$$\int_{S_{ij}} |\nabla\varphi^{(ij)}(\mathbf{x})|^2 d\mathbf{x} = (t_i - t_j) \int_{S_j} \frac{\partial\varphi^{(ij)}}{\partial\mathbf{n}}(\mathbf{x}) d\mathbf{x}.$$

Then,

$$\int_{S_j} \frac{\partial\varphi^{(ij)}}{\partial\mathbf{n}}(\mathbf{x}) d\mathbf{x} = \frac{1}{t_i - t_j} \int_{S_{ij}} |\nabla\varphi^{(ij)}(\mathbf{x})|^2 d\mathbf{x}. \qquad (3.66)$$

We denoted by $\varphi^{\pm 1}(\mathbf{x})$ solution of the problem (3.50) for $t_i = 1/2$, $t_j = -1/2$ (see Definition 10). It means, in particular, that $\varphi^{\pm 1}(\mathbf{x})$ is the solution of the problem (3.50) for two bodies subjected to the difference of potentials equal to 1. Then,

$$\nabla\varphi^{(ij)}(\mathbf{x}) = (t_i - t_j)\nabla\varphi^{\pm 1}(\mathbf{x}), \qquad (3.67)$$

By virtue of (3.67), equality (3.66) takes the form

$$\int_{S_j} \frac{\partial\varphi^{(ij)}}{\partial\mathbf{n}} d\mathbf{x} = (t_i - t_j) \int_{S_{ij}} |\nabla\varphi^{\pm 1}(\mathbf{x})|^2 d\mathbf{x}. \qquad (3.68)$$

Definition 12. *The quantity*

$$C^{S_{ij}} = \int_{S_{ij}} |\nabla\varphi^{\pm 1}(\mathbf{x})|^2 d\mathbf{x} \qquad (3.69)$$

is referred to the capacity of the pair of the sets S_i and S_j (or D_i and D_j) with respect to the domain S_{ij}.

We note the following equality

$$C^{S_{ij}} = \int_{S_{ij}} |\nabla\varphi^{\pm 1}(\mathbf{x})|^2 d\mathbf{x} = \int_{S_j} \frac{\partial\varphi^{\pm 1}}{\partial\mathbf{n}}(\mathbf{x}) d\mathbf{x} \qquad (3.70)$$

takes place.

The capacities $C_{ij}^{(2)}$ (3.51) and $C^{S_{ij}}$ (3.69) are different. The capacity $C^{S_{ij}}$ can be treated as the capacity of a capacitor with the layers S_i and S_j under the condition that the electric field does not leave domain S_{ij}. Although, it is physically impossible, the definition is mathematically correct. The capacity $C^{S_{ij}}$ depends on the choice of the domain S_{ij}.

We calculated the flux $\int_{S_i} \dfrac{\partial \varphi^{(ij)}}{\partial \mathbf{n}}(\mathbf{x})d\mathbf{x}$ coming to the body D_i through one specific channel S_{ij}. Since the body D_i has several neighbors, generally, the integral in (3.53) is the total flux over all channels (which, by construction, do not intersect each other) that leads to D_i. Thus, to construct the trial function, we must specify the fluxes in each channel (the set of the local fluxes) in such a manner that the flux balance holds for all bodies simultaneously (globally). For this, we use the solution of the finite-dimensional network model (3.35) and (3.36). We consider the numbers (which have the meaning of fluxes in the network model)

$$p_{ij} = C^{(2)}_{ij}(t^{net}_i - t^{net}_j), \tag{3.71}$$

where $C^{(2)}_{ij}$ is the capacity of the pair of the bodies D_i and D_j in R^n (they are the coefficients of the network model, see (3.35)) and $\{t^{net}_i, i \in K\}$ are determined from solution of the network problem (3.35) and (3.36). In accordance with (3.35), p_{ij} satisfy the balance conditions for all bodies simultaneously. Condition (3.53) will be satisfied, if the flux in each channel S_{ij} satisfies the equality

$$\int_{S_i} \frac{\partial \varphi^{(ij)}}{\partial \mathbf{n}}(\mathbf{x})d\mathbf{x} = p_{ij}. \tag{3.72}$$

We note that this trick works for the bodies belonging to the alive net $K = CH(1)$, only, because for other bodies the quantities p_{ij} are not determined.

We assume that in the channel S_{ij} the function $\varphi(\mathbf{x})$ has the form

$$\varphi(\mathbf{x}) = \lambda_{ij}\varphi^{\pm 1}(\mathbf{x}). \tag{3.73}$$

Substituting (3.73) in (3.72), we obtain that λ_{ij} satisfies the condition

$$\int_{S_i} \lambda_{ij}\frac{\partial \varphi^{\pm 1}}{\partial \mathbf{n}}(\mathbf{x})d\mathbf{x} = \lambda_{ij}\int_{S_i} \frac{\partial \varphi^{\pm 1}}{\partial \mathbf{n}}(\mathbf{x})d\mathbf{x} = p_{ij}.$$

With regard for (3.70), this is equivalent to the following equality

$$\lambda_{ij}C^{S_{ij}} = p_{ij}. \tag{3.74}$$

Substituting (3.71) in (3.74), we obtain the equality

$$\lambda_{ij} = \frac{C^{(2)}_{ij}}{C^{S_{ij}}}(t^{net}_i - t^{net}_j). \tag{3.75}$$

If the trial function in the channel S_{ij} has the form (3.73), condition (3.53) is satisfied for

$$\mathbf{v}(\mathbf{x}) = \nabla \varphi(\mathbf{x}). \tag{3.76}$$

Note that for function (3.76) equality (3.52) is satisfied (in accordance with Lemma 3, (3.52) satisfied for $\nabla\varphi^{\pm 1}(\mathbf{x})$, then it is satisfied for $\lambda_{ij}\nabla\varphi^{\pm 1}(\mathbf{x})$). The condition (3.54) is satisfied for the function (3.76) since it is the solution of the problem (3.50). Thus, the function (3.76) under condition (3.75) can be taken as a trial function.

Now, we calculate integrals in (3.28) for the trial function $\mathbf{v}(\mathbf{x})$ (3.76) under the conditions (3.73), (3.75).

Computation of the volume integral from (3.28)

We call bodies closely placed if the distance between the neighbors tends to zero as $\delta \to 0$. We have defined the trial function $\mathbf{v}(\mathbf{x})$ by the equality (3.55).

Since $\mathbf{v}(\mathbf{x}) = 0$ outside the channels, the volume integrals are calculated over the channels, only. First, we calculate the integral over one channel S_{ij}. For function (3.76) under the conditions (3.73), (3.75)

$$\int_{S_{ij}} |\lambda_{ij}\nabla\varphi^{\pm 1}(\mathbf{x})|^2 dx = \lambda_{ij}^2 \int_{S_{ij}} |\nabla\varphi^{\pm 1}(\mathbf{x})|^2 dx = \lambda_{ij}^2 C^{S_{ij}}. \tag{3.77}$$

With regard for (3.75), the right-hand side of (3.77) is equal to

$$\lambda_{ij}^2 C^{S_{ij}} = \left[\frac{C_{ij}^{(2)}}{C^{S_{ij}}}\right]^2 (t_i^{net} - t_j^{net})^2 C^{S_{ij}} = \frac{\left[C_{ij}^{(2)}\right]^2}{C^{S_{ij}}}(t_i^{net} - t_j^{net})^2. \tag{3.78}$$

Since the channels do not intersect one other, the integral $-\dfrac{1}{2}\displaystyle\int_Q \mathbf{v}^2(\mathbf{x})dx$ is equal to sum over all channels between the bodies belonging to the network $K = CH(1)$. Using (3.78), we obtain

$$-\frac{1}{2}\int_Q \mathbf{v}^2(\mathbf{x})dx = -\frac{1}{2}\sum_{S_{ij}\in NCH(1)} \frac{\left[C_{ij}^{(2)}\right]^2}{C^{S_{ij}}}(t_i^{net} - t_j^{net})^2. \tag{3.79}$$

Here, index S_{ij} in the sum indicates that the summation is performed over the channels (i.e., one channel contributes one summand); $NCH(1)$ denotes the channels between the bodies belonging to the network $K = CH(1)$. Writing (3.79), we take into account that under construction of the function $\mathbf{v}(\mathbf{x})$, only the bodies belonging to nets (chains) $CH(1)$ are connected by channels. Outside the mentioned channels $\mathbf{v}(\mathbf{x}) = 0$.

When the summation over channels in (3.79) is replaced by summation over the indices, each channel contributes two summands. The result is

$$-\frac{1}{2}\int_Q \mathbf{v}^2(\mathbf{x})dx = -\frac{1}{4}\sum_{i\in K}\sum_{j\in N_i} \frac{\left[C_{ij}^{(2)}\right]^2}{C^{S_{ij}}}(t_i^{net} - t_j^{net})^2. \tag{3.80}$$

The factor $1/4$ appears for the reason of double counting the channels (see Section 3.3 for the detailed explanation).

Computation of the boundary integral from (3.28)

Now, we consider electrostatic channel S_{1j} between the body D_1 and the neighbor bodies D_j ($j \in N_1$). We introduce function $\varphi(\mathbf{x})$ in the channel S_{1j} as

$$\varphi(\mathbf{x}) = \frac{\varphi_1^0 + t_j^{net}}{2} + (\varphi_1^0 - t_j^{net})\varphi^{\pm 1}(\mathbf{x}), \tag{3.81}$$

and define trial function $\mathbf{v}(\mathbf{x})$ as

$$\mathbf{v}(\mathbf{x}) = \begin{cases} (\varphi_1^0 - t_j^{net})\nabla\varphi^{\pm 1}(\mathbf{x}) \text{ in } S_{1j}, \\ 0 \text{ outside } S_{1j} \end{cases} \tag{3.82}$$

This function is divergence free in all domain Q.

For the function $\varphi^{\pm 1}(\mathbf{x})$ the following equality takes place, see (3.69):

$$\int_{S_1} \frac{\partial \varphi^{\pm 1}}{\partial \mathbf{n}}(\mathbf{x})d\mathbf{x} = \int_{S_{1j}} |\nabla\varphi^{\pm 1}(\mathbf{x})|^2 d\mathbf{x}. \tag{3.83}$$

Here S_1 is the base of the channel S_{1j}, which borders with the body D_1.

From (3.82), (3.83), we obtain

$$\int_{S_1} \mathbf{v}(\mathbf{x})\mathbf{n}d\mathbf{x} = (\varphi_1^0 - t_j^{net})\int_{S_{1j}} |\nabla\varphi^{\pm 1}(\mathbf{x})|^2 d\mathbf{x}. \tag{3.84}$$

In accordance with the definition

$$\int_{S_{1j}} |\nabla\varphi^{\pm 1}(\mathbf{x})|^2 d\mathbf{x} = C^{S_{1j}}.$$

Then (3.84) is equal to the flux

$$p_{1j} = (\varphi_1^0 - t_j^{net})C^{S_{1j}} \tag{3.85}$$

from the body D_j to the body D_1 through the channel S_{1j}.

The formula (3.85) gives the flux coming to the body D_1 through the single channel S_{1j}. The sum of fluxes over all channels entering D_1 (i.e., the total flux to the body D_1). We denote this sum by P^+ and write

$$P^+ = \sum_{j \in N_1} C^{S_{1j}}(\varphi_1^0 - t_j^{net}).$$

The boundary integral from (3.28) over ∂D_1 is equal to $\varphi_1^0 P^+$. The boundary integral from (3.28) over ∂D_2 can be computed in the similar way and it is equal to $\varphi_2^0 P^-$, where

$$P^- = \sum_{j \in N_2} C^{S_{2j}}(\varphi_2^0 - t_j^{net}),$$

N_1 and N_2 are indices of the neighbors of the bodies D_1 and D_2 correspondingly.

Thus, both integrals in (3.28) (volume and boundary) are calculated and we obtain the following estimate

$$\mathcal{E} \geq -\frac{1}{4} \sum_{i \in K} \sum_{j \in N_i} \frac{\left[C_{ij}^{(2)} \right]^2}{C^{S_{ij}}} (t_i^{net\cdot} - t_j^{net})^2 + \varphi_1^0 P^+ + \varphi_2^0 P^-. \tag{3.86}$$

3.4.4. Refined upper-sided estimate

In this section, we refine the general upper-sided estimate (3.16). In a domain between neighboring bodies (in the channel, but now it is a cylindrical channel), the trial functions are taken equal to the electrostatics solution. Outside the channels (where the fluxes are physically insignificant), the trial functions are taken from the condition that their derivatives are bounded.

Cylindrical channel and trial function

We consider a cylindrical channel R_{ij}, see Fig. 3.11. The channel R_{ij} has a shape of circular cylinder connecting two bodies.

We define a trial function in the channel R_{ij} as

$$\varphi(\mathbf{x}) = \frac{t_i + t_j}{2} + (t_i - t_j)\varphi^{\pm 1}(\mathbf{x}). \tag{3.87}$$

where $\varphi^{\pm 1}(\mathbf{x})$ is solution of electrostatic problem (3.50) and $\{t_i,\ i = 1, \ldots, N\}$ are arbitrary real numbers. It means that the trial function (3.87) is a restriction of the electrostatic solution to the channel R_{ij}. We investigate the integral $\int_{R_{ij}} |\nabla \varphi(\mathbf{x})|^2 dx$ for the cylindrical channel S_{ij} and trial function defined by (3.87).

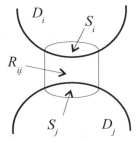

Figure 3.11. *Cylindrical channel between bodies.*

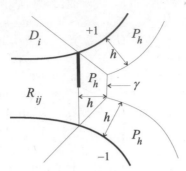

Figure 3.12. *A covering neighborhood P_h of bodies and channel between bodies.*

Note that

$$\varphi(\mathbf{x}) = t_i \text{ on } D_i \text{ and } \varphi(\mathbf{x}) = t_j \text{ on } D_j. \tag{3.88}$$

It follows from the definitions of the function $\varphi^{\pm 1}(\mathbf{x})$ (see Definition 10).

For the function (3.87), we have for the channel R_{ij}

$$\int_{R_{ij}} |\nabla \varphi(\mathbf{x})|^2 d\mathbf{x} = (t_i - t_j)^2 \int_{R_{ij}} |\nabla \varphi^{\pm 1}(\mathbf{x})|^2 d\mathbf{x}.$$

Construction of a trial function outside the channel R_{ij}

We demonstrate that the function (3.87) can be continued from the cylindrical channel R_{ij} to whole domain Q in such a way that it will belong to the functional space V (3.10) and its derivatives are restricted uniformly over the positions of the bodies.

Definition 13. *We say that a quantity (function) f depending of the parameter $\delta \to 0$ is separated from zero uniformly as $\delta \to 0$ if there exists a number $a > 0$ such that $f \geq a$ for all $\delta \to 0$.*

The diameter of the channel R_{ij} is fixed. Outside the channel R_{ij} the distance between the bodies D_i and D_j is separated from zero uniformly as $\delta \to 0$. Also, distances between different channels are separated from zero uniformly as $\delta \to 0$. Outside the channels, we can construct a neighborhood P_h of the body D_i and channel R_{ij} which width h does not depend on $\delta \to 0$, see Fig. 3.12. We call this neighborhood the covering if one side of the neighborhood contacts the body D_i and the channel R_{ij} and another side does not contact the body D_i nor other bodies. A typical fragment of the covering neighborhood P_h is shown in Fig. 3.12. As $\delta \to 0$, the neighborhood P_h passes to its limit (corresponding to $\delta = 0$) position. The covering channel also passes to a limit position and saves the properties indicated above for all $\delta \to 0$.

In the domain P_h, we can construct a function $\psi^\delta(\mathbf{x})$ that takes specified values on the bodies D_i and D_j and on the boundary of the channel R_{ij} and vanishes on the surface γ shown at Fig. 3.12. The question consists of the possibility to construct a function $\psi^\delta(\mathbf{x})$ subjected to the additional condition that its derivatives are bounded uniformly over $\delta \to 0$.

Since, as $\delta \to 0$, domain P_h tends to its limit (corresponding to $\delta = 0$) position, function $\psi^\delta(\mathbf{x})$ with the required properties can be constructed if derivatives of its boundary value (i.e., value of function $\psi^\delta(\mathbf{x})$ on the boundary of the domain P_h) is uniformly bounded, as $\delta \to 0$.

We know that

$$\psi^\delta(\mathbf{x}) = \frac{1}{2} \text{ at boundary of domain } D_i,$$

$$\psi^\delta(\mathbf{x}) = 0 \text{ at } \gamma,$$

and on the boundary of the channel R_{ij},

$$\psi^\delta(x) = \frac{t_i + t_j}{2} + (t_i - t_j)\varphi^{\pm 1}(\mathbf{x}). \tag{3.89}$$

We have no more *a priori* information about the function $\varphi^\delta(\mathbf{x})$ except it is the solution of the problem (3.50). Nevertheless, this is sufficient to obtain the required estimates. To do that we need some estimates for solution of equations of elliptic type in "interior domain" and near a "smooth boundary". Such estimates are known, see, e.g. [105, 260, 266, 321]. We formulate results by Ladizenskaiy and Ural'tseva [200, Chapter 15] as applied to the problem under consideration.

Lemma 4. *[200]. Let $u(\mathbf{x})$ be a solution of Dirichlet problem for Laplace equation in a domain Ω. Let $S_1 \subseteq \partial\Omega$ be a part of the boundary of domain Ω and S_1 is a surface of class C^2. Boundary condition*

$$u(\mathbf{x}) = \varphi(\mathbf{x})$$

is set in S_1, with function $\varphi(\mathbf{x}) \in C^2(S_1)$. Let $\Omega_1 \subseteq \Omega$ be a subset of Ω, such that the distance between Ω_1 and $\partial\Omega \backslash S_1$ is $\rho > 0$.

Then, $\|\nabla u\|_{C^1(\Omega_1)}$ can be estimated from above by a constant, which depends only on $|u|_{L_2(\Omega)}$, $|\varphi|_{C^2(S_1)}$, ρ and smoothness of the surface S_1.

Formulating Lemma 4, we use the obvious fact that a function $u(\mathbf{x}) \in C^2(S_1)$ can be continued from a surface S_1 of C^2 class into all domain Ω in such a way that the norm $\|u\|_{C^2(\Omega)}$ is bounded by quantity *const* $\|\nabla u\|_{C^2(S_1)}$, where *const* $< \infty$ does not depend on the function $u(\mathbf{x})$.

Note 1. *The smoothness (the class C^m) of a surface is determined by the smoothness of functions used for the unbending the surface [200]. In Lemma 4, the unbending function for surface S_1 belongs to C^2.*

Note 2. *With regard to principle of maximum harmonic function, we can change $\|u\|_{L_2(\Omega)}$ for $\|u\|_{(\partial\Omega)}$ in Lemma 4.*

We introduce the following function

$$\varphi(\mathbf{x}) = \begin{cases} \dfrac{t_i + t_j}{2} + (t_i - t_j)\varphi^{\pm 1}(\mathbf{x}) \text{ in } R_{ij}, \\[2mm] \psi^\delta(\mathbf{x}) \text{ in } P_h, \\[2mm] 0 \text{ outside } R_{ij} \bigcup P_h, \end{cases} \tag{3.90}$$

The last two lines in (3.90) describe the continuation of the function (3.87) from the channel R_{ij} to whole domain Q. By construction, the function (3.90) belongs to the set V and can be used as a trial function in (3.16).

For (3.90), we write

$$I(\varphi) = \frac{1}{2} \int_Q |\nabla \varphi(\mathbf{x})|^2 d\mathbf{x} =$$

$$= \frac{1}{4} \sum_{k=1}^{C} \sum_{i \in CH(k)} \sum_{j \in N_i} (t_i - t_j)^2 C^{R_{ij}} + \int_{P_h} |\nabla \psi^\delta(\mathbf{x})|^2 d\mathbf{x}, \tag{3.91}$$

where C is the total numbers of chains (nets) and $CH(k)$ denotes the k-th chine (net).

The quantity

$$C^{R_{ij}} = \int_{R_{ij}} |\nabla \varphi^{\pm 1}(\mathbf{x})|^2 dx \tag{3.92}$$

is capacity of the bodies D_i and D_j with respect to the cylindrical channel R_{ij}.

Taking into account (3.16), (3.91) and (3.92), we obtain the following estimate:

$$\mathcal{E} \leq \frac{1}{4} \sum_{k=1}^{C} \sum_{i \in CH(k)} \sum_{j \in N_i} (t_i - t_j)^2 C^{R_{ij}} + \int_{P_h} |\nabla \psi^\delta(\mathbf{x})|^2 dx, \tag{3.93}$$

It is necessary estimate the last integral in the right-hand part of (3.93). We consider the function $\psi^\delta(\mathbf{x})$ in the domain P_h. The boundary value of this function is known exactly except its value on the boundary of the channel R_{ij}. On R_{ij}, this function is equal to

$$\frac{t_i + t_j}{2} + (t_i - t_j)\varphi^{\pm 1}(\mathbf{x})$$

but we do not know the exact value of $\varphi^{\pm 1}(\mathbf{x})$. We investigate the function $\varphi^{\pm 1}(\mathbf{x})$ on the boundary of the channel R_{ij}. Consider a part of boundary of the channel R_{ij} and of domain D_i, see Fig. 3.13. Denote this surface by r. It is possible to surround the surface r with a neighborhood O_z of thickness $z > 0$ not depending on δ, see

Fig. 3.13. On the boundary of O_z, with regard to principle of maximum for Laplace equation,

$$|\psi^\delta(\mathbf{x})| \leq \frac{1}{2}.$$

Then, due to Lemma 4, $|\nabla\varphi^{\pm 1}(\mathbf{x})|$ is bounded on the boundary of the channel R_{ij} uniformly with respect to δ then, there exists a function $\psi^\delta(\mathbf{x})$, which is equal to $\varphi^{\pm 1}(\mathbf{x})$ at the boundary of the channel R_{ij} and has the first derivatives bounded in P_h uniformly over $\delta \to 0$. Then, taking into account that

$$|\nabla\psi^\delta(\mathbf{x})| = |(t_i - t_j)\nabla\varphi^{\pm 1}(\mathbf{x})|$$

is bounded in P_h and the domain P_h is finite, we obtain that the integral

$$\int_{P_h} |\nabla\psi^\delta(\mathbf{x})|^2 dx \leq C, \tag{3.94}$$

where $C < \infty$ does not depend on δ.

3.5. Completion of proof of the theorems

According to the estimates (3.86) and (3.93),

$$\mathcal{E} \geq -\frac{1}{4} \sum_{i \in CH(1)} \sum_{j \in N_i} \frac{\left[C_{ij}^{(2)}\right]^2}{C^{S_{ij}}} (t_i^{net} - t_j^{net})^2 + \varphi_1^0 P^+ + \varphi_2^0 P^-, \tag{3.95}$$

$$\mathcal{E} \leq \frac{1}{4} \sum_{k=1}^{C} \sum_{i \in CH(k)} \sum_{j \in N_i} C^{R_{ij}} (t_i^{net} - t_j^{net})^2 + \int_{P_h} |\nabla\psi^\delta(\mathbf{x})|^2 dx. \tag{3.96}$$

We note that the estimates (3.93), (3.94) are valid for arbitrary $\{t_i, i = 1, \ldots, N\}$. In particular, they are valid for the solution $\{t_i^{net}, i \in K\}$ of the network problem (3.35) and (3.36). For our aim, we will use the estimates (3.93) and (3.94) in the form (3.96).

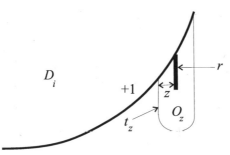

Figure 3.13. *Fragment of the covering neighborhood presented in Fig. 3.12.*

Lemma 5. *The following equality holds for the solution $\{t_i^{net}, i \in K\}$ of the network problem (3.35) and (3.36)*

$$\varphi_1^0 P^+ + \varphi_2^0 P^- = \frac{1}{2} \sum_{i \in K} \sum_{j \in N_i} C_{ij}^{(2)} (t_i^{net} - t_j^{net})^2, \qquad (3.97)$$

here $K = CH(1)$ is the alive net.

Formula (3.97) differs from the usual Green formula by the factor $1/2$. It appears for the same reason that $1/4$ appears in the discrete formula for energy presented above in Section 3.3.

Proof. We write

$$\frac{1}{2} \sum_{i \in K} \sum_{j \in N_i} C_{ij}^{(2)} (t_i^{net} - t_j^{net})^2 = \qquad (3.98)$$

$$= \frac{1}{2} \sum_{i \in K} \sum_{j \in N_i} C_{ij}^{(2)} (t_i^{net} - t_j^{net})(t_i^{net} - t_j^{net}) =$$

$$= \frac{1}{2} \sum_{i \in K} t_i^{net} \left[\sum_{j \in N_i} C_{ij}^{(2)} (t_i^{net} - t_j^{net}) \right] - \frac{1}{2} \sum_{i \in K} t_j^{net} \left[\sum_{i \in N_i} C_{ij}^{(2)} (t_i^{net} - t_j^{net}) \right].$$

The first sum in the square brackets in the right-hand side of (3.98) is equal to zero due to (3.35) for $i \in I = \{3, ..., N\}$. Then only terms, which correspond to D_1 and D_2 remain

$$\frac{1}{2} \sum_{i=1,2} t_i^{net} \sum_{j \in N_i} C_{ij}^{(2)} (t_i^{net} - t_j^{net}). \qquad (3.99)$$

We consider the last sum in (3.98). Note that the summation is carried out over all ribs of the net K. Taking into account that $C_{ij} = C_{ji}$, we can change the order of summation and rewrite the last sum in (3.98) as

$$\frac{1}{2} \sum_{j \in K} t_j^{net} \left[\sum_{i \in N_j} C_{ij}^{(2)} (t_i^{net} - t_j^{net}) \right].$$

For the reasons presented above, it is also equal to (3.99).

Note that $t_1^{net} = \varphi_1^0$ and $t_2^{net} = \varphi_2^0$ in (3.99). Hence (3.98) is exactly equal to $\varphi_1^0 P^+ + \varphi_2^0 P^-$. **Proof is completed**.

From (3.95) and (3.97) we obtain the following lower-sided estimate

$$\mathcal{E} \geq \frac{1}{2} \sum_{i \in K} \sum_{j \in N_i} C_{ij}^{(2)} \left(-\frac{1}{2} \frac{\left[C_{ij}^{(2)} \right]^2}{C^{S_{ij}}} + 1 \right) (t_i^{net} - t_j^{net})^2. \qquad (3.100)$$

From (3.96) and (3.100), we obtain the following two-sided estimate:

$$\frac{1}{2} \sum_{i \in K} \sum_{j \in N_i} C_{ij}^{(2)} \left(-\frac{1}{2} \frac{\left[C_{ij}^{(2)} \right]^2}{C^{S_{ij}}} + 1 \right) (t_i^{net} - t_j^{net})^2 \leq \mathcal{E} \leq \qquad (3.101)$$

$$\leq \frac{1}{4} \sum_{k=1}^{C} \sum_{i \in CH(k)} \sum_{j \in N_i} C^{S_{ij}} (t_i^{net} - t_j^{net})^2 + \int_{P_h} |\nabla \psi^\delta(\mathbf{x})|^2 d\mathbf{x}.$$

3.5.1. Theorem about NL zones

We consider two bodies (domains) D_i and D_j in R^n and a channel K between bodies. Here K implies electrostatic channel S_{ij} or cylindrical channel R_{ij} (we do not use channels of other types here). Recall that we denote $\delta_{ij} = \delta_{ij}(\delta)$ the distance between the bodies D_i and D_j.

We denote

$$\mathcal{J} = \{(i, j) : \delta_{ij}(\delta) \to 0 \text{ as } \delta \to 0\}$$

the pairs of the bodies, which touch one another in the limit positions (near touching bodies as $\delta \to 0$),

$$K = \bigcup_{(i,j) \in \mathcal{J}} K_{ij}$$

the system of channels between the near touching bodies (here K_{ij} means S_{ij} or R_{ij}). We denote by D the system of the near touching bodies.

Definition 14. *We refer quantities (functions) f and g asymptotically equivalent as $\delta \to 0$ and write $f \sim g$ as $\delta \to 0$, if*

$$\frac{f}{g} \to 1 \text{ as } \delta \to 0.$$

Theorem 1. *For the bodies and channels described above, the energy outside the channels K*

$$\int_{R^n \setminus (K \bigcup D)} |\nabla \varphi^{\pm 1}(\mathbf{x})|^2 d\mathbf{x} \leq C, \qquad (3.102)$$

where $C < \infty$ does not depend on δ.

Proof. We consider two almost touching as $\delta \to 0$ (touching in those limit positions, as $\delta = 0$) bodies D_i and D_j. The consideration of moving bodies creates some technical difficulties in obtaining of estimates. However, the fact that the bodies move to the known limit positions (position of touching) reduces these difficulties considerably.

Estimation of the energy outside the sphere $\{|\mathbf{x}| = R\}$

We surround two approaching (moving to their limit positions) bodies D_i and D_j by a sphere

$$S_R = \{\mathbf{x} : |\mathbf{x}| = R\}$$

of fixed radius R. The bodies D_i and D_j stay inside the sphere S_R for all $\delta \to 0$. Prove that the energy outside the sphere S_R is bounded uniformly with respect to $\delta \to 0$.

Case $n = 3$

On the sphere S_R by virtue of the principle of maximum for Laplace equation,

$$|\varphi^{\pm 1}(\mathbf{x})| \leq \frac{1}{2}$$

irrespective of the positions of the bodies D_i and D_j. We write Poisson's integral (see, e.g., [94]),

$$\varphi^{\pm 1}(\mathbf{x}) = \frac{1}{4\pi R} \int_{|\mathbf{y}|=R} \frac{R^2 - |\mathbf{x}|^2}{|\mathbf{x} - \mathbf{y}|^2} \varphi_R(\mathbf{y}) d\mathbf{y}, \tag{3.103}$$

where $\varphi_R(\mathbf{y})$ means the value of the function $\varphi^{\pm 1}(\mathbf{y})$ on the sphere

$$S_R = \{\mathbf{y} : |\mathbf{y}| = R\}.$$

From (3.103), it follows, see [94], that

$$|\varphi^{\pm 1}(\mathbf{x})| \leq \frac{C}{|\mathbf{x}|^2}, \tag{3.104}$$

as $\mathbf{x} \to 0$, where $C < \infty$ does not depend on position of D_i and D_j.

Differentiating (3.103) (for $|\mathbf{y}| \geq R$ it is possible to change the order of differentiation and integration operations in (3.103), see, e.g., [94]), we have

$$\frac{\partial \varphi^{\pm 1}}{\partial x_i}(\mathbf{x}) = \frac{1}{4\pi R} \int_{|\mathbf{y}|=R} \frac{\partial}{\partial x_i} \frac{R^2 - |\mathbf{x}|^2}{|\mathbf{x} - \mathbf{y}|^2} \varphi_R(\mathbf{y}) d\mathbf{y}. \tag{3.105}$$

With regard to

$$|\varphi_R(\mathbf{y})| \leq 1 \text{ on } S_R = \{\mathbf{x} : |\mathbf{x}| = R\},$$

we obtain from (3.105) that

$$\left| \frac{\partial \varphi^{\pm 1}}{\partial x_i}(\mathbf{x}) \right| \leq \frac{1}{4\pi R} \int_{|\mathbf{y}|=R} \left| \frac{\partial}{\partial x_i} \frac{R^2 - |\mathbf{x}|^2}{|\mathbf{x} - \mathbf{y}|^2} \right| d\mathbf{y}. \tag{3.106}$$

For $|\mathbf{x}| \geq R$ and $|\mathbf{y}| \leq R$, we have

$$\frac{\partial}{\partial x_i} \frac{R^2 - |\mathbf{x}|^2}{|\mathbf{x} - \mathbf{y}|^2} =$$

$$= \frac{-2x_i|\mathbf{x} - \mathbf{y}|^2 - 2(R^2 - |\mathbf{x}|^2)(x_i - y_i)}{|\mathbf{x} - \mathbf{y}|^4} \sim$$

$$\sim \frac{|\mathbf{y}|^2 + |\mathbf{y}|}{|\mathbf{y}|^4} \sim \frac{1}{|\mathbf{y}|^2}$$

as $\mathbf{y} \to \infty$.

Then, it follows from this estimate and (3.106) that

$$|\nabla \varphi^{\pm 1}(\mathbf{x})| \leq \frac{C}{R^2}, \tag{3.107}$$

where $C < \infty$ does not depend on the positions of D_i and D_j during their approach as $\delta \to 0$.

Applying Green's formula to the spherical layer $R \leq |\mathbf{x}| \leq \rho$, we obtain with regard for (3.104) and (3.107)

$$\int_{R \leq |\mathbf{y}| \leq \rho} |\nabla \varphi^{\pm 1}(\mathbf{x})|^2 d\mathbf{x} = \tag{3.108}$$

$$= \int_{|\mathbf{y}|=R} \varphi^{\pm 1}(\mathbf{x}) \frac{\partial \varphi^{\pm 1}}{\partial \mathbf{n}}(\mathbf{x}) d\mathbf{x} + \int_{|y|=\rho} \varphi^{\pm 1}(\mathbf{x}) \frac{\partial \varphi^{\pm 1}}{\partial \mathbf{n}}(\mathbf{x}) dx \leq$$

$$\leq \int_{|\mathbf{y}|=R} \frac{C}{RR^2} d\mathbf{x} + \int_{|y|=\rho} \frac{C}{\rho \rho^2} d\mathbf{x} = 4\pi R^2 \frac{C}{R^3} + 4\pi \rho^2 \frac{C}{\rho^3},$$

where $C < \infty$ does not depend on the position of D_i and D_j when these bodies approach each other. When $\rho \to \infty$, it follows from (3.108) that

$$\int_{R \leq |\mathbf{x}|} |\nabla \varphi^{\pm 1}(\mathbf{x})|^2 d\mathbf{x} \leq C_1,$$

where $C_1 < \infty$ does not depend on δ.

Case $n = 2$

In this case from the Poisson integral it follows that

$$|\varphi^{\pm 1}(\mathbf{x})| \leq \frac{C}{|\mathbf{x}|}, \tag{3.109}$$

as $\mathbf{x} \to 0$, where $C < \infty$ does not depend on position of D_i and D_j and (3.107) holds for $n = 2$.

Applying Green's formula to the cylindrical layer $R \leq |\mathbf{x}| \leq \rho$, we obtain with regard for (3.107) and (3.109)

$$\int_{R \leq |\mathbf{y}| \leq \rho} |\nabla \varphi^{\pm 1}(\mathbf{x})|^2 d\mathbf{x} = 4\pi R \frac{C}{R^2} + 4\pi \rho \frac{C}{\rho^2} < \infty. \qquad (3.110)$$

Estimate of energy inside the sphere $\{|\mathbf{x}| = R\}$

Now, we consider set

$$M = S_R \backslash (D_i \bigcup D_j \bigcup K_{ij})$$

– the sphere S_R without the bodies (domain) D_i and D_j and the channel K. We can cover set M with a finite set of domains $\{O_s, s = 1, \ldots, S\}$, which do not depend on the position of the domains D_i and D_j. When the bodies D_i and D_j approach to each other, the domains $\{O_s, s = 1, \ldots, S\}$ can overlap one another. This overlapping can only increase the energy computed over O_s, $s = 1, \ldots, S$. Thus, the sum of energies over $\{O_s, s = 1, \ldots, S\}$ is an upper-sided estimate for the energy in M. A typical domain O_s is shown in Fig. 3.14.

Domain O_s is an inner subdomain of domain Q_s^+, where domain Q_s^+ does not depend on the position of the bodies D_i and D_j. The domain Q_s^+ follows the body with which it is associated. The function $\varphi^{\pm 1}(\mathbf{x})$ and the boundaries of the body (domain) D_i and the domain O_s satisfy condition of Lemma 4. Namely, the following equalities are satisfied for the function $\varphi^{\pm 1}(\mathbf{x})$:

$$\varphi^{\pm 1}(\mathbf{x}) = \frac{1}{2} \text{ on } D_i,$$

$$|\nabla \varphi^{\pm 1}(\mathbf{x})| \leq \frac{1}{2} \text{ on the boundary of domain } Q_s^+.$$

The boundary the boundary of domain Q_s^+ can be taken as smooth as necessary and Q_s^+ can be taken in such a way that the distance between the boundary of

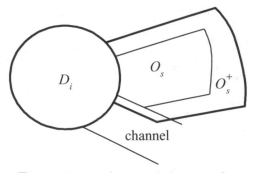

Figure 3.14. *A typical domain O_s.*

domain Q_s^+ and boundary of domain O_s is isolated from zero uniformly with respect to δ. Then, by virtue of Lemma 4, $|\nabla\varphi^{\pm 1}(\mathbf{x})|$ is bounded in O_s by a value, which does not depend on position of D_i and D_j.

Thus, the integral of function $|\nabla\varphi^{\pm 1}(\mathbf{x})|^2$ over $M = S_R \backslash (D_i \bigcup D_j \bigcup R_{ij})$ is bounded uniformly with respect to δ. **Proof is completed.**

3.5.2. Theorem about asymptotic equivalence of the capacities

Now we prove that the asymptotic equivalence of capacities $C_{ij}^{(2)}$, $C^{S_{ij}}$ and $C^{R_{ij}}$ as $\delta \to 0$ follows from Theorem 1 under condition (3.1).

Theorem 2. *If the bodies D_i and D_j are neighbors and the condition*

$$C_{ij}^{(2)} \to \infty \quad as \ \delta \to 0 \tag{3.111}$$

holds then the capacities $C_{ij}^{(2)}$, $C^{S_{ij}}$ and $C^{R_{ij}}$ are asymptotically equivalent:

$$C_{ij}^{(2)} \sim C^{S_{ij}} \sim C^{R_{ij}} \tag{3.112}$$

as $\delta \to 0$.

Proof. The asymptotic equivalence $C_{ij}^{(2)} \sim C^{R_{ij}}$ follows from the equality

$$C_{ij}^{(2)} = C^{R_{ij}} + \int_{R^n \backslash (R_{ij} \bigcup D_i \bigcup D_j)} |\nabla\varphi^{\pm 1}(\mathbf{x})|^2 d\mathbf{x}, \tag{3.113}$$

which follows from the definitions of the quantities $C_{ij}^{(2)}$ and $C^{R_{ij}}$ and the boundedness of integral

$$\int_{R^n \backslash (R_{ij} \bigcup D_i \bigcup D_j)} |\nabla\varphi^{\pm 1}(\mathbf{x})|^2 d\mathbf{x}$$

uniformly over $\delta \to 0$ (with regard to Theorem 1).

For the bodies D_i and D_j, integral

$$\int_{R^n \backslash (D_i \bigcup D_j)} |\nabla\varphi^{\pm 1}(\mathbf{x})|^2 d\mathbf{x}$$

is equal to capacity $C_{ij}^{(2)}$ of these bodies in R^n. Under condition (3.111)

$$\int_{R^n \backslash (D_i \bigcup D_j)} |\nabla\varphi^{\pm 1}(\mathbf{x})|^2 d\mathbf{x} \to \infty$$

as $\delta \to 0$.

With regard to Theorem 1, integral

$$\int_{R^n \backslash (R_{ij} \bigcup D_i \bigcup D_j)} |\nabla\varphi^{\pm 1}(\mathbf{x})|^2 d\mathbf{x}$$

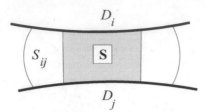

Figure 3.15. *Cylinder* $\mathbf{S} \subset S_{ij}$.

is bounded uniformly with respect to $\delta \to 0$. Then $C^{R_{ij}} \to \infty$ as $\delta \to 0$.

Dividing (3.113) by $C^{R_{ij}}$, we obtain

$$\frac{C_{ij}^{(2)}}{C^{R_{ij}}} = 1 + \frac{\int_{R^n \backslash (R_{ij} \bigcup D_i \bigcup D_j)} |\nabla \varphi^{\pm 1}(\mathbf{x})|^2 dx}{C^{R_{ij}}} \to 1 \qquad (3.114)$$

as $\delta \to 0$.

Now we prove that

$$C_{ij}^{(2)} \sim C^{S_{ij}}$$

as $\delta \to 0$.

The channel S_{ij} is not a cylindrical channel. By construction its width is separated from zero uniformly for all positions of D_i and D_j as $\delta \to 0$. Then, it is possible to inscribe into S_{ij} a cylinder $\mathbf{S}$, which width does not depend on the positions of D_i and D_j as $\delta \to 0$, see Fig. 3.15.

We consider the capacity

$$C^{\mathbf{S}} = \int_{\mathbf{S}} |\nabla \varphi^{\pm 1}(\mathbf{x})|^2 dx$$

with respect to this cylinder $\mathbf{S}$. The following inequalities hold

$$C^{\mathbf{S}} \leq C^{S_{ij}} \leq C_{ij}^{(2)}, \qquad (3.115)$$

because the capacities $C^{\mathbf{S}}$, $C^{S_{ij}}$, $C_{ij}^{(2)}$ are computed by integration of the function $|\nabla \varphi^{\pm}(\mathbf{x})|^2 \geq 0$ over the domains $S \subset S_{ij} \subset R^n \backslash (D_i \bigcup D_j)$.

For cylindrical domain $\mathbf{S}$ Theorem 2 is valid. By virtue of this,

$$C^{\mathbf{S}} \sim C_{ij}^{(2)}$$

as $\delta \to 0$.

With regard to this equivalence, it follows from (3.115) that

$$C^{S_{ij}} \sim C_{ij}^{(2)}$$

as $\delta \to 0$.

Proof is completed.

We define the energy $\mathcal{E}_d$ in the finite-dimensional network introduced above (see Section 3.3) as

$$\mathcal{E}_d = \frac{1}{4} \sum_{i \in K} \sum_{j \in N_i} C_{ij}^{(2)} (t_i^{net} - t_j^{net})^2, \tag{3.116}$$

where $\{t_i^{net}, i \in K\}$ is a solution of the discrete network problem (3.35) and (3.36). The multiplier $\dfrac{1}{4}$ in formula (3.116) instead of the usual multiplier $\dfrac{1}{2}$ is a result of double counting of the channels in the formula (3.116).

3.5.3. Theorem about network approximation

The pair capacities depend on the distances between the bodies. We assume that all the distances between the bodies forming chains (nets) have the same order when $\delta \to 0$.

Lemma 6. *Let all capacities $C_{ij}^{(2)}$ of neighboring bodies belonging to an alive network have the same order $f(\delta) \to \infty$ as $\delta \to 0$. Then the energy $\mathcal{E}_d$ (3.116) in the network has the order $f(\delta)$ as $\delta \to 0$, and*

$$\sum_{k=1}^{C} \sum_{i \in CH(k)} \sum_{j \in N_i} C^{R_{ij}} (t_i - t_j)^2 = \tag{3.117}$$

$$= \sum_{i \in K} \sum_{j \in N_i} C^{R_{ij}} (t_i^{net} - t_j^{net})^2 + O(1) \ \text{as} \ \delta \to 0,$$

here $K = CH(1)$ is an alive net and $\{t_i, i = 1, \ldots, N\}$ are determined from (3.2)–(3.8).

Formula (3.117) saves its validity if one changes $C^{R_{ij}}$ for any of the asymptotically equivalent capacities indicated in (3.112).

Proof. The condition of Lemma 6 means that there exist $0 < c, C < \infty$ such that

$$cf(\delta) \le C_{ij}^{(2)} \le Cf(\delta) \ \text{as} \ \delta \to 0,$$

for any neighboring bodies.

Consider the alive network (indices $i \in K$) with the following characteristics (conductivities) of edges:

$$H_{ij} = \begin{cases} cf(\delta) \ \text{if} \ C_{ij}^{(2)} > 0, \\[2mm] 0 \ \text{if} \ C_{ij}^{(2)} = 0. \end{cases}$$

The equality $C_{ij}^{(2)} = 0$ takes place for not neighboring bodies.

For the quadratic form with the coefficients $C_{ij}^{(2)}$ and H_{ij} the following inequality takes place:

$$\sum_{i \in K} \sum_{j \in N_i} C_{ij}^{(2)} (z_i - z_j)^2 \geq \sum_{i \in K} \sum_{j \in N_i} H_{ij} (z_i - z_j)^2 \qquad (3.118)$$

for any $\mathbf{z} \in R^N$.

Let $\{t_i, i \in I\}$ and $\{T_i, i \in I\}$ be solutions of problem (3.35), (3.36) and solution of the similar problem with coefficients (the edge's conductivities) H_{ij}, correspondingly. The last problem has the form,

$$f(\delta) \sum_{j \in N_i} H_{ij} (T_i - T_j) = 0, i \in K, \qquad (3.119)$$

$$T_1 = \varphi_1^0, T_2 = \varphi_2^0.$$

The energy in the network $\{\mathbf{x}_i, C_{ij}^{(2)}; i, j \in K\}$ is greater than the energy in the network $\{\mathbf{x}_i, H_{ij}; i, j \in K\}$. Really, these energies are minimums of the functions

$$\sum_{i \in K} \sum_{j \in N_i} C_{ij}^{(2)} (z_i - z_j)^2 \qquad (3.120)$$

and

$$\delta \sum_{i \in K} \sum_{j \in N_i} H_{ij} (z_i - z_j)^2, \qquad (3.121)$$

correspondingly.

With regard to (3.118), we have

$$\sum_{i \in K} \sum_{j \in N_i} C_{ij}^{(2)} (t_i^{net} - t_j^{net})^2 \geq \sum_{i \in K} \sum_{j \in N_i} H_{ij} (t_i^{net} - t_j^{net})^2 \geq \qquad (3.122)$$

$$\geq c f(\delta) \sum_{i \in K} \sum_{j \in N_i} (t_i^{net} - t_j^{net})^2.$$

The solution of problem (3.119) does not depend on $f(\delta)$ (because dividing equation (3.119) by $f(\delta)$, we arrive at a problem, which does not depend on $f(\delta)$. The left-hand side of (3.122) is $4\mathcal{E}_d$, the right-hand side of (3.122) has the form $f(\delta)const$ with $const < \infty$ not depending on δ_{ij}.

By virtue of (3.122) the energy of alive net

$$\mathcal{E}_d = \frac{1}{4} \sum_{i \in K} \sum_{j \in N_i} C_{ij}^{(2)} (t_i^{net} - t_j^{net})^2 \leq C_1 f(\delta), \qquad (3.123)$$

where $C_1 > 0$.

Now we prove that it is possible to take off summation over the insulated bodies and insulated chains from the sum

$$\sum_{k=1}^{C} \sum_{i \in CH(k)} \sum_{j \in N_i} C^{R_{ij}} (t_i - t_j)^2 \qquad (3.124)$$

with accuracy $O(1)$. To prove this, we take

$$t_i = t_i^{net} \text{ for } i \in K = CH(1)$$

and

$$t_i = const_k \text{ for } i \in K = CH(k) \ (k > 1).$$

We note that $t_i = const_k$ for $i \in K = CH(k)$ is the solution of the finite-dimensional network problem for insulated net $CH(k) \ (k > 1)$. The network problem for insulated net $CH(k) \ (k > 1)$ has form

$$\sum_{j \in N_i} C^{(2)}_{i(k)j} (t_{i(k)} - t_j), \ i(k) \in CH(k) \ (k > 1). \qquad (3.125)$$

Thus, the sums over insulated nets $CH(k) \ (k > 1)$ give no contribution to the sum (3.124).

For two bodies, which do not belong to the same net $CH(k) \ (k \geq 1)$, the following cases are possible:
1. the bodies belong to different nets,
2. one body belongs to a net, the other body is an insulated body,
3. both of the bodies are insulated.

For all the cases listed

$$C^{R_{ij}} \leq C^{(2)}_{ij}$$

and $C^{(2)}_{ij}$ corresponding to insulated bodies are restricted uniformly over $\delta \to 0$. These capacities are restricted uniformly by virtue of Lemma 2. The quantities $\{t_i, i \in I\}$ are restricted uniformly over $\delta \to 0$. Then the terms corresponding to the cases listed above are restricted uniformly over $\delta \to 0$ (have order $O(1) \ll f(\delta) \to \infty$ as $\delta \to 0$).

The last proposition of the lemma follows directly from the Definition 14. **Proof is completed**.

Theorem 3. *Let capacities $C^{(2)}_{ij}$ of all neighboring bodies forming alive network have the same order $f(\delta) \to \infty$ as $\delta \to 0$. Then energy of the collection of the bodies determined by (3.14)*

$$\mathcal{E} \to \infty \text{ as } \delta \to 0.$$

The leading term of the energy $\mathcal{E}$ (3.14) of the continuum problem (3.3)–(3.8) is asymptotically equivalent, as $\delta \to 0$, to the energy

$$\mathcal{E}_d = \frac{1}{4} \sum_{i \in K} \sum_{j \in N_i} C_{ij}^{(2)} (t_i^{net} - t_j^{net})^2 \qquad (3.126)$$

of the of network model:

$$\mathcal{E} \sim \mathcal{E}_d, \qquad (3.127)$$

where $\{t_i^{net}, i \in K\}$ is solution of the network model (3.35) and (3.36) and $C_{ij}^{(2)}$ means the capacity of the pair of the bodies D_i and D_j in R^n, $K = CH(1)$ is the alive net.

Proof. With regard to (3.101),

$$\frac{1}{2} \sum_{i \in K} \sum_{j \in N_i} C_{ij}^{(2)} \left(-\frac{1}{2} \frac{\left[C_{ij}^{(2)} \right]^2}{C^{S_{ij}}} + 1 \right) (t_i^{net} - t_j^{net})^2 \leq \qquad (3.128)$$

$$\leq \mathcal{E} \leq \frac{1}{4} \sum_{k=1}^{C} \sum_{i \in CH(k)} \sum_{j \in N_i} C^{R_{ij}} (t_i^{net} - t_j^{net})^2 + \int_{P_h} |\nabla \psi^\delta(\mathbf{x})|^2 d\mathbf{x}.$$

The sum in the right-hand part (3.128) involves more items than the right-hand part (3.128). By virtue of Lemma 6, we can save in the right-hand part (3.128) the leading term only and rewrite (3.128) as

$$\frac{1}{2} \sum_{i \in K} \sum_{j \in N_i} C_{ij}^{(2)} \left(-\frac{1}{2} \frac{\left[C_{ij}^{(2)} \right]^2}{C^{S_{ij}}} + 1 \right) (t_i^{net} - t_j^{net})^2 \leq \qquad (3.129)$$

$$\leq \mathcal{E} \leq \frac{1}{4} \sum_{i \in K} \sum_{j \in N_i} C^{R_{ij}} (t_i^{net} - t_j^{net})^2 + \int_{P_h} |\nabla \psi^\delta(\mathbf{x})|^2 d\mathbf{x} + O(1).$$

where $\{t_i^{net}, i \in K\}$ is the solution of the network model, which does not depend on δ.

Integral

$$\int_{P_h} |\nabla \psi^\delta(\mathbf{x})|^2 d\mathbf{x} < C,$$

where $C < \infty$ does not depend on δ, see (3.94).

Let us tend δ to zero. By virtue of Theorem 2,

$$\frac{C_{ij}^{(2)}}{C^{S_{ij}}} \to 1, \quad \frac{C^{R_{ij}}}{C_{ij}^{(2)}} \to 1. \qquad (3.130)$$

By virtue of the first limit in (3.130) we have

$$-\frac{1}{2}\frac{C_{ij}^{(2)}}{C^{S_{ij}}} + 1 = \frac{1}{2} + o(1), \tag{3.131}$$

where $O(\delta) \to 0$ as $\delta \to 0$.

From the second limit in (3.130) it follows that

$$C^{R_{ij}} = C_{ij}^{(2)} + C_{ij}^{(2)} o(1). \tag{3.132}$$

Then, we can write (3.129) in the form

$$\frac{1}{4}\sum_{i \in K}\sum_{j \in N_i} C_{ij}^{(2)}(t_i^{net} - t_j^{net})^2 + 2\max||C_{ij}^{(2)}||o(1) \leq \tag{3.133}$$

$$\leq \mathcal{E} \leq \frac{1}{4}\sum_{i \in K}\sum_{j \in N_i} C_{ij}^{(2)}(t_i^{net} - t_j^{net})^2 + 2\max||C_{ij}^{(2)}||o(1) +$$

$$+ \int_{P_h} |\nabla\psi^\delta(\mathbf{x})|^2 d\mathbf{x} + O(1).$$

Here we take into account that $\varphi_1^0 \leq t_i \leq \varphi_2^0$ for $i \in K$ by virtue of Lemma 1 and $\varphi_1^0 \leq t_i \leq \varphi_2^0$ for other (insulated) bodies by virtue of Lemma 2.

In accordance with the condition of Theorem 1, all capacities $C_{ij}^{(2)}$ have the same order $f(\delta) \to \infty$, as $\delta \to 0$. Then, by virtue of Lemma 6, quantity

$$\mathcal{E}_d = \sum_{i \in K}\sum_{j \in N_i} C_{ij}^{(2)}(t_i^{net} - t_j^{net})^2 \tag{3.134}$$

has the order $f(\delta) \to \infty$, as $\delta \to 0$.

Dividing both pats of (3.133) by (3.134), we obtain

$$1 + 2\frac{\max||C_{ij}^{(2)}||}{\frac{1}{4}\sum_{i \in K}\sum_{j \in N_i} C_{ij}^{(2)}(t_i^{net} - t_j^{net})^2}o(1) \leq \tag{3.135}$$

$$\leq \frac{\mathcal{E}}{\frac{1}{4}\sum_{i \in K}\sum_{j \in N_i} C_{ij}^{(2)}(t_i^{net} - t_j^{net})^2} \leq$$

$$\leq 1 + 2\frac{\max||C_{ij}^{(2)}||}{\frac{1}{4}\sum_{i \in K}\sum_{j \in N_i} C_{ij}^{(2)}(t_i^{net} - t_j^{net})^2}o(1)+$$

$$+ \frac{\displaystyle\int_{P_h} |\nabla \psi^\delta(\mathbf{x})|^2 d\mathbf{x} + O(1)}{\dfrac{1}{4} \sum_{i \in K} \sum_{j \in N_i} C_{ij}^{(2)} (t_i^{net} - t_j^{net})^2}.$$

Integral $\displaystyle\int_{P_h} |\nabla \psi^\delta(\mathbf{x})|^2 d\mathbf{x}$ is bounded uniformly with respect to $\delta \to 0$, see (3.94). In the ratio

$$\frac{\max \|C_{ij}^{(2)}\|}{\dfrac{1}{4} \sum_{i \in K} \sum_{j \in N_i} C_{ij}^{(2)} (t_i^{net} - t_j^{net})^2}$$

the nominator and the denominetor have the same order by virtue of Lemma 6. Then both the left-hand and the right-hand parts of inequality (3.135) tend to 1 as $\delta \to 0$. As a result, we obtain from (3.135) that

$$\frac{\mathcal{E}}{\dfrac{1}{4} \sum_{i \in K} \sum_{j \in N_i} C_{ij}^{(2)} (t_i^{net} - t_j^{net})^2} \to 1$$

as $\delta \to 0$, or

$$\mathcal{E} \sim \frac{1}{4} \sum_{i \in K} \sum_{j \in N_i} C_{ij}^{(2)} (t_i^{net} - t_j^{net})^2 = \mathcal{E}_d \qquad (3.136)$$

as $\delta \to 0$. **Proof is completed.**

3.5.4. Asymptotic behavior of capacity of a network

Now, we can solve the problem described in Section 1.4 for a system of closely placed bodies. We consider an alive net of closely placed bodies, Fig. 3.16 (left). If the difference of potentials applied to bodies D_1 and D_2 is equal to unity ($\varphi_2^0 - \varphi_1^0 = 1$), the capacity C^{net} of the system of bodies is equal to the double energy of the electric field corresponding to the system of bodies:

$$C^{net} = 2\mathcal{E}.$$

In the frameworks of Theorem 3, the leading term of the energy $\mathcal{E}$ (3.14) of the continuum problem (3.3)–(3.8) is asymptotically equivalent, as $\delta \to 0$, to the energy $\mathcal{E}_d$ (3.126) of the finite-dimensional network model (3.35). Then, the capacity C^{net} of system of bodies is asymptotically equivalent to $\mathcal{E}_d$:

$$C^{net} \sim 2\mathcal{E}_d, \qquad (3.137)$$

as $\delta \to 0$. We note that $\mathcal{E}_d$ is equal to the capacity of a net of capacitors.

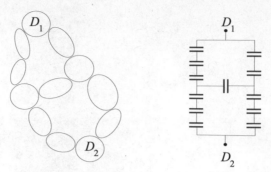

Figure 3.16. *An alive net of closely placed bodies and the corresponding network of capacitors.*

Fig. 3.16 illustrates this result. The capacity of the net of the bodies shown in the left side of the figure is equivalent to the capacity of the system of capacitors shown in the right side of the figure. Their capacitors are equal to $C_{ij}^{(2)}$, where $C_{ij}^{(2)}$ means the capacity of the neighbor bodies in R^n ($n = 2, 3$).

The sufficient condition of the equivalence of the capacities of the networks shown in Fig. 3.16 is the following: the capacities $C_{ij}^{(2)}$ of all neighbor bodies forming an alive network have the same order $f(\delta) \to \infty$ as $\delta \to 0$. The necessary condition sounds as $C_{ij}^{(2)} \to \infty$ as $\delta \to 0$ for the neighbor bodies. Note that under these conditions, the insulated bodies (if exist) do not influence the asymptotic (3.137).

3.5.5. Asymptotic of the total flux through network

Now we consider an asymptotic behavior of total flux (total current in the case of the electroconductivity problem) in the network under consideration.

Multiply the equation (3.3) by $\varphi(\mathbf{x})$ and integrate by parts. Then

$$0 = -\int_Q |\nabla\varphi(\mathbf{x})|^2 dx + \sum_{i=1}^{N} \int_{\partial D_i} \frac{\partial\varphi}{\partial\mathbf{n}}(\mathbf{x})\varphi(\mathbf{x})dx. \qquad (3.138)$$

For the integrals over the boundaries of the bodies, we use (3.4) and (3.6) to obtain

$$\int_{\partial D_i} \frac{\partial\varphi}{\partial\mathbf{n}}(\mathbf{x})\varphi(\mathbf{x})dx = t_i \int_{\partial D_i} \frac{\partial\varphi}{\partial\mathbf{n}}dx \quad (i \in K), \qquad (3.139)$$

$$\int_{\partial D_i} \frac{\partial\varphi}{\partial\mathbf{n}}(\mathbf{x})\varphi(\mathbf{x})dx = \varphi_i^0 \int_{\partial D_i} \frac{\partial\varphi}{\partial\mathbf{n}}dx \quad (i = 1, 2).$$

By virtue of (3.5), the integrals with indices $i \in K$ in (3.139) are equal to zero, and (3.138) becomes

$$\int_Q |\nabla\varphi(\mathbf{x})|^2 dx = \varphi_1^0 \int_{\partial D_1} \frac{\partial\varphi}{\partial\mathbf{n}}dx + \varphi_2^0 \int_{\partial D_2} \frac{\partial\varphi}{\partial\mathbf{n}}dx. \qquad (3.140)$$

Now, we multiply the equation (3.3) by 1 and integrate by parts. Then

$$0 = \sum_{i=1}^{N} \int_{\partial D_i} \frac{\partial \varphi}{\partial \mathbf{n}}(\mathbf{x}) d\mathbf{x}. \tag{3.141}$$

By virtue of (3.5), the integrals with indices $i \in K$ are equal to zero, and we obtain from (3.141) the following equality

$$\int_{\partial D_1} \frac{\partial \varphi}{\partial \mathbf{n}} d\mathbf{x} = - \int_{\partial D_2} \frac{\partial \varphi}{\partial \mathbf{n}} d\mathbf{x}. \tag{3.142}$$

The physical meaning of (3.142) is that the flux in (from) the bodies D_1 and D_2, which are poles of the network, are equal in absolute value and have opposite signs. The absolute value of (3.142) is called the total flux through the network.

Combining (3.140) and (3.142), we obtain the following formula:

$$(\varphi_1^0 - \varphi_2^0) \int_{\partial D_1} \frac{\partial \varphi}{\partial \mathbf{n}}(\mathbf{x}) d\mathbf{x} = \int_Q |\nabla \varphi(\mathbf{x})|^2 d\mathbf{x} = 2\,\mathcal{E}, \tag{3.143}$$

where $\varphi(\mathbf{x})$ is the solution of the problem (3.3)–(3.8).

Lemma 7. *For the solution $\{t_i^{net}, i \in K\}$ of the network problem (3.35) and (3.36)*

$$P^+ - P^- = 0, \tag{3.144}$$

here $K = CH(1)$ is the alive net.

To prove the lemma, it is sufficient to summarize all fluxes in p_{ij} (3.37) and then take into account the balance equations (3.35). We note that the difference in the formulas (3.142) and (3.144) in the difference of definitions of the fluxes in / from bodies D_1 and D_2 in continuum and discrete models.

From Lemma 7 and equality (3.97), we obtain

$$(\varphi_1^0 - \varphi_2^0)P^+ = -(\varphi_1^0 - \varphi_2^0)P^- = \frac{1}{2} \sum_{i \in K} \sum_{j \in N_i} C_{ij}^{(2)} (t_i^{net} - t_j^{net})^2 = 2\,\mathcal{E}_d. \tag{3.145}$$

The left-hand side of (3.143) is equal to the total flux through the network multiplied by $(\varphi_1^0 - \varphi_2^0) \neq 0$. Combining the equalities (3.143) and (3.145) with Theorem 3 and Lemma 7, we arrive at the conclusion that the flux through the continuum network is approximated by the flux through the discrete network:

$$\int_{\partial D_1} \frac{\partial \varphi}{\partial \mathbf{n}} d\mathbf{x} \sim P^+, \tag{3.146}$$

$$\int_{\partial D_2} \frac{\partial \varphi}{\partial \mathbf{n}} d\mathbf{x} \sim -P^-,$$

as $\delta \to 0$.

We note that the minus sign in the second formula (3.146) is a result of definitions of fluxes in continuum and network models.

3.6. Some consequences of the theorems about NL zones and network approximation

Now, we discuss some consequences of physical value following from the results obtained above.

3.6.1. Dykhne experiment and energy localization

We refer to "Dykhne experiment" as the experiment arranged by Dykhne to measure the conductivity of a disordered collection of bodies. The aim of experiment was the testing of well-known Keller-Dykhne formula for effective conductivity of planar random composite material [110, 167]. In the experiment, a number of conducting spheres were placed short distances from one another. After that a voltage was applied to the system. It is reported [257] that the result of that experiment was the inflammation of the system.

It follows from Theorem 3 that under condition (3.102) the total energy has the order of $C_{ij}^{(2)}$. For spheres of the radii R

$$C_{ij}^{(2)} = -\pi R \ln \delta \to \infty \text{ as } \delta \to 0,$$

thus, condition (3.1) is satisfied.

Then the total energy of the electric field in the system of the spheres

$$\int_Q |\nabla \varphi(\mathbf{x})|^2 d\mathbf{x} \to \infty \text{ as } \delta \to 0.$$

Theorem 1 states that the energy outside channels is finite when $\delta \to 0$. Thus, large amount of energy is localized in the system of channels and only a finite energy is distributed outside the channels. Then the energy accumulated in the channels is asymptotically equal to the total energy of the system of the bodies $\int_Q |\nabla \varphi(\mathbf{x})|^2 d\mathbf{x}$.

Denote the total measure (area if $n = 2$, volume if $n = 3$) of the channels by $M_{channels}$. The width the channels has the order of characteristic distance $\delta << 1$ between the closely placed bodies. Then the measure $M_{channels}$ of the channels is small as compared with the characteristic dimension V of the domain occupied by the collection of the bodies. Then the density of energy in the channels, estimated as

$$\frac{\int_Q |\nabla \varphi(\mathbf{x})|^2 d\mathbf{x}}{M_{channels}},$$

is greater than the average density of energy

$$\frac{\int_Q |\nabla \varphi(\mathbf{x})|^2 d\mathbf{x}}{|Q|}.$$

We find that in the Dykhne experiment most of the energy must be accumulated in domains of small measure (in the channels). The growth of the local energy in the channels can lead to failure of channels, in particular, it can lead to the inflammation of the channels. Note that the inflammation, as a result of the concentration of energy, can arise in a system not subjected to large differences of potentials. The authors made a conjecture that the Dykhne experiment fixed a manifestation of the energy localization effect.

3.6.2. Explanation of Tamm shielding effect

We present here a fragment from the Tamm book [353] (the first edition of the book published in 1927). On page 59 Tamm described the classical shielding effect. Then, Tamm discussed capacity of a collection of bodies and wrote: *"If none of two (or more) conductors forms a closed system confining the remaining conductors, then the capacitance of this system of conductors (of this capacitor) will practically be independent of the arrangement of the surrounding bodies only if the dimension of the conductors are great in comparison with the distance between them. Only under this condition, the space between the plates* [closely placed surfaces of conductors are treated as plates of a capacitor] *be shielded to a considerable extent, if not completely, by the plates themselves from the action of an external field."*

The phenomenon described by I.E. Tamm corresponds to the case when two alive bodies D_1 and D_2 are closely placed and other bodies are not close to these bodies, see Fig. 3.5 (a). In this case we have a chain consisting of the bodies D_1 and D_2 and insulated system consisting of the remaining bodies.

The capacity $\mathcal{C}_{12}$ of the pair of the bodies D_1 and D_2 considered as elements of a system of bodies (i.e., in the presence of other bodies) can be calculated as the coefficient in the formula

$$\mathcal{E} = \frac{1}{2}\mathcal{C}_{12}(\varphi_1^0 - \varphi_2^0)^2, \tag{3.147}$$

where $\mathcal{E}$ is the energy of the electric field corresponding to the system of bodies under consideration.

In accordance with Theorem 3, the energy $\mathcal{E}$ of the collection of these bodies under condition

$$C_{12}^{(2)} \to \infty \text{ as } \delta \to 0 \tag{3.148}$$

is asymptotically equivalent to the energy E of the network formed by two alive bodies D_1 and D_2:

$$\mathcal{E} \sim \mathcal{E}_d = \frac{1}{4}[C_{12}^{(2)}(\varphi_1^0 - \varphi_2^0)^2 + C_{21}^{(2)}(\varphi_2^0 - \varphi_1^0)] = \tag{3.149}$$

$$= \frac{1}{2}C_{12}^{(2)}(\varphi_1^0 - \varphi_2^0)^2$$

as $\delta \to 0$. Here we use the symmetry of capacities $C_{12}^{(2)} = C_{21}^{(2)}$.

Comparing the formulas (3.147) and (3.149), we conclude that

$$\mathcal{C}_{12} \sim C_{12}^{(2)} \tag{3.150}$$

as $\delta \to 0$.

Formula (3.150) means that the capacity of two closely placed bodies, considered as elements of a system of bodies, does not depend on other bodies, which are situated not closely to these two bodies. It expresses the matter of the Tamm shielding effect. Note that the effect takes place under condition (3.148), which was not mentioned in any form in the book by I.E. Tamm.

·   We now give an explanation of the physics of the shielding effect from the point of view of the asymptotic theory presented above. In the case of unbounded increasing of the capacity with approach the bodies (i.e., under the condition (3.148)), the energy localization in the small size channel between the bodies (it follows from Theorem 1). The flux in the channel becomes greater than the flux outside the channel. Thus, the flux in the channel protects itself from the effect of external fields by its large magnitude. This form of self-protection does not require any material shielding. So, the mechanism of Tamm shielding effects differs from the classical shielding, where a material shield is a necessary element of the phenomenon. It also differs from the mechanism originally described by I.E. Tamm [353, 354]. Tamm shielding effect is a localization phenomenon.

3.7. Capacity of a pair of bodies dependent on shape

The problem of dependence of transport properties of systems of bodies on the individual geometry of the bodies is a practically important problem, which has attracted attention of numerous scientists, see, e.g., [37, 127, 116, 126, 271]. Various techniques were developed to relate transport property of a system of bodies and individual shape of the bodies. Our approach relates transport property of a system of bodies to capacity of pairs of neighbor bodies $C_{ij}^{(2)}$. It is known that the capacity $C_{ij}^{(2)}$ is strongly related to the shape of the bodies (often it is said that capacity has geometrical characteristics). So, our approach naturally relates transport property of a system of bodies and individual shape of the bodies.

Previously, we did not use any information about the specific shape of the bodies. We only assumed that the boundaries of the bodies are piecewise smooth (we used it to write formulas like integration by parts). Also, we do not directly use the dimension of the problem. The network approximation theorem was proved in the terms of capacity $C_{ij}^{(2)}$ under condition (3.1):

$$C_{ij}^{(2)} \to \infty \text{ as } \delta \to 0.$$

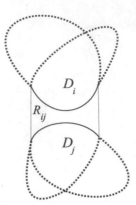

Figure 3.17. *Variations of the parts of the domains outside the channel R_{ij} (drawn in dotted lines) do not influence the asymptotic behavior of capacity $C_{ij}^{(2)}$.*

The capacities $C_{ij}^{(2)}$ depend both on the geometry of bodies and dimension of the problem. Then, the shape of bodies and dimension of the problem influence the network model through the capacities $C_{ij}^{(2)}$.

Present values of capacities $C_{ij}^{(2)}$ for pairs of bodies of various shapes for various dimensions.

Asymptotic behavior of capacity is determined by the local geometry of bodies

As follows from Theorem 1, asymptotic behavior of capacity of two closely placed bodies is determined by the shape of bodies in the region near the point of about touching of the bodies. Really, we can compute the capacity $C^{R_{ij}}$ with respect to the cylindrical channel R_{ij} between the bodies, see Fig. 3.17.

Computing the capacity with respect to the cylindrical channel, we ignore all information about the geometry of body outside R_{ij} and only use information about fragments $R_{ij} \bigcap D_i$ and $R_{ij} \bigcap D_i$ of these bodies, only. By virtue of Theorem 2, the capacity $C^{R_{ij}}$ is asymptotically equivalent to the classical capacity $C_{ij}^{(2)}$ of the bodies D_i and D_j in R^n (and on the contrary).

This conclusion is valid under the condition

$$C_{12}^{(2)} \to \infty \text{ as } \delta \to 0.$$

Then, computing the asymptotic behavior of capacity, we can take into account the shape of the bodies in the region near the point of about touching of the bodies and vary the shape of the bodies outside this region as is suitable for our computations.

3.7.1. Capacity of the pair cone–plane

In this section, we consider a pair cone–plane and demonstrate the boundedness of the capacity $C_{cp}^{(2)}$ of this pair R^3 as $\delta \to 0$. As above, we assume the dielectric constant of medium outside the bodies equal to unity. In the general case, it is necessary to multiply the formulas for computation of capacity by the dielectric constant.

We consider a pair of bodies which have in the near touching region the cone-plane geometry, see Fig. 3.18. We denote domain occupied by the cone-like body by C and domain with planar top by

$$\Pi = \{(x, y, 0) : (x, y) \in R^2\}.$$

The upper bound for the capacity of the two bodies $C_{cp}^{(2)}$ follows from (3.16) and it has the form

$$C_{cp}^{(2)} \leq \frac{1}{2} \int_{R^3 \backslash (C \bigcup \Pi)} |\nabla \varphi(\mathbf{x})|^2 d\mathbf{x} \text{ for any } \varphi(\mathbf{x}) \in V, \qquad (3.151)$$

where

$$V = \{\varphi(\mathbf{x}) \in H^1(R^3 \backslash (C \bigcup \Pi)) : \varphi(\mathbf{x}) = 1/2 \text{ on } S, \varphi(\mathbf{x}) = -1/2 \text{ on } \Pi\}. \ (3.152)$$

We consider cylindrical channel S_{cp} of radius R and its neighborhood P_h, in which one can define a test function that vanishes outside P_h and has uniformly bounded derivatives uniformly with respect to δ in P_h. Thus, to verify the boundedness of capacity, it is enough to make sure of the boundedness in the channel S_{cp} (see Fig. 3.18) of the right side of (3.151) for some test function from V, i.e., we have to consider the integral

$$I_S(\varphi) = \frac{1}{2} \int_{S_{cp}} |\nabla \varphi(\mathbf{x})|^2 d\mathbf{x} \qquad (3.153)$$

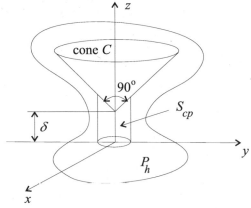

Figure 3.18. *Scheme of cone–plane pair for the three-dimensional case.*

with the conditions (3.152).

We assume $\varphi(\mathbf{x})$ to be linear in z:

$$\varphi(\mathbf{x}) = -\frac{1}{2} + \frac{z}{\delta + \sqrt{x^2 + y^2}}$$

(we consider a circular cone with a vertex angle of 90°, see Fig. 3.18). The function $\varphi(\mathbf{x})$ depends on all the variables and has three nonzero partial derivatives:

$$\frac{\partial \varphi}{\partial x} = \frac{\dfrac{xz}{\sqrt{x^2 + y^2}}}{(\delta + \sqrt{x^2 + y^2})^2} = \frac{\dfrac{zr\cos\theta}{r}}{(\delta + r)^2} = \frac{z\cos\theta}{(\delta + r)^2},$$

$$\frac{\partial \varphi}{\partial y} = \frac{\dfrac{yz}{\sqrt{x^2 + y^2}}}{(\delta + \sqrt{x^2 + y^2})^2} = \frac{\dfrac{zr\sin\theta}{r}}{(\delta + r)^2} = \frac{z\sin\theta}{(\delta + r)^2},$$

$$\frac{\partial \varphi}{\partial z} = \frac{1}{\delta + \sqrt{x^2 + y^2}} = \frac{1}{\delta + r},$$

where (r, θ, z) are cylindrical coordinates.

Calculate $\int_{S_{cp}} |\nabla\varphi|^2 d\mathbf{x}$ for the above function. In cylindrical coordinates we have

$$\int_{S_{cp}} \left(\frac{\partial\varphi}{\partial x}(\mathbf{x})\right)^2 d\mathbf{x} = \int_0^R \int_0^{r+\delta} \left(\frac{\partial\varphi}{\partial x}\right)^2 r\, dr\, dz \leq$$

$$\leq \int_0^R \frac{r\, dr}{(\delta + r)^4} \int_0^{r+\delta} z^2 r\, dz = \int_0^R \frac{r\, dr}{(\delta + r)^4} \frac{(r+\delta)^3}{4} < \frac{R}{3},$$

which takes into account that $r < r + \delta$. The precise value of R is immaterial. It is essential that it can be chosen to be independent on δ. The integral of $\left|\dfrac{\partial\varphi}{\partial y}(\mathbf{x})\right|^2$ can be estimated similarly. We have

$$\int_S \left|\frac{\partial\varphi}{\partial z}(\mathbf{x})\right|^2 d\mathbf{x} = \int_0^R \frac{r\, dr}{(r+\delta)^2} \int_0^{r+\delta} dz =$$

$$= \int_0^R \frac{r\, dr}{(r+\delta)^2}(r+\delta) = \int_0^R \frac{r\, dr}{r+\delta} \leq R,$$

which takes into account that $r < r + \delta$.

Consequently, the capacity $C_{cp}^{(2)}$ is bounded uniformly with respect to δ (in particular, as $\delta \to 0$).

3.7.2. Capacity of the pair angle–line

In this section, we consider that the two-dimensional pair angle–line (or angle–half-line) demonstrate the boundedness of the capacity $C_{al}^{(2)}$ in R^2 for this pair.

We consider a pair of bodies having in the near touching region angle–line geometry, see Fig. 3.19. The lower bound for the capacity C_{al} of the pair of bodies follows from (3.28) and it has the form

$$C_{al}^{(2)} \geq -\frac{1}{2} \int_{R^3 \backslash (A \bigcup \Pi)} |\mathbf{v}(\mathbf{x})|^2 d\mathbf{x} + \int_{\partial A \bigcup \partial \Pi} \varphi^0(\mathbf{x}) \mathbf{v}(\mathbf{x}) \mathbf{n} d\mathbf{x}, \qquad (3.154)$$

$$\text{div}\mathbf{v} = 0;$$

where $\varphi^0(\mathbf{x}) = 1$ for the angle A and $\varphi^0(\mathbf{x}) = -1$ for the line Π. For the 90° angle, as in Fig. 3.19, we consider cylindrical channel S_{al} and introduce the function

$$\mathbf{v}(\mathbf{x}) = \mathbf{v}(x) = \begin{cases} \dfrac{\mathbf{e}_2}{\delta + x} & \text{for } 0 < x < x_0, \\ 0 & \text{for } x \leq 0 \text{ and } x \geq x_0. \end{cases} \qquad (3.155)$$

We note that $\mathbf{v}(x) \neq 0$ at $0 < x < x_0$ and $\text{div}\mathbf{v}(\mathbf{x}) = 0$ throughout the region. Really, in the channel

$$\text{div}\mathbf{v}(\mathbf{x}) = 0 + \frac{\partial}{\partial y}\left(\frac{1}{\delta + x}\right) = 0.$$

Note that $\mathbf{v}(x)$ is piecewise smooth but it does not lead to δ-function because $v_1(x) = 0$ and we differentiate $v_1(x)$ in y.

The normal vector for the angle shown in Fig. 3.19 is

$$\mathbf{n} = \left(\frac{1}{\sqrt{2}}, -\frac{1}{\sqrt{2}}\right).$$

The normal vector for the line is $\mathbf{n} = (0, 1)$.

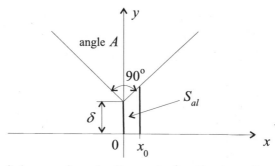

Figure 3.19. *Scheme of angle–line pair for the two-dimensional case.*

For (3.155) the right-hand side of the inequality (3.154) takes the form

$$\frac{1}{2} \int_0^{x_0} dx \int_0^{x+\delta} \frac{1}{\delta + x} dy - \int_0^{x_0} \left(1 + \frac{1}{\sqrt{2}} \sqrt{2}\right) dx, \qquad (3.156)$$

where the appearance of the factor $\sqrt{2}$ is due to the dependence

$$ds = \sqrt{dx^2 + dy^2} = \sqrt{2} dx.$$

Integrating (3.156), we obtain

$$\frac{x_0}{2} - \left(-\frac{1}{2} + 2\right) \int_0^{x_0} \frac{dx}{\delta + x} = \frac{x_0}{2} - \frac{3}{2} \ln(\delta + x) \Big|_0^{x_0} =$$

$$= \frac{x_0}{2} - \frac{3}{2} \ln(\delta + x_0) - \frac{3}{2} \ln(\delta) \to \infty$$

as $\delta \to 0$.

Consequently, the capacity $C_{al}^{(2)} \to \infty$ when $\delta \to 0$.

3.7.3. Other examples

In this section, we present values of capacity for some pairs of bodies of various geometries. The formulas presented are derived using Keller's approach, see Section 1.2.3.

Three-dimensional cases

1). Pair sphere–sphere of radii R. In this case [340]

$$C^{(2)} = -\pi R \ln \delta \to \infty \text{ as } \delta \to 0.$$

It justifies the result of Keller [166] for periodic system of spheres. The condition (3.1) is satisfied, thus network models exist for composites filled with spherical particles and systems of bodies.

2). Pair sphere–sphere (different radii).

We consider two spheres of radii R_1 and R_2. Using Keller method [166], we write local flux in cylindrical coordinates as

$$\mathbf{v}(r) = \left(0, 0, \frac{1}{\delta + (\rho_1 + \rho_2)\frac{r^2}{2}}\right),$$

where

$$\rho_1 = \frac{1}{R_1}, \quad \rho_2 = \frac{1}{R_2}$$

are curvatures of the spheres. Then the total flux between the spheres is

$$2\pi \int_0^\infty \frac{r\,dr}{\delta + (\rho_1 + \rho_2)\dfrac{r^2}{2}} = 2\pi \frac{1}{\rho_1 + \rho_2} \ln\left(\delta + (\rho_1 + \rho_2)\frac{r^2}{2}\right)\Big|_0^\infty.$$

This integral diverges "at infinity" similarly to integral (1.48). If we restrict the interval of integration from above, we obtain that the leading term is

$$-2\pi \frac{1}{\rho_1 + \rho_2} \ln\delta = -2\pi \frac{R_1 R_2}{R_1 + R_2} \ln\delta. \tag{3.157}$$

Now we derive this formula from an accurate formula for capacity of two spheres, which has the following form [340]:

$$C = -\frac{4\pi R_1 R_2}{R_1 + R_2 + \delta}\mathrm{sh}\alpha \sum_{n=1}^\infty \mathrm{cosech}(n\alpha),$$

where α is a parameter introduced by the equality [340]

$$\mathrm{ch}\alpha = \frac{(R_1 + R_2 + \delta)^2 - R_1^2 - R_2^2}{2R_1 R_2}.$$

We can write

$$C = -\frac{4\pi R_1 R_2}{R_1 + R_2 + \delta}\mathrm{sh}\alpha \sum_{n=1}^\infty \mathrm{cosech}(n\alpha) = \tag{3.158}$$

$$= \frac{4\pi R_1 R_2}{R_1 + R_2 + \delta}\mathrm{sh}\alpha \sum_{n=1}^\infty \frac{1}{e^{n\alpha} - e^{-n\alpha}} =$$

$$= -\frac{4\pi R_1 R_2}{R_1 + R_2 + \delta}\mathrm{sh}\alpha \sum_{n=1}^\infty \frac{2e^{-n\alpha}}{1 - e^{2n\alpha}} \approx$$

$$\approx -\frac{4\pi R_1 R_2}{R_1 + R_2 + \delta} \sum_{n=1}^\infty \frac{e^{-n\alpha}}{n}$$

and for small δ

$$\mathrm{ch}\alpha = \frac{(R_1 + R_2 + \delta)^2 - R_1^2 - R_2^2}{2R_1 R_2} \approx 1 + \frac{R_1 + R_2}{2R_1 R_2}\delta. \tag{3.159}$$

Taking into account the equality

$$\frac{e^{-n\alpha}}{n} = -\int_{\alpha}^{\infty} e^{-n\alpha} d\alpha,$$

we write the right-hand part of (3.158) as

$$\frac{4\pi R_1 R_2}{R_1 + R_2 + \delta} \sum_{n=1}^{\infty} \int_{\alpha}^{\infty} e^{-n\alpha} d\alpha = \tag{3.160}$$

$$= \frac{4\pi R_1 R_2}{R_1 + R_2 + \delta} \int_{\alpha}^{\infty} \frac{e^{-\alpha}}{1 - e^{-\alpha}} d\alpha =$$

$$= \frac{4\pi R_1 R_2}{R_1 + R_2 + \delta} \ln(1 - e^{-\alpha}).$$

For arbitrary α,

$$\mathrm{ch}\,\alpha \approx 1 + \frac{\alpha^2}{2},$$

then

$$\alpha \approx \sqrt{2(\mathrm{ch}\,\alpha - 1)}.$$

Taking into account (3.159), we obtain

$$\alpha \approx \sqrt{\frac{R_1 + R_2}{2R_1 R_2}\delta}.$$

Substituting this expression for α in (3.160), we obtain

$$C \approx -\frac{4\pi R_1 R_2}{R_1 + R_2} \ln \alpha \approx \tag{3.161}$$

$$\approx -\frac{4\pi R_1 R_2}{R_1 + R_2} \ln \sqrt{\frac{R_1 + R_2}{2R_1 R_2}\delta} =$$

$$= -\frac{2\pi R_1 R_2}{R_1 + R_2} \ln \left(\frac{R_1 + R_2}{2R_1 R_2}\delta \right).$$

The right-hand sides of (3.157) and (3.161) are asymptotically equivalent as $\delta \to 0$.

3) Pair plane–plane (classical planar capacitor)

$$C_{pp}^{(2)} = \frac{S}{\delta},$$

where S means the area of plane surface. The condition (3.1) is satisfied.

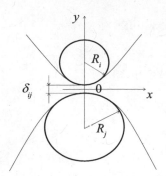

Figure 3.20.   Two neighbor disks and the tangential parabolas.

4) Pair of paraboloids of 4-th power. We consider surfaces described by the equation

$$z = \pm \frac{1}{2}(r^4 + \delta),$$

where $r = \sqrt{x^2 + y^2}$ and the distance between surfaces is equal to δ. For this case

$$C_{PP}^{(2)} = \int_0^\infty \frac{r\,dr}{\delta + r^4} = \frac{1}{\sqrt{\delta}} \arctan\left(\frac{r^2}{\delta}\right)\Bigg|_0^\infty = \frac{\pi}{\sqrt{\delta}} \to \infty \text{ as } \delta \to 0.$$

The condition (3.1) is satisfied, thus network models exist.

Two-dimensional cases

1). Pair disk–disk of radii R. In this case (we assume circular disk radius R)

$$C_{dd}^{(2)} = \pi\sqrt{\frac{R}{\delta}} \to \infty \text{ as } \delta \to 0.$$

The condition (3.111) is satisfied, a network model exists. This case leads to result of [41] for disordered system of disks.

2). Pair disk–disk of different radii. We employ Keller's method to derive an approximate formula for the flux between two disks (the i-th and the j-th) of arbitrary radii R_i and R_j placed at the distance δ_{ij}, see Fig. 3.20. We approximate the disks by the tangential parabolas

$$y = \frac{\delta_{ij}}{2} + \rho_i \frac{x^2}{2}, \tag{3.162}$$

and

$$y = -\left(\frac{\delta_{ij}}{2} + \rho_j \frac{x^2}{2}\right),$$

where

$$\rho_i = \frac{1}{R_i}, \quad \rho_j = \frac{1}{R_j} \tag{3.163}$$

are curvatures of the disks equal to the curvatures of the tangential paraboloids (3.162) in the points of tangency.

The distance between the parabolas is equal to

$$H(x) = \delta_{ij} + (\rho_i + \rho_j)\frac{x^2}{2}. \tag{3.164}$$

The local flux $\mathbf{v}(\mathbf{x})$ between the disks is of the form

$$\mathbf{v}(x) = \left(0, \frac{t_i - t_j}{H(x)}\right), \tag{3.165}$$

that is we assume that it is proportional to the difference in temperature or potential and inversely proportional to the distance between the disks.

Then the total (integral) flux between the disks can be calculated as follows:

$$q_{ij} = (t_i - t_j)\int_{-\infty}^{\infty} \frac{dx}{\delta + (\rho_i + \rho_j)\dfrac{x^2}{2}} =$$

$$= \frac{t_i - t_j}{\left(\dfrac{\rho_i + \rho_j}{2}\right)^{1/2}\delta_{ij}^{1/2}} \arctan\left[\left(\frac{\rho_i + \rho_j}{2\delta_{ij}}\right)^{1/2} x\right]\Bigg|_{-\infty}^{\infty}. \tag{3.166}$$

We write in (3.166) the limits of integration $-\infty$ and ∞. It is not a good choice, because the flux (3.165) is determined only between disks and we change the limits of integration $-\infty$ and ∞ for $-R_i$ and R_i or $-R_j$ and R_j, see Section 1.2.3. In the case under consideration there is no difference if we integrate from $-\infty$ to ∞ or from $-R_i$ to R_i or from $-R_j$ to R_j because the integral (3.166) converges at infinity and its asymptotic behavior with respect to δ_{ij} is the same for any choice of the limit of integration. Note that for spheres (the case analyzed in [166]) the corresponding integral diverges.

If the sum of the curvatures $\rho_i + \rho_j$ is not small, and the distance δ_{ij} between the disks is small, then the value of arctangent in the right-hand side (3.166) is approximately π. Taking into account (3.163), we obtain

$$q_{ij} = C_{ij}^{(2)}(t_i - t_j), \tag{3.167}$$

where

$$C_{ij}^{(2)} = \pi\frac{\sqrt{\dfrac{2R_iR_j}{R_i + R_j}}}{\sqrt{\delta_{ij}}}. \tag{3.168}$$

Note that capacity $C_{ij}^{(2)}$ is equal to the specific flux (i.e., flux per unit difference of potential). Formula (3.167) can be derived from the exact solution of the Laplace equation for two disks as the leading term in asymptotic in δ_{ij}. This exact solution (computation of capacity of two cylinders) can be found in [340].

3) Pair polyhedron–polyhedron. In two-dimensional case, two polyhedra in the general position are similar to the pair angle–line. In two-dimensional case the condition (3.111) is satisfied for both smooth and not smooth bodies, thus in two-dimensional case a network model exists for every shape of filled particles.

Numerical experiments predicts that other orders of capacity $C^{(2)}$ are possible. The capacity of a pair of smooth bodies tends to infinity as the bodies draw together, while the order in δ depends on the shape of the bodies.

Comparison of capacity for systems of disks and polyhedra

Compare capacity $C_{dd}^{(2)}$ of the pair disk–disk and capacity $C_{al}^{(2)}$ of the pair angle–line in two-dimensional case. The pair angle–line is a model of local shape for polyhedral bodies. The pair disk–disk is a model of local shape for smooth bodies. We have,

$$\frac{C_{dd}^{(2)}}{C_{al}^{(2)}} \cong \frac{\delta^{-\frac{1}{2}}}{\ln \delta} \to \infty \text{ as } \delta \to 0.$$

It means that disks are more effective than polyhedrons if we want to increase the effective capacity of a system.

In this section, presentation was given in the terms of electrostatics. The problems like (3.3)–(3.8) arise in thermoconductivity, elasticity and other disciplines, where the concept of capacity traditionally is not used. Thus, it is possible to expand the concept of capacity to thermoconductivity, elasticity and other disciplines.

3.7.4. Transport properties of systems of smooth and angular bodies

As was noted, the problem of influence of shape of bodies on the overall properties of a system of bodies or particles filled composite attracted attention of many researchers, see, e.g., [37, 126, 127, 384]. Our approach based on the notion of capacity, solves the problem of influence of shape of bodies on the overall transport properties of a system of closely placed bodies automatically. As an example, we would like to present an answer to the question: "Which system demonstrates higher transport properties, smooth or angular bodies?" as it follows from the theory developed in this chapter.

In the three-dimensional case, two polyhedra in the general position are similar to the pair cone–plane in the near touching region. Thus, for them $C^{(2)}$ is restricted as $\delta \to 0$. The condition (3.111) is not satisfied, thus network models do not approximate the original continuum problem for polyhedron-shaped bodies. This example is interesting for practice. Often powders are obtained by using melting technology. The melting technology usually produced polyhedron-shaped particles. There exist technologies producing smooth particles (sphere- or spheroid-shaped). In accordance with our computations, it is impossible to produce a system (composite material) with high transport properties on the basis of polyhedron-shaped particles, even packing such particles densely. Our computations predict the system of bodies (composite material) with high transport properties can be produced if smooth bodies / particles are used.

Chapter 4

NETWORK APPROXIMATION FOR POTENTIALS OF CLOSELY PLACED BODIES

In the previous chapter, we considered a boundary-value problem, which describes distribution of electric field around a system of perfectly conducting bodies. It was demonstrated that the original continuum problem can be approximated as the interparticle distance $\delta \to 0$ by a finite dimensional network problem in the sense of closeness of total flux, energy or capacity (these quantities are expressed one through the others).

All the characteristics mentioned above are integral in nature. For the continuum problem they all are expressed in the form of integrals over a domain, for the discrete problem — in the form of a sum over the bodies forming the system.

We put the question: *"Do the total flux, energy and capacity exhaust the characteristics of continuum model which can be approximated with the corresponding network model?"*

In this chapter, we demonstrate that additions to this list are potentials of the bodies. To the best knowledge of the authors, the problem of closeness of potentials determined from the continuum problem and the potential of nodes determined from the corresponding network problem was not discussed before [174, 186]. At the same time, it is an interesting and important problem from both a mathematical and engineering point of view. As was explained in Section 2.5.3, this problem arises as a non-trivial problem when one analyzes non-periodic arrays of bodies. For a periodic array of bodies, the difference of potentials of bodies is equal to the period of the array multiplied by the value of the overall electric field. If bodies are symmetric (disks, spheres, cubes, etc.), it is possible compute those potentials. For a periodic array of non-symmetric bodies such computation usually is impossible in a simple way.

Again condition (3.1) will play a keystone role for the existence of the network approximation.

4.1. Formulation of the problem of approximation of potentials of bodies

We consider a system of non-overlapping and not touching domains $\{D_i, i = 1, \ldots, N\}$ $\subset R^n$ ($n = 2, 3$) occupied by the perfectly conducting bodies. As above, we denote

$$Q = R^n \setminus \bigcup_{i=1}^{N} D_i$$

space outside the bodies.

We consider the boundary-value problem

$$\Delta \varphi = 0 \text{ in } Q; \tag{4.1}$$

$$\varphi(\mathbf{x}) = t_i \text{ on } D_i, i \in I = \{3, \ldots, N\}; \tag{4.2}$$

$$\int_{\partial D_i} \frac{\partial \varphi}{\partial n}(\mathbf{x}) d\mathbf{x} = 0, i \in I; \tag{4.3}$$

$$\varphi(\mathbf{x}) = \varphi_1^0 \text{ on } D_1, \varphi(\mathbf{x}) = \varphi_2^0 \text{ on } D_2; \tag{4.4}$$

$$\varphi(\mathbf{x}) \leq const \text{ as } |\mathbf{x}| \to \infty \ (n = 2); \tag{4.5}$$

$$\varphi(\mathbf{x}) \to 0 \text{ as } |\mathbf{x}| \to \infty \ (n = 3). \tag{4.6}$$

As above, unknowns in (4.1)–(4.6) are function $\varphi(\mathbf{x})$ and real numbers $\{t_i, i \in I\}$.

The boundary-value problem (4.1)–(4.6) is equivalent to the minimization problem

$$I(\varphi) = \frac{1}{2} \int_Q |\nabla \varphi(\mathbf{x})|^2 d\mathbf{x} \to \min, \tag{4.7}$$

considered on the set of functions

$$V = \Big\{ \varphi(\mathbf{x}) \in H^1(Q) : \varphi(\mathbf{x}) = \varphi_1^0 \text{ on } D_1, \varphi(\mathbf{x}) = \varphi_2^0 \text{ on } D_2;$$

$$\varphi(\mathbf{x}) = t_i \text{ on } D_i, i \in I;$$

$$|\varphi(\mathbf{x})| \leq const < \infty \text{ as } |\mathbf{x}| \to \infty \ (n = 2), \varphi(\mathbf{x}) \to 0 \text{ as } |\mathbf{x}| \to \infty \ (n = 3) \Big\}.$$

where $\{t_i, i \in I\}$ are not fixed constants (generally, they are different in different bodies).

The energy $\mathcal{E}$ corresponding to the problem (4.1)–(4.6) is introduced as (see Definition 6)

$$\mathcal{E} = I(\varphi^{cont}) = \min_{\varphi \in V} I(\varphi). \tag{4.8}$$

We denote $\varphi^{cont}(\mathbf{x})$ and $\{t_i^{cont}, i \in I\}$ the function and values of the variables $\{t_i, i \in I\}$ determined from solution of the boundary-value problem (4.1)–(4.6). In Chapter 3, we paid attention to the asymptotic behavior of the energy $\mathcal{E}$ (4.8). In this chapter, we investigate the asymptotic behavior of the potentials $\{t_i^{cont}, i \in I\}$ of the bodies $\{D_i, i \in I\}$ as the interparticle distance $\delta \to 0$.

The heuristic network problem corresponding to the continuum problem (4.1)–(4.6) was constructed in Section 3.3. It is associated with the network (graph)

$$\mathbf{G} = \{\mathbf{x}_i, C_{ij}^{(2)}; i, j \in K\} \qquad (4.9)$$

with the nodes $\{\mathbf{x}_i, i \in K\}$ corresponding to the bodies $\{D_i, i \in K\}$ forming the alive network and specific transport characteristics $C_{ij}^{(2)}$ equal to the capacity of the bodies D_i and D_j, K are indices of the bodies forming the network.

Definition 15. *The bodies $\{D_i, i = 1, \ldots, N\}$ are called uniformly dense packing if the distances between any neighboring bodies satisfy the condition*

$$\delta_{ij} = d_{ij}\delta \ (i \in K, \ j \in N_i),$$

where $0 < c \le d_{ij} \le C < \infty$ as $\delta \to 0$.

The condition of uniformly dense packing means that distances between any neighboring bodies from $\{D_i, i = 1, \ldots, N\}$ tend to zero uniformly with respect to δ. It also means that the bodies $\{D_i, i = 1, \ldots, N\}$ forms one alive network (in other words there are no insulated bodies in the case under consideration). We restrict our consideration to the case of the uniformly dense packing of bodies in order to be concentrated on the principle idea of asymptotic analysis of the problem and not digress for the issues related to topology of sets of bodies. Since under the condition of uniform dense packing, the bodies form a unique alive network, we have

$$\mathbf{G} = CH(1) = K = \{1, 2, ..., N\}. \qquad (4.10)$$

The fluxes arising in the network satisfy the Kirchhoff equations for the interior nodes of the network $\mathbf{G}$:

$$\sum_{j \in N_i} C_{ij}^{(2)}(t_i - t_j) = 0, \ i \in I \qquad (4.11)$$

(N_i means the set of nodes of the network $\mathbf{G}$ adjacent to the $i-$th node) supplemented with the boundary conditions on the nodes of the network:

$$t_1 = \varphi_1^0, \ t_2 = \varphi_2^0. \qquad (4.12)$$

We denote the solution of the finite-dimensional network problem (4.11) and (4.12) by $\{t_i^{net}, i \in K\}$.

The quantities
$$\mathbf{t}^{cont} = \{t_i^{cont}, i \in K\},$$

which are real potentials determined from solution of the continuum boundary-value problem (4.1)–(4.6), and
$$\mathbf{t}^{net} = \{t_i^{net}, i \in K\}$$

which are potentials of the network nodes determined from solution of the finite-dimensional problem (4.11) and (4.12), are not equal, generally. At the same time, numerical experiments [174] (see also Chapter 2) provide us with strong argument in favor of the coincidence of the potentials of the bodies determined from the solution of the continuum problem and the potentials of the nodes of the network for a system of dense packed bodies.

In this chapter, we present proof of the asymptotic coincidence of the potentials $\mathbf{t}^{cont}$ and $\mathbf{t}^{net}$ under the condition of uniformly dense packing of the bodies $\{D_i, i \in K\}$ and condition (3.1). Namely, we prove that

$$|\mathbf{t}^{cont} - \mathbf{t}^{net}| \to 0$$

as $\delta \to 0$.

As follows from Chapter 3, see formula (3.127), under the condition of uniformly dense packing of bodies, the energies corresponding to the original continuum problem (4.1)–(4.6) and network problem (4.11) and (4.12) asymptotically equivalent as $\delta \to 0$:

$$\mathcal{E} \sim \mathcal{E}_d, \tag{4.13}$$

where energy $\mathcal{E}$ is defined by (4.8) and

$$\mathcal{E}_d = \frac{1}{4} \sum_{i \in K} \sum_{j \in N_i} C_{ij}^{(2)} (t_i^{net} - t_j^{net})^2$$

is the energy corresponding to the network problem (4.11) and (4.12).

As was noted above, see formula (4.8), $\mathcal{E} = I(\varphi^{cont})$, where the functional $I(\varphi)$ is defined by (4.7).

In accordance with Definition 14, (4.13) means that

$$\frac{I(\varphi^{cont})}{\mathcal{E}_d} = 1 + o(1) \tag{4.14}$$

as $\delta \to 0$. We use the standard notation $o(1)$ (1.60).

From (4.14) it follows that

$$\left| I(\varphi^{cont}) - \mathcal{E}_d \right| = o(1) E_d \tag{4.15}$$

as $\delta \to 0$.

4.2. Proof of the network approximation theorem for potentials

This section contains the three-step proof of the approximation theorem for the potentials of bodies determined from the solution of the continuum and the network problems.

4.2.1. An auxiliary boundary-value problem

We consider the following boundary-value problem:

$$\Delta\varphi = 0 \text{ in } Q; \tag{4.16}$$

$$\varphi(\mathbf{x}) = t_i^{net} \text{ on } D_i, i \in I = \{3, \dots, N\}; \tag{4.17}$$

$$\varphi(\mathbf{x}) = \varphi_1^0 \text{ on } D_1, \varphi(\mathbf{x}) = \varphi_2^0 \text{ on } D_2; \tag{4.18}$$

$$|\varphi(\mathbf{x})| \leq const < \infty \text{ as } |\mathbf{x}| \to \infty \ (n = 2), \tag{4.19}$$

$$\varphi(\mathbf{x}) \to 0 \text{ as } |\mathbf{x}| \to \infty \ (n = 3), \tag{4.20}$$

where $\mathbf{t}^{net} = \{t_i^{net}, i \in I\}$ is solution of the network problem (4.11) and (4.12). Since $\{t_i^{net}, i \in I\}$ in (4.16)–(4.20) are known, this is a classical electrostatic problem for a system of bodies, see Section 1.4. We denote solution of the problem (4.16)–(4.20) by $\varphi^{net}(\mathbf{x})$.

The problem (4.16)–(4.20) is outer Dirichlet problem in which potentials of all bodies $\{D_i, i = 1, \dots, N\}$ are known. Note that we give the bodies $\{D_i, i \in I\}$ values of the potentials determined from solution of the network problem (4.11) and (4.12). Note that the original problem (4.1)–(4.6) is not a Dirichlet problem.

Function $\varphi^{net}(\mathbf{x})$ is solution of the minimization problem

$$I(\varphi) = \frac{1}{2} \int_Q |\nabla\varphi(\mathbf{x})|^2 d\mathbf{x} \to \min \tag{4.21}$$

on the set of functions

$$V^{net} = \Big\{ \varphi(\mathbf{x}) \in H^1(Q) : \varphi(\mathbf{x}) = t_i^{net} \text{ on } D_i, i \in I; \tag{4.22}$$

$$|\varphi(\mathbf{x})| \leq const < \infty \text{ as } |\mathbf{x}| \to \infty \ (n = 2), \ \varphi(\mathbf{x}) \to 0 \text{ as } |\mathbf{x}| \to \infty \ (n = 3) \Big\}.$$

Now, write the dual problem corresponding to (4.22).

We denote

$$W = \{\mathbf{v}(\mathbf{x}) \in L_2(Q) : \operatorname{div}\mathbf{v}(\mathbf{x}) \in L_2(Q)\}.$$

Lemma 8. *The problem dual with respect to (4.21) has the form*

$$J(\mathbf{v}) \to \max, \quad \mathbf{v}(\mathbf{x}) \in W, \tag{4.23}$$

where

$$J(\mathbf{v}) = -\frac{1}{2} \int_Q \mathbf{v}^2(\mathbf{x}) dx + \sum_{i=1}^{N} t_i^{net} \int_{\partial D_i} \mathbf{v}(\mathbf{x})\mathbf{n} dx + \tag{4.24}$$

$$+ \varphi_1^0 \int_{\partial D_1} \mathbf{v}(\mathbf{x})\mathbf{n}\mathbf{x} + \varphi_2^0 \int_{\partial D_2} \mathbf{v}(\mathbf{x})\mathbf{n} dx$$

and function $\mathbf{v}(\mathbf{x}) \in W$ *satisfies the condition*

$$|\mathbf{v}(\mathbf{x})| \le \frac{const}{|\mathbf{x}|^2} \quad as \quad |\mathbf{x}| \to \infty. \tag{4.25}$$

Proof. By using the Legendre transform (3.18), we write (4.8) in the form

$$\mathcal{E} = \frac{1}{2} \min_{\varphi \in V} \int_Q |\nabla \varphi(\mathbf{x})|^2 dx = \tag{4.26}$$

$$= \frac{1}{2} \min_{\varphi \in V} \max_{\mathbf{v} \in W} \int_Q \left(\mathbf{v}(\mathbf{x})\nabla \varphi(\mathbf{x}) - \frac{1}{2}\mathbf{v}^2(\mathbf{x}) \right) dx =$$

$$= \frac{1}{2} \max_{\mathbf{v} \in W} \min_{\varphi \in V} \int_Q \left(\mathbf{v}(\mathbf{x})\nabla \varphi(\mathbf{x}) - \frac{1}{2}\mathbf{v}^2(\mathbf{x}) \right) dx.$$

Since the quadratic functional $I(\varphi)$ satisfies the conditions of Proposition 5.2 Chapter III from [113], we can interchange max and min in (4.26).

Since the term $\frac{1}{2}\mathbf{v}^2(\mathbf{x})$ in (4.26) does not depend on $\nabla\varphi(\mathbf{x})$, we can write (4.26) in the form

$$\mathcal{E} = \frac{1}{2} \max_{\mathbf{v} \in W} \left(\min_{\varphi \in V} \int_Q \mathbf{v}(\mathbf{x})\nabla \varphi(\mathbf{x}) - \frac{1}{2} \int_Q \mathbf{v}^2(\mathbf{x}) \right) dx. \tag{4.27}$$

Integrate by parts in the first integral in (4.27), we obtain

$$\mathcal{E} = \frac{1}{2} \max_{\mathbf{v} \in W} \left[\min_{\varphi \in V} \left(-\int_Q \operatorname{div}\mathbf{v}(\mathbf{x})\varphi(\mathbf{x}) dx + \sum_{i=1}^{N} \int_{\partial D_i} \varphi(\mathbf{x})\mathbf{v}(\mathbf{x})\mathbf{n} dx \right) - \tag{4.28} \right.$$

$$\left. -\frac{1}{2} \int_Q \mathbf{v}^2(\mathbf{x}) dx \right].$$

Here we use that for $\varphi(\mathbf{x}) \in V$ the integral at infinity

$$\int_{|\mathbf{x}|=\rho} \mathbf{v}(\mathbf{x})\frac{\partial\varphi}{\partial\mathbf{n}}(\mathbf{x})dx \to 0 \text{ as } \rho \to 0,$$

due to (4.25).

We know values of $\varphi(\mathbf{x})$ on $\{\partial D_i, i = 1, ..., N\}$ (see the boundary condition (4.17) and (4.18)). Thus, we can write (4.28) as follows

$$\mathcal{E} = \frac{1}{2}\max_{\mathbf{v}\in W}\left[\min_{\varphi\in V}\left(-\int_Q \text{div}\mathbf{v}(\mathbf{x})\varphi(\mathbf{x})dx\right) + \right. \tag{4.29}$$

$$+\sum_{i\in I} t_i^{net}\int_{\partial D_i} \mathbf{v}(\mathbf{x})\mathbf{n}dx +$$

$$\left.+\varphi_1^0\int_{\partial D_1}\mathbf{v}(\mathbf{x})\mathbf{n}\mathbf{x} + \varphi_2^0\int_{\partial D_2}\mathbf{v}(\mathbf{x})\mathbf{n}dx - \frac{1}{2}\int_Q \mathbf{v}^2(\mathbf{x})dx\right].$$

The first term in (4.29) is linear in $\varphi(\mathbf{x})$ and it is equal to $-\infty$ with the exception of the case when $\text{div}\mathbf{v}(\mathbf{x}) \neq 0$. As a result, we arrive at (4.24). **The proof is completed.**

Lemma 9. *The following inequality holds*

$$\left|I(\varphi^{net}) - \mathcal{E}_d\right| = o(1)\mathcal{E}_d \tag{4.30}$$

as $\delta \to 0$, where $I(\varphi^{net})$ is the energy corresponding to the problem (4.16)-(4.20) (value of Dirichlet integral for the function $\varphi^{net}(\mathbf{x})$).

Proof. Function $\varphi^{net}(\mathbf{x})$ is solution of the minimization problem (4.21) on the set V^{net} (4.22).

The condition $\mathbf{v}(\mathbf{x}) \in W$ makes it possible to determine the trace of the function $\mathbf{v}(\mathbf{x})\mathbf{n}$ on the surfaces $\{\partial D_i, i = 1, ..., N\}$ as element of functional space $H^{-1/2}(Q)$ and apply the formula of integration by parts to the functions under consideration [113, 212].

It is known (see, e.g., [113, 212]) that for solutions of the problem (4.16)–(4.20) and (4.23) the following equalities have place:

$$\mathbf{v}(\mathbf{x}) = \nabla\varphi^{net}(\mathbf{x}) \tag{4.31}$$

and

$$I(\varphi^{net}) = J(\mathbf{v}). \tag{4.32}$$

From (4.21), (4.23) and (4.31), (4.32), we obtain the following two-sided estimate for $I(\varphi^{net})$:

$$-\frac{1}{2}\int_Q \mathbf{v}^2(\mathbf{x})dx + \sum_{i\in I} t_i \int_{\partial D_i} \mathbf{v}(\mathbf{x})\mathbf{n}dx + \tag{4.33}$$

$$+\varphi_1^0 \int_{\partial D_1} \mathbf{v}(\mathbf{x})\mathbf{n}dx + \varphi_2^0 \int_{\partial D_2} \mathbf{v}(\mathbf{x})\mathbf{n}dx \le$$

$$\le I(\varphi^{net}) \le \frac{1}{2}\int_Q |\nabla\varphi(\mathbf{x})|^2 dx$$

for any $\varphi(\mathbf{x}) \in V^{net}$ and any $\mathbf{v}(\mathbf{x}) \in W$ which satisfies (4.25).

In Section 3.4.3, we constructed trial function $\mathbf{v}(\mathbf{x}) \in W$, such that
1) $\operatorname{div}\mathbf{v}(\mathbf{x}) = 0$ in Q (see formula (3.52) and Lemma 3);
2) $\mathbf{v}(\mathbf{x}) = 0$ outside channels (see formula (3.55)), thus the condition (4.25) is satisfied;
3) equations (3.53) are satisfied, i.e.,

$$\int_{\partial D_i} \mathbf{v}(\mathbf{x})\mathbf{n}dx = 0, \ i \in I;$$

$$\varphi_1^0 \int_{\partial D_1} \mathbf{v}(\mathbf{x})\mathbf{n}dx + \varphi_2^0 \int_{\partial D_2} \mathbf{v}\mathbf{n}dx = 0.$$

In Section 3.4.4, we constructed trial function $\varphi(\mathbf{x})$, such that

$$\varphi(\mathbf{x}) \in V \tag{4.34}$$

and, see formula (3.88),

$$\varphi(\mathbf{x}) = t_i^{net} \text{ on } D_i. \tag{4.35}$$

With these trial functions Theorem 3 was proved. Thus, for the mentioned functions $\varphi(\mathbf{x})$ and $\mathbf{v}(\mathbf{x})$

$$|J(\mathbf{v}) - I(\varphi)| = o(1)\mathcal{E}_d, \tag{4.36}$$

$$|I(\varphi) - \mathcal{E}_d| = o(1)\mathcal{E}_d.$$

From (4.34) and (4.35) it follows that $\varphi(\mathbf{x}) \in V^{net}$. Then we can use the functions $\varphi(\mathbf{x})$ and $\mathbf{v}(\mathbf{x})$ constructed in Sections 3.4.3 and 3.4.4 as trial functions in (4.33) and from (4.33)–(4.36) we obtain (4.30). **The proof is completed.**

From (4.15) and (4.30), we obtain that

$$\left|I(\varphi^{cont}) - I(\varphi^{net})\right| = o(1)\mathcal{E}_d \tag{4.37}$$

as $\delta \to 0$.

We note that by virtue of (4.15), (4.30), it is possible to change $\mathcal{E}_d$ for $I(\varphi^{cont})$ or $I(\varphi^{net})$ in right-hand side of (4.37).

Now, we prove that from (4.37) the closeness of the potentials of bodies in the continuum problem and potential of nodes in the network model as $\delta \to 0$ follows. Namely, it holds the following theorem.

Theorem 4. *Network approximation theorem for potentials of nodes.*

Let the bodies $\{D_i, i = 1, \ldots, N\}$ satisfy the condition of uniformly dense packing and capacities of any pair of the neighbor bodies satisfy the condition

$$C_{ij}^{(2)} \to \infty \ as \ \delta \to 0 \ (i \in K, j \in N_i). \tag{4.38}$$

Then

$$|\mathbf{t}^{cont} - \mathbf{t}^{net}| \to 0 \tag{4.39}$$

as $\delta \to 0$, where $\mathbf{t}^{cont} = \{t_i^{cont}, i \in K\}$ are the values of the potentials of the perfectly conducting bodies determined from solution of the problem (4.1)–(4.6) and $\mathbf{t}^{net} = \{t_i^{net}, i \in K\}$ are the potentials of the nodes determined from solution of the network problem (4.11) and (4.12).

The next sections contain the proof of Theorem 4. The idea of the proof is the following. Under condition (4.38), the quantities $I(\varphi^{cont})$ and $I(\varphi^{net})$ have the order of $C_{ij}^{(2)}$ and (4.37) means asymptotic closeness of $I(\varphi^{cont})$ and $I(\varphi^{net})$. Generally, from the closeness of the energies, it does not follow the closeness of solutions. In the considered case, in addition, $\varphi^{net}(\mathbf{x}) \in V$ and function $\varphi^{cont}(\mathbf{x})$ gives the minimal value to the functional $I(\varphi)$ on V. We will demonstrate that under this additional condition the closeness of the energies leads to the closeness of solutions.

4.2.2. An auxiliary estimate for the energies

Lemma 10. *For any $\varphi(\mathbf{x}) \in V$ satisfying the conditions*

$$|\varphi(\mathbf{x})| \leq C \ , \ |\nabla \varphi(\mathbf{x})| \leq \frac{C}{|\mathbf{x}|^2} \ as \ |\mathbf{x}| \to \infty \ (n = 2), \tag{4.40}$$

$$|\varphi(\mathbf{x})| \leq \frac{C}{|\mathbf{x}|} \ , \ |\nabla \varphi(\mathbf{x})| \leq \frac{C}{|\mathbf{x}|^2} \ as \ |\mathbf{x}| \to \infty \ (n = 3),$$

where $C < \infty$, the equality

$$\int_Q |\nabla \varphi^{cont}(\mathbf{x}) - \nabla \varphi(\mathbf{x})|^2 d\mathbf{x} = \int_Q |\nabla \varphi(\mathbf{x})|^2 d\mathbf{x} - \int_Q |\nabla \varphi^{cont}(\mathbf{x})|^2 d\mathbf{x}$$

holds.

We note that the solutions of problems (4.1)–(4.6) and (4.16)–(4.20) satisfy the conditions (4.40) for the corresponding dimension n (the proof can be found in [94, 366]).

Proof. We take a function $\psi(\mathbf{x}) \in V$, satisfying condition (4.40) and the additional condition

$$\psi(\mathbf{x}) = 0 \text{ on } D_1 \text{ and } D_2; \tag{4.41}$$

and consider the following integral depending on the real parameter t:

$$F(t) = \int_Q |\nabla \varphi^{cont}(\mathbf{x}) + t\nabla\psi(\mathbf{x})|^2 d\mathbf{x}. \tag{4.42}$$

The derivative of $F(t)$ with respect to the parameter t is

$$\frac{dF}{dt}(t) = 2\int_Q (\nabla\varphi^{cont}(\mathbf{x}) + t\nabla\psi(\mathbf{x}))\nabla\psi(\mathbf{x})d\mathbf{x} = \tag{4.43}$$

$$= 2\int_Q \nabla\varphi^{cont}(\mathbf{x})\nabla\psi(\mathbf{x})d\mathbf{x} + 2t\int_Q |\nabla\psi(\mathbf{x})|^2 d\mathbf{x}.$$

We consider the first integral in the right-hand side of (4.43). Integrating it by parts, we obtain

$$\int_Q \nabla\varphi^{cont}(\mathbf{x})\nabla\psi(\mathbf{x})d\mathbf{x} = -\int_Q \Delta\varphi^{cont}(\mathbf{x})\psi(\mathbf{x})d\mathbf{x} + \tag{4.44}$$

$$+ \sum_{i=3}^{N} \int_{\partial D_i} \frac{\partial\varphi^{cont}}{\partial\mathbf{n}}(\mathbf{x})\psi(\mathbf{x})d\mathbf{x} +$$

$$+ \int_{\partial D_1} \frac{\partial\varphi^{cont}}{\partial\mathbf{n}}(\mathbf{x})\psi(\mathbf{x})d\mathbf{x} + \int_{\partial D_2} \frac{\partial\varphi^{cont}}{\partial\mathbf{n}}(\mathbf{x})\psi(\mathbf{x})d\mathbf{x}.$$

Here we use the integral at infinity

$$\int_{|\mathbf{x}|=\rho} \frac{\partial\varphi^{cont}}{\partial\mathbf{n}}(\mathbf{x})\psi(\mathbf{x})d\mathbf{x} \to 0 \text{ as } \rho \to 0,$$

due to (4.40). For the both cases $n = 2$ and $n = 3$ it is estimated from above by the quantity

$$c_1\frac{C^2}{\rho} \to 0 \text{ as } \rho \to 0.$$

Here $c_1 = 2\pi$ if $n = 2$ and $c_1 = 4\pi$ if $n = 3$.

The first integral in the right-hand side (4.44) is equal to zero due to (4.1).

For the second integral in the right-hand side of (4.44), taking into account that $\psi(\mathbf{x}) \in V$, we have

$$\int_{\partial D_i} \frac{\partial \varphi^{cont}}{\partial \mathbf{n}}(\mathbf{x})\psi(\mathbf{x})d\mathbf{x} = \tilde{t}_i \int_{\partial D_i} \frac{\partial \varphi^{cont}}{\partial \mathbf{n}}(\mathbf{x})d\mathbf{x} = 0 \ (i \in I),$$

where $\tilde{t}_i$ is the value of the function $\psi(\mathbf{x})$ on ∂D_i (since the function $\psi(\mathbf{x}) \in V$, it takes constant values on ∂D_i). The last equality follows from (4.3) (we recall that $\varphi^{cont}(\mathbf{x})$ is the solution of the problem (4.1)–(4.6)).

By virtue of (4.41), the third and the fourth integrals in the right-hand side of (4.44) are equal to zero.

As a result, (4.43) becomes

$$\frac{dF}{dt}(t) = 2t \int_Q |\nabla\psi(\mathbf{x})|^2 d\mathbf{x}. \tag{4.45}$$

Integrating (4.45) with respect to t from 0 to 1, we obtain (note that the function $\psi(\mathbf{x})$ does not depend on t)

$$F(1) - F(0) = \int_Q |\nabla\psi(\mathbf{x})|^2 d\mathbf{x}. \tag{4.46}$$

Due to the definition of the function $F(t)$ (4.42),

$$F(1) = \int_Q |\nabla\varphi^{cont}(\mathbf{x}) + \nabla\psi(\mathbf{x})|^2 d\mathbf{x}, \tag{4.47}$$

$$F(0) = \int_Q |\nabla\varphi^{cont}(\mathbf{x})|^2 d\mathbf{x}.$$

Now we take

$$\psi(\mathbf{x}) = \varphi(\mathbf{x}) - \varphi^{cont}(\mathbf{x}), \tag{4.48}$$

where $\varphi(\mathbf{x}) \in V$ and satisfies (4.40). Note that $\varphi^{cont}(\mathbf{x}) \in V$ by definition of the function $\varphi^{cont}(\mathbf{x})$ and satisfies (4.40). The function $\varphi^{cont}(\mathbf{x})$ satisfies (4.40) because it is the solution of the Laplace equation with the conditions (4.6) (the proof can be found in [94, 366]). Then the function $\psi(\mathbf{x})$ defined by (4.48) satisfies (4.40) and (4.41). For this function $\psi(\mathbf{x})$, (4.47) gives

$$F(1) = \int_Q |\nabla\varphi(\mathbf{x})|^2 d\mathbf{x},$$

$$F(0) = \int_Q |\nabla\varphi^{cont}(\mathbf{x})|^2 d\mathbf{x}$$

and (4.46) gives

$$\int_Q |\nabla\varphi(\mathbf{x})|^2 dx - \int_Q |\nabla\varphi^{cont}(\mathbf{x})|^2 dx = \int_Q |\nabla(\varphi(\mathbf{x}) - \varphi^{cont}(\mathbf{x}))|^2 dx. \qquad (4.49)$$

Proof is completed.

4.2.3. Estimate of difference of solutions of the original problem and the auxiliary problem

We consider the function

$$\psi(\mathbf{x}) = \varphi^{net}(\mathbf{x}) - \varphi^{cont}(\mathbf{x}), \qquad (4.50)$$

which is the difference of solutions of the problems (4.1)–(4.6) and the problem (4.16)–(4.20) and estimate the Dirichlet integral of this function.

The function $\psi(\mathbf{x})$ is solution of the following problem (see problems (4.1)–(4.6) and (4.16)–(4.20), which define the functions $\varphi^{cont}(\mathbf{x})$ and $\varphi^{net}(\mathbf{x})$):

$$\Delta\psi = 0 \text{ in } Q; \qquad (4.51)$$

$$\psi(\mathbf{x}) = \tau_i \text{ on } D_i, i \in I; \qquad (4.52)$$

$$\psi(\mathbf{x}) = 0 \text{ on } D_1 \text{ and } D_2; \qquad (4.53)$$

$$\psi(\mathbf{x}) \leq const \text{ as } |\mathbf{x}| \to \infty \ (n = 2); \qquad (4.54)$$

$$\psi(\mathbf{x}) \to 0 \text{ as } |\mathbf{x}| \to \infty \ (n = 3); \qquad (4.55)$$

where $\{\tau_i, i \in I\}$ are introduced as the differences

$$\tau_i = t_i^{cont} - t_i^{net}.$$

The quantity

$$I(\psi) = \frac{1}{2}\int_Q |\nabla\psi(\mathbf{x})|^2 dx = \frac{1}{2}\int_Q |\nabla(\varphi^{net}(\mathbf{x}) - \varphi^{cont}(\mathbf{x}))|^2 dx = I(\varphi^{cont} - \varphi^{net})$$

is the energy corresponding to the problem (4.51)–(4.55).

From (4.49) with $\varphi(\mathbf{x}) = \varphi^{net}(\mathbf{x})$, it follows that

$$\left| \int_Q |\nabla\varphi^{net}(\mathbf{x})|^2 dx - \int_Q |\nabla\varphi^{cont}(\mathbf{x})|^2 dx \right| = 2I(\varphi^{cont} - \varphi^{net}) \qquad (4.56)$$

or

$$\left| I(\varphi^{net}) - I(\varphi^{cont}) \right| = I(\varphi^{cont} - \varphi^{net}). \qquad (4.57)$$

Definition 16. *For two nonzero quantities a and b of the same order of magnitude, the number*

$$\frac{|a-b|}{|a|} \approx \frac{|a-b|}{|b|}$$

is called the relative difference of a and b.

We explain an idea of the next step of the proof in physics terms. In accordance with (4.37), the relative difference of the energies $I(\varphi^{cont})$ and $I(\varphi^{net})$ is small as compared with $\mathcal{E}_d$. At the same time, this is greater than $I(\psi)$ divided by $I(\varphi^{cont})$ or $I(\varphi^{net})$. We investigate $I(\psi)$, which has the physical meaning of the energy corresponding to the electrostatic problem (4.51)–(4.55). The physics intuition based on the idea of the effect of the energy localization (see Chapter 3) suggests that the energy $I(\psi)$ takes the value of the order of $C_{ij}^{(2)}$ if even one of the quantities τ_i is not small. The unique case when the energy $I(\psi)$ may be not large as compared with $I(\varphi^{cont})$ and $I(\varphi^{net})$ is the case when all τ_i, $i \in I$ are small. Now, we present the formal calculations.

Due to (4.37) and (4.49)

$$I(\varphi^{cont} - \varphi^{net}) = \left| I(\varphi^{cont}) - I(\varphi^{net}) \right| = o(1)\mathcal{E}_d \qquad (4.58)$$

as $\delta \to 0$.

Now, we prove that

$$I(\varphi^{cont} - \varphi^{net}) \geq C \min_{i,j} |\tau_i - \tau_j| \mathcal{E}_d \qquad (4.59)$$

as $\delta \to 0$, where the minimum is taken over the indices i, j of the neighbor nodes of the network and $0 < C < \infty$ does not depend on $\{\tau_i, i \in I\}$.

We use the low-sided estimate similar to (4.33) for the solution $\psi(\mathbf{x}) = \varphi^{net}(\mathbf{x}) - \varphi^{cont}(\mathbf{x})$ of the problem (4.51)–(4.55) and write

$$I(\varphi^{cont} - \varphi^{net}) \geq -\frac{1}{2} \int_Q \mathbf{v}^2(\mathbf{x})d\mathbf{x} + \sum_{i \in I} \tau_i \int_{\partial D_i} \mathbf{v}(\mathbf{x})\mathbf{n}d\mathbf{x}, \qquad (4.60)$$

where $\mathbf{v}(\mathbf{x}) \in W$ and $|\mathbf{v}(\mathbf{x})| \leq \dfrac{const}{|\mathbf{x}|^2}$ as $|\mathbf{x}| \to \infty$.

We use the notion of electrostatic channel which was introduced in Section 3.4.2, see Definition 11. In the case under consideration, the electrostatic channel between two bodies D_i and D_j is associated with the following electrostatics problem in R^3 for the bodies D_i and D_j with the potentials $\tau_i \neq \tau_j$:

$$\Delta\varphi^{(ij)} = 0 \text{ in } R^3\backslash(D_i \bigcup D_j); \qquad (4.61)$$

$$\varphi^{(ij)} = \tau_i \text{ on } D_i, \varphi^{(ij)} = \tau_j \text{ on } D_j; \qquad (4.62)$$

$$|\varphi^{(ij)}(\mathbf{x})| \leq const \text{ as } |\mathbf{x}| \to \infty \ (n = 2); \qquad (4.63)$$

$$|\varphi^{(ij)}(\mathbf{x})| \to 0 \text{ as } |\mathbf{x}| \to \infty \ (n = 3). \qquad (4.64)$$

We consider the electrostatic channel S_{ij} between the neighbor bodies D_i and D_j, see Fig. 3.9. We choose the channel S_{ij} narrow enough so that it does not intersect similar channels between other bodies. Such choice is always possible under the condition of uniformly dense packing when the neighboring bodies are placed closely to one another. We consider the function

$$\mathbf{v}(\mathbf{x}) = \begin{cases} (\tau_i - \tau_j)\nabla\varphi^{\pm 1}(\mathbf{x}) \text{ in channel } S_{ij}, \\ 0 \text{ outside channel } S_{ij}, \end{cases} \tag{4.65}$$

where $\varphi^{\pm 1}(\mathbf{x})$ is the solution of problem (4.61)–(4.64) for the case $\tau_i = 1/2$ and $\tau_j = -1/2$ (see Definition 10).

It is clear that gradients of the functions $\varphi^{ij}(\mathbf{x})$ and $\varphi^{\pm 1}(\mathbf{x})$ are related as follows:

$$\nabla\varphi^{(ij)}(\mathbf{x}) = (\tau_i - \tau_j)\nabla\varphi^{\pm 1}(\mathbf{x}). \tag{4.66}$$

Since the channel is an electrostatic one,

$$\mathbf{v}(\mathbf{x})\mathbf{n} = 0 \text{ on } \partial S_{ij}^{lat}, \tag{4.67}$$

where ∂S_{ij}^{lat} denotes the lateral boundary of the channel S_{ij}, see Fig. 3.10. From (4.67) and (4.61), it follows that the function $\mathbf{v}(\mathbf{x})$ (4.65) satisfies the condition

$$\mathrm{div}\mathbf{v}(\mathbf{x}) = 0$$

in whole domain Q (see Section 3.4.3 for detailed proof). By definition (4.65), $\mathbf{v}(\mathbf{x})=0$ outside the channel S_{ij}. Then condition

$$|\mathbf{v}(\mathbf{x})| \leq \frac{const}{|\mathbf{x}|^2} \text{ as } |\mathbf{x}| \to 0$$

is satisfied evidently.

Thus, the function $\mathbf{v}(\mathbf{x})$ (4.65) can be used in (4.60) as a trial function. For the function (4.65), the first integral in the right-hand side of (4.60) is

$$-\frac{1}{2}\int_Q \mathbf{v}^2(\mathbf{x})d\mathbf{x} = -\frac{1}{2}\int_{S_{ij}} \mathbf{v}^2(\mathbf{x})d\mathbf{x} = \tag{4.68}$$

$$= -\frac{1}{2}(\tau_i - \tau_j)^2\int_K |\nabla\varphi^{\pm 1}(\mathbf{x})|^2 d\mathbf{x}.$$

For the function (4.65), the second integral in the right-hand side of (4.60) is

$$\sum_{k \in I} \tau_k \int_{\partial D_k} \mathbf{v}(\mathbf{x})\mathbf{n}d\mathbf{x} = \tau_i \int_{\partial D_i} \mathbf{v}(\mathbf{x})\mathbf{n}d\mathbf{x} + \tau_j \int_{\partial D_j} \mathbf{v}(\mathbf{x})\mathbf{n}d\mathbf{x}.$$

As a result of our choice of the function (4.65) and channel S_{ij}, the flux through the top S_i and bottom S_j, see Fig. 3.9, surfaces of the channel S_{ij} are equal to zero and

$$\int_{S_i} \mathbf{v}(\mathbf{x})\mathbf{n}dx = -\int_{S_j} \mathbf{v}(\mathbf{x})\mathbf{n}dx.$$

Then

$$\sum_{k \in I} \tau_k \int_{\partial D_k} \mathbf{v}(\mathbf{x})\mathbf{n}dx = (\tau_i - \tau_j)\int_{S_i} \mathbf{v}(\mathbf{x})\mathbf{n}dx. \tag{4.69}$$

By virtue of the first line of (4.65) and the equality

$$\int_{S_i} \nabla\varphi^{\pm}(\mathbf{x})\mathbf{n}dx = \int_{S_{ij}} |\nabla\varphi^{\pm}(\mathbf{x})|^2 dx$$

the right-hand part of (4.69) is equal to

$$(\tau_i - \tau_j)^2 \int_{S_{ij}} |\nabla\varphi^{\pm 1}(\mathbf{x})|^2 dx. \tag{4.70}$$

By virtue of (4.68) and (4.70), the right-hand side of (4.60) is equal to

$$\frac{1}{2}(\tau_i - \tau_j)^2 \int_{S_{ij}} |\nabla\varphi^{\pm 1}(\mathbf{x})|^2 dx$$

and we have that

$$I(\varphi^{cont} - \varphi^{net}) \geq \frac{1}{2}(\tau_i - \tau_j)^2 \int_{S_{ij}} |\nabla\varphi^{\pm 1}(\mathbf{x})|^2 dx. \tag{4.71}$$

Under the condition (4.38), the magnitude of the integral $\int_{S_{ij}} |\nabla\varphi^{\pm 1}(\mathbf{x})|^2 dx$ for a pair of bodies has the order of $C_{ij}^{(2)}$, see Theorem 2. Due to this we obtain

$$I(\varphi^{net} - \varphi^{cont}) = I(\varphi^{net} - \varphi^{cont}) \geq cC_{ij}^{(2)}(\tau_i - \tau_j)^2, \tag{4.72}$$

where $0 < c < \infty$ does not depend on $\{\tau_i\}$.

Combining (4.72) with (4.56)–(4.58), we obtain

$$C_{ij}^{(2)}(\tau_i - \tau_j)^2 \leq o(1)\mathcal{E}_d$$

or

$$(\tau_i - \tau_j)^2 = o(1) \frac{\mathcal{E}_d}{C_{ij}^{(2)}} \tag{4.73}$$

for the neighboring bodies.

Under the condition of the uniformly dense packing for bodies $\{D_i, i = 1, \dots, N\}$ and condition (4.38) for the capacities of any pair of the neighbor bodies, $\mathcal{E}_d$ has the order of $C_{ij}^{(2)}$ as $\delta \to 0$, see Section 3.5.2. Then

$$\frac{\mathcal{E}_d}{C_{ij}^{(2)}} < const < \infty \text{ as } \delta \to 0 \ (i, j \in 1, \dots, N). \tag{4.74}$$

From (4.73) and (4.74), we have

$$|\tau_i - \tau_j| = o(1). \tag{4.75}$$

For the bodies D_1 and D_2 we have $\tau_1 = 0$ and $\tau_2 = 0$. The graph $\mathbf{G}$ (4.9) is connected, see (4.10). From this note and (4.75) it follows that $|\tau_i| = o(1)$ for all $i = 1, \dots, N$, or $|t_i^{cont} - t_i^{net}| = o(1) \to 0$ as $\delta \to 0$. From this, (4.39) follows. **Proof is completed**.

4.3. The speed of convergence of potentials for a system of circular disks

As follows from (4.73)

$$|\tau_i - \tau_j| = \sqrt{o(1) \frac{\mathcal{E}_d}{C_{ij}^{(2)}}}. \tag{4.76}$$

It was demonstrated in [41] that for circular disks the order of $o(1)$ in (4.58) is

$$o(1) = \sqrt{\frac{\delta}{R}}, \tag{4.77}$$

R is the radius of the disks.

From (4.76) and (4.77), we have

$$|\tau_i - \tau_j| = \sqrt{o(1) \frac{\mathcal{E}_d}{C_{ij}^{(2)}}} = \sqrt{\sqrt{\frac{\delta}{R}} \frac{\mathcal{E}_d}{C_{ij}^{(2)}}}. \tag{4.78}$$

Since $\mathcal{E}_d \sim C_{ij}^{(2)}$ as $\delta \to 0$, we obtain from (4.78) that

$$|\tau_i - \tau_j| \sim \sqrt[4]{\frac{\delta}{R}}, \tag{4.79}$$

as $\delta \to 0$.

From (4.73) and (4.79) we have that the speed of convergence of the energies (capacity) is $\sqrt{\dfrac{\delta}{R}}$ and the speed of convergence of the potentials is $\sqrt[4]{\dfrac{\delta}{R}}$.

For small $\dfrac{\delta}{R}$

$$\sqrt{\frac{\delta}{R}} < \sqrt[4]{\frac{\delta}{R}}. \tag{4.80}$$

It means that speed of convergence of potentials is slower than the speed of convergence of the energies (capacity). Numerical computations presented in Section 2.5 demonstrated slow convergence of potentials. Thus, we observe the agreement between the mathematical and numerical predictions of speed of convergence.

Finalizing this chapter, we emphasize again the role of condition (3.1). The network approximation theorem for potentials (as well as the network approximation theorem for energies) was proved under that condition (see condition (4.38) in the formulation of Theorem 4). The authors cannot predict asymptotic behavior of a system of closely placed bodies under the condition $\delta \to 0$ when the capacities $C_{ij}^{(2)}$ are restricted as $\delta \to 0$. We recall that such a situation takes place for a system of polyhedron-shaped bodies for the dimension $n = 3$.

Chapter 5

ANALYSIS OF TRANSPORT PROPERTIES OF HIGHLY FILLED CONTRAST COMPOSITES USING THE NETWORK APPROXIMATION METHOD

In this chapter, we consider particle-filled composite material. In previous chapters, we kept the electrostatic interpretation of the considered transport problem, but now we treat the transport problem as the problem of electrical conduction, see Table 1 in Preface. In highly filled composite materials the filling particles are closely packed. Then there is a reason to apply the theory developed above to highly filled contrast composites.

5.1. Modification of the network approximation method as applied to particle-filled composite materials

Although the transport problems for composite material and system of bodies have many similarities, they also have essential differences. We emphasize the following (main from our point of view) differences between the problems considered in this chapter and Chapters 3 and 4:

• Considering a system of bodies (see Chapter 3), we formulate the problem in R^n ($n = 2, 3$). As a result, we arrive at the outer boundary-value problem. Considering a composite material, we deal with a finite specimen or construction and arrive at an inner boundary-value problem.

• For composite materials, the potentials are specified not for two or more small number of predetermined bodies / particles, but for a boundary of domain. For the bodies / particles belonging to (touching) the boundary of domain, the boundary

139

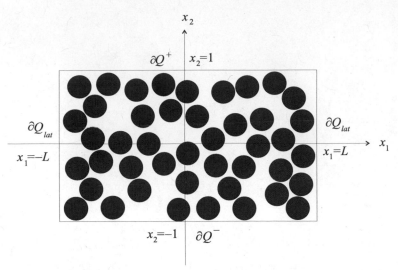

Figure 5.1. *A domain filled with randomly distributed particles (illustration is given for the dimension $n = 2$).*

conditions are formulated similarly to Chapter 3 (i.e., values of potential are known for these bodies / particles). For other parts of the boundary, the boundary condition is formulated for the matrix.

• For composite materials, the object of interest, usually, is not the capacity but the effective transport characteristic of the composite, which is the total flux (the sum of fluxes through all parts of the boundary, particles and matrix). Usually, the flux through the matrix is the largest portion of the total flux.

The problems listed above do not arise when we analyze a system of bodies.

In this chapter, we introduce the notion of effective conductivity of high-contrast composite filled with closely placed particles and then describe a number of tricks, which allow us to modify the methods from Chapters 3 and 4 for composite materials of the type mentioned. We do not present the proofs of the theorems in this chapter and do not present a complete theory of contrast composite materials. The proofs can be found in [183, 185]. A complete theory of contrast composite materials will be presented in another book.

5.1.1. Formulation of the problem

Consider a mathematical model of a particle-filled composite material. We introduce the domain

$$\Pi = [-L, L]^{n-1} \times [-1, 1]$$

occupied by the composite material depicted in Fig. 5.1 ($n = 2, 3$ is dimension of the problem).

We denote by $\{D_i, i = 1, ..., N\}$ the bodies which model the particles, where N is the total number of particles. The particles do not necessarily form any periodic array. The only restriction we impose is that the domains $\{D_i, i = 1, ..., N\}$ do not overlap and do not touch one another. Then

$$Q = \Pi \setminus \bigcup_{i=1}^{N} D_i$$

is the matrix (the perforated domain Π).

We denote by

$$\partial Q^+ = \{\mathbf{x} \in R^n : x_n = 1\}$$

and

$$\partial Q^- = \{\mathbf{x} \in R^n : x_n = -1\}$$

the top and bottom boundary of the domain Q and we denote

$$\partial Q_{lat} = \partial Q \setminus (\partial Q^+ \bigcup \partial Q^-)$$

the lateral (left and right in the case under consideration) boundaries of the domain Q, see Fig. 5.1.

On the boundaries ∂Q^+ and ∂Q^- we stay the following boundary conditions:

$$\varphi(\mathbf{x}) = 1 \text{ on } \partial Q^+, \ \varphi(\mathbf{x}) = -1 \text{ on } \partial Q^-$$

corresponding to applied potential of ± 1 to the top (∂Q^+) and bottom (∂Q^-) boundaries, respectively.

On ∂Q_{lat} we set the boundary conditions (in this chapter, ε implies the electrical conductivity, see Table 1)

$$\varepsilon(\mathbf{x}) \frac{\partial \varphi}{\partial \mathbf{n}} = 0 \text{ on } \partial Q_{lat}.$$

The thickness of the domain Π in the Ox_n-axis direction is equal to two. The difference of the potential applied to the top and bottom boundaries ∂Q^+ and ∂Q^- of the domain is also equal to two. Thus, the average difference of potential per unit length in the Ox_n-axis direction is equal to one. In other words, with our choice of the domain Π and boundary conditions on the boundaries ∂Q^+ and ∂Q^- we consider a composite material subject to an applied external field (strength of electric field in the case under consideration) of unit value.

We introduce the functional space

$$V = \Big\{ \varphi(\mathbf{x}) \in H^1(Q) : \varphi(\mathbf{x}) = t_i \text{ on } D_i, i = 1, ..., N;$$

$$\varphi(\mathbf{x}) = 1 \text{ on } \partial Q^+, \ \varphi(\mathbf{x}) = -1 \text{ on } \partial Q^- \Big\}. \tag{5.1}$$

In the definition (5.1) $\{t_i,\ i \in I\}$ are arbitrary constants for the particles, which do not touch Q^+ or Q^-. The set of indices of such particles is denoted by I. It is possible that some particles touch the boundaries Q^+ or Q^-. Such particles have potentials 1 or -1, correspondingly.

The functional space (5.1) corresponds to perfectly conducting particles, which occupy finite domain Q. The difference between the functional spaces corresponding to finite and infinite domains becomes clear if seen compared to definitions (5.1) and (3.10).

We consider the following variational problem

$$I(\varphi) \to \min, \ \varphi(\mathbf{x}) \in V, \tag{5.2}$$

where

$$I(\varphi) = \frac{1}{2} \int_Q \varepsilon(\mathbf{x})|\nabla\varphi(\mathbf{x})|^2 dx.$$

Here $\varepsilon(\mathbf{x})$ means the local transport characteristic (electric conductivity in the case under consideration). For composite material, it may depend on the spatial variable (the case of inhomogeneous matrix).

The problem (5.2) is equivalent to the following boundary-value problem (the Euler–Lagrange equations):

$$\frac{\partial}{\partial x_i}\left(\varepsilon(\mathbf{x})\frac{\partial\varphi}{\partial x_i}\right) = 0 \text{ in } Q; \tag{5.3}$$

$$\varphi(\mathbf{x}) = t_i \text{ on } \partial D_i, \ i \in I; \tag{5.4}$$

$$\int_{\partial D_i} \varepsilon(\mathbf{x})\frac{\partial\varphi}{\partial\mathbf{n}}d\mathbf{x} = 0; \tag{5.5}$$

$$\varphi(\mathbf{x}) = 1 \text{ on } \partial Q^+, \ \varphi(\mathbf{x}) = -1 \text{ on } \partial Q^-; \tag{5.6}$$

$$\varepsilon(\mathbf{x})\frac{\partial\varphi}{\partial\mathbf{n}} = 0 \text{ on } \partial Q_{lat}. \tag{5.7}$$

Here $\mathbf{n}$ is the outward unit normal to the boundaries of the corresponding domains (Q or D_i).

Condition (5.5) is obtained by integration by parts in the Euler–Lagrange equation, which corresponds to the functional (5.2), where the condition (5.4) was taken into account. This condition means that the total flux of the current entering each particle D_i is zero. Condition (5.7) means that the lateral boundaries of the domain Π are insulated.

The unknowns in the problem (5.3)–(5.7) (like in problem (3.3)–(3.8)) are function $\varphi(\mathbf{x})$ in domain Q and real numbers $\{t_i, i \in I\}$ (potentials of the particles $\{D_i, i \in I\}$, I is the set indices of bodies, which are not subjected to direct application of voltage.

The problems (5.2) and (5.3)–(5.7) have (the same) unique solution. The proof is similar to the proof presented in Section 3.2.1.

5.1.2. Effective conductivity of the composite material

The effective conductivity can be defined in several equivalent forms. One of the definitions was presented in Section 1.1. There exists other ways for computation of total effective conductivity of the composite material.

We start with a definition which is natural from the physical point of view. We define first the total flux D_{total} through a cross-section of the domain (for example, through the upper boundary ∂Q^+) as

$$D_{total} = \int_{\partial Q^+} \varepsilon(\mathbf{x})\frac{\partial \varphi}{\partial \mathbf{n}}(\mathbf{x})d\mathbf{x}.$$

We will need an expression for this flux in terms of the value of the energy integral $I(\varphi)$, which is commonly used in homogenization theory as a definition of the effective conductivity [30]. For the sake of completeness, we show here the equivalence of these two definitions.

Multiply the equation (5.3) by $\varphi(\mathbf{x})$ and integrate by parts. Then

$$0 = -\int_Q \varepsilon(\mathbf{x})|\nabla\varphi(\mathbf{x})|^2 d\mathbf{x} + \int_{\partial Q^+} \varepsilon(\mathbf{x})\frac{\partial \varphi}{\partial \mathbf{n}}(\mathbf{x})\varphi(\mathbf{x})d\mathbf{x} + \qquad (5.8)$$

$$+ \int_{\partial Q^-} \varepsilon(\mathbf{x})\frac{\partial \varphi}{\partial \mathbf{n}}(\mathbf{x})\varphi(\mathbf{x})d\mathbf{x} + \sum_{i=1}^N \int_{\partial D_i} \varepsilon(\mathbf{x})\frac{\partial \varphi}{\partial \mathbf{n}}(\mathbf{x})\varphi(\mathbf{x})d\mathbf{x}.$$

For the integrals over the boundaries of the particles, we use (5.4) and (5.5) to obtain

$$\int_{\partial D_i} \varepsilon(\mathbf{x})\frac{\partial \varphi}{\partial \mathbf{n}}(\mathbf{x})\varphi(\mathbf{x})d\mathbf{x} = t_i \int_{\partial D_i} \varepsilon(\mathbf{x})\frac{\partial \varphi}{\partial \mathbf{n}}d\mathbf{x} = 0. \qquad (5.9)$$

Furthermore,

$$\int_{\partial Q^+} \varepsilon(\mathbf{x})\frac{\partial \varphi}{\partial \mathbf{n}}(\mathbf{x})\varphi(\mathbf{x})d\mathbf{x} + \int_{\partial Q^-} \varepsilon(\mathbf{x})\frac{\partial \varphi}{\partial \mathbf{n}}(\mathbf{x})\varphi(\mathbf{x})d\mathbf{x} =$$

$$= \int_{\partial Q^+} \varepsilon(\mathbf{x})\frac{\partial \varphi}{\partial \mathbf{n}}(\mathbf{x})d\mathbf{x} - \int_{\partial Q^-} \varepsilon(\mathbf{x})\frac{\partial \varphi}{\partial \mathbf{n}}(\mathbf{x})d\mathbf{x} \qquad (5.10)$$

due to (5.6).

Multiply the equation (5.3) by 1 and integrate the result by parts, we obtain

$$0 = \int_{\partial Q^+} \varepsilon(\mathbf{x}) \frac{\partial \varphi}{\partial \mathbf{n}}(\mathbf{x}) d\mathbf{x} + \int_{\partial Q^-} \varepsilon(\mathbf{x}) \frac{\partial \varphi}{\partial \mathbf{n}}(\mathbf{x}) d\mathbf{x} -$$

$$- \sum_{i=1}^{N} \int_{\partial D_i} \varepsilon(\mathbf{x}) \frac{\partial \varphi}{\partial \mathbf{n}}(\mathbf{x}) d\mathbf{x}.$$

The last integral is equal to zero due to (5.5). Then

$$\int_{\partial Q^+} \varepsilon(\mathbf{x}) \frac{\partial \varphi}{\partial \mathbf{n}}(\mathbf{x}) d\mathbf{x} + \int_{\partial Q^-} \varepsilon(\mathbf{x}) \frac{\partial \varphi}{\partial \mathbf{n}}(\mathbf{x}) d\mathbf{x} = 0. \tag{5.11}$$

The physical meaning of (5.11) is that the total flux through the low boundary ∂Q^- is equal to the total flux through the upper boundary ∂Q^+.

From (5.11) and (5.10) we obtain

$$\int_{\partial Q^+} \varepsilon(\mathbf{x}) \frac{\partial \varphi}{\partial \mathbf{n}}(\mathbf{x}) \varphi(\mathbf{x}) d\mathbf{x} + \int_{\partial Q^-} \varepsilon(\mathbf{x}) \frac{\partial \varphi}{\partial \mathbf{n}}(\mathbf{x}) \varphi(\mathbf{x}) d\mathbf{x} = 2 \int_{\partial Q^+} \varepsilon(\mathbf{x}) \frac{\partial \varphi}{\partial \mathbf{n}}(\mathbf{x}) d\mathbf{x}. \tag{5.12}$$

Combining (5.3), (5.8), (5.9) and (5.12), we finally obtain the following formula:

$$\int_{\partial Q^+} \varepsilon(\mathbf{x}) \frac{\partial \varphi}{\partial \mathbf{n}}(\mathbf{x}) d\mathbf{x} = \frac{1}{2} \int_Q \varepsilon(\mathbf{x}) |\nabla \varphi(\mathbf{x})|^2 d\mathbf{x}, \tag{5.13}$$

where $\varphi(\mathbf{x})$ is the solution of the problem (5.3)–(5.7) (or equivalently, it solves the problem (5.2)).

The left-hand side of (5.13) is equal to the total flux D_{total} through the upper boundary ∂Q^+. The right-hand side of (5.13) is the energy

$$\mathcal{E} = \frac{1}{2} \int_Q \varepsilon(\mathbf{x}) |\nabla \varphi(\mathbf{x})|^2 d\mathbf{x}. \tag{5.14}$$

In accordance with (5.13) and (5.14) we have

$$D_{total} = \mathcal{E}.$$

We first define the effective conductivity for the continuum problem and then construct its approximation in the corresponding discrete problem.

Definition 17. *The total flux D_{total} through the boundary ∂Q^+ divided by the measure $|\partial Q^+|$ of the boundary ∂Q^+*

$$\widehat{\varepsilon} = \frac{D_{total}}{|\partial Q^+|} = \frac{1}{|\partial Q^+|} \int_{\partial Q^+} \varepsilon(\mathbf{x}) \frac{\partial \varphi}{\partial \mathbf{n}}(\mathbf{x}) d\mathbf{x} \tag{5.15}$$

is called the effective (overall) conductivity.

In the case under consideration, the measure $|\partial Q^+|$ of ∂Q^+ is equal to $4L^2$ (the area of ∂Q^+) in the three-dimensional case and it is equal to $2L$ (the length of interval ∂Q^+) in the two-dimensional case.

Due to (5.13) we have the following formulas for $\widehat{\varepsilon}$:

$$\widehat{\varepsilon} = \frac{1}{2|\partial Q^+|} \int_Q \varepsilon(\mathbf{x})|\nabla\varphi(\mathbf{x})|^2 d\mathbf{x}, \tag{5.16}$$

and

$$\widehat{\varepsilon} = \frac{1}{|\partial Q^+|} \int_{\partial Q^+} \varepsilon(\mathbf{x})\frac{\partial\varphi}{\partial\mathbf{n}}(\mathbf{x})d\mathbf{x}, \tag{5.17}$$

where $\varphi(\mathbf{x})$ is the solution of the problem (5.3)–(5.7) (or the equivalent extremal problem (5.2)).

Also we can write

$$\widehat{\varepsilon} = \frac{1}{2|\partial Q^+|} \min_{\varphi \in V} \int_Q \varepsilon(\mathbf{x})|\nabla\varphi(\mathbf{x})|^2 d\mathbf{x}. \tag{5.18}$$

The equality (5.18) can be understood as an extremal principle for the effective conductivity. Similar extremal principles are known for homogenized characteristics of solid bodies [246, 390, 391], plates and beams [182].

Composite material with homogeneous matrix

If material of the matrix is homogeneous, then

$$\varepsilon(\mathbf{x}) = \varepsilon = const > 0. \tag{5.19}$$

The corresponding problem is obtained by changing $\varepsilon(\mathbf{x})$ for ε in the problem (5.1) and (5.2), as well as in the problem (5.3)–(5.7) and formulas (5.16)–(5.18).

Transition to dimensionless problem

Observing the equations (5.3)–(5.7) and formulas (5.16)–(5.18) with $\varepsilon(\mathbf{x}) = \varepsilon$, we find that all the equations involving ε are homogeneous (the free terms are zero). Under condition (5.19) we can divide these equations by ε and, as a result, eliminate ε from the boundary problem (5.3)–(5.7). But we cannot not eliminate ε from the formulas (5.16)–(5.18) for the computation of effective conductivity $\widehat{\varepsilon}$ (5.15) and energy $\mathcal{E}$ (5.14).

Then the minimization problem (5.2) can be written as follows:

$$I(\varphi) \to \min, \quad \varphi(\mathbf{x}) \in V, \tag{5.20}$$

where

$$I(\varphi) = \frac{1}{2} \int_Q |\nabla\varphi(\mathbf{x})|^2 d\mathbf{x}. \tag{5.21}$$

The boundary-value problem (5.3)–(5.7) takes the form

$$\Delta\varphi = 0 \text{ in } Q; \tag{5.22}$$

$$\varphi(\mathbf{x}) = t_i \text{ on } \partial D_i, \ i \in I; \tag{5.23}$$

$$\int_{\partial D_i} \frac{\partial\varphi}{\partial\mathbf{n}}(\mathbf{x})d\mathbf{x} = 0, \ i \in I; \tag{5.24}$$

$$\varphi(\mathbf{x}) = 1 \text{ on } \partial Q^+, \ \varphi(\mathbf{x}) = -1 \text{ on } \partial Q^-; \tag{5.25}$$

$$\frac{\partial\varphi}{\partial\mathbf{n}}(\mathbf{x}) = 0 \text{ on } \partial Q_{lat}. \tag{5.26}$$

The formulas (5.16)–(5.18) for the computation of effective conductivity $\widehat{\varepsilon}$ take the form

$$\widehat{\varepsilon} = \frac{\varepsilon}{2|\partial Q^+|} \int_Q |\nabla\varphi(\mathbf{x})|^2 d\mathbf{x}, \tag{5.27}$$

$$\widehat{\varepsilon} = \frac{\varepsilon}{|\partial Q^+|} \int_{\partial Q^+} \frac{\partial\varphi}{\partial\mathbf{n}}(\mathbf{x})d\mathbf{x}, \tag{5.28}$$

$$\widehat{\varepsilon} = \frac{\varepsilon}{2|\partial Q^+|} \min_{\varphi\in V} \int_Q |\nabla\varphi(\mathbf{x})|^2 d\mathbf{x}. \tag{5.29}$$

5.1.3. Modeling particle-filled composite materials using the Delaunay–Voronoi method. The notion of pseudo-particles

Now we describe a trick, which allows us to modify the method developed in the previous chapters for finite domains (i.e., for composite materials). The complete analysis of the problem of network modeling as applied to transport in particle-filled composite materials can be found in [38, 41, 176, 183, 184, 185].

The general idea of modeling disordered materials using the Delaunay–Voronoi method was presented in Section 3.2.3. An additional problem in our consideration arises due to the external boundary, which represents the particle–wall interactions.

The Delaunay–Voronoi method also works for finite domains, see Fig. 5.2. We note that in finite domain, it partitions the boundary of the domain. Namely, consider a particle, which lies near the boundary of a rectangular domain described

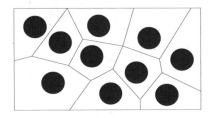

Figure 5.2. *Voronoi tessellations in a bounded domain.*

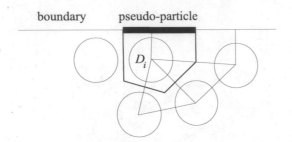

Figure 5.3. *Pseudo-particle corresponding to the near-boundary particle D_i, Voronoi cells and edges of the Delaunay graph.*

above, see Fig. 5.3. Then a face of its Voronoi cell coincides with a part of the boundary of this domain. We call this part of the boundary a *pseudo-particle*. The construction of the Delaunay graph for the problem (5.22)–(5.26) has some special features due to the boundary conditions (5.25), (5.26). The particle–wall interaction at the boundaries $\partial Q^+ \bigcup \partial Q^-$ is taken into account by the corresponding construction of the Delaunay graph. We incorporate such interactions using edges of the Delaunay graph, which connect centers of near boundary particles with corresponding pseudo-particles. These edges are chosen to be perpendicular to the boundary, see Fig. 5.3, which reflects the physics of the particle–wall interactions.

This construction allows us to not distinguish between the particles and pseudo-particles, which is a very convenient way to incorporate the boundary effects. Also, when local fluxes are computed one can treat any pseudo-particle as a particle of infinite radius $R = \infty$ – zero curvature, correspondingly.

As a result of the addition of pseudo-particles to the set of original particles, the number of particles (now we assume that particle means both an original particle and a pseudo-particle) becomes $k \geq N$ (N is the number of original particles). Fig. 5.4 displays the system of pseudo-particles (corresponding particles are displayed in Fig. 5.1) and scheme of potentials applied to pseudo-particles.

Since the asymptotic behavior of the problem under consideration is determined by the local geometry of the particles (original particles and a pseudo-particles), see Section 3.7, we can change the size of pseudo-particles (compare Fig. 5.3 and Fig. 5.4).

5.1.4. Heuristic network model for highly filled composite material

As above, see Chapters 3 and 4, the network model is associated with the weighted graph:

$$\mathbf{G} = \{\mathbf{x}_i, C_{ij}^{(2)}; \ i,j = 1, ..., K\}. \tag{5.30}$$

The reference points of the particles $\mathbf{x}_i$ are the vertices of this graph, and the weights $C_{ij}^{(2)}$ are assigned to the corresponding edges. But now, the term particles implies both the original (real) particles and pseudo-particles.

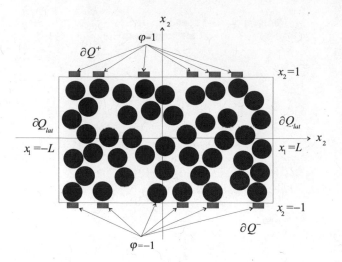

Figure 5.4. *System of particles and pseudo-particles corresponding to Fig. 5.1 and scheme of potentials applied (illustration is given for the dimension $n = 2$).*

We denote S^+ and S^- the set of vertices corresponding to the pseudo-particles and particles which touch the boundaries ∂Q^+ and ∂Q^-, correspondingly. Note that all pseudo-particles belong to the set $S^+ \bigcup S^-$. The remaining vertices, whose indices are denoted by

$$I = \{i = 1, ..., N\} \setminus (S^+ \bigcup S^-),$$

form the set of internal vertices of the network.

The energy localization is a local effect. It takes place for particles placed in a restricted domain just as for bodies in whole space, see Section 3.7 and numerical examples in Chapter 2. Then the discrete representation of the Dirichlet energy integral (5.21) is, as above (see Chapter 3),

$$\frac{1}{4} \sum_{i,j=1}^{K} C_{ij}^{(2)} (t_i - t_j)^2,$$

where the factor $\dfrac{1}{4}$ appears due to the fact that we count each channel twice (see comment after formula (3.116)). Thus, the discrete version of the minimization problem (5.20) is given by

$$\frac{1}{4} \sum_{i,j=1}^{K} C_{ij}^{(2)} (t_i - t_j)^2 \rightarrow \min, \qquad (5.31)$$

which corresponds to (5.20), where the minimum is taken over all t_i which satisfy

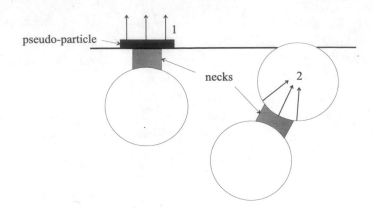

Figure 5.5. *Flux through the boundary: 1 – flux through matrix (into pseudo-particle), 2 – flux into boundary-touching particle.*

the conditions

$$t_i = 1 \text{ for } i \in S^+, \tag{5.32}$$

$$t_i = -1 \text{ for } i \in S^-.$$

The equality (5.32) is a discrete version of the boundary condition (5.24).

It is clear that the minimization problem (5.31), (5.32) is equivalent to solving the algebraic system

$$\sum_{j \in N_i} C_{ij}^{(2)}(t_i - t_j) = 0, \ i \in I, \tag{5.33}$$

where N_i stands for the set of neighbors of the vertex $\mathbf{x}_i$.

Finally, we write a discrete formula for the effective conductivity, which corresponds to (5.28):

$$\widehat{\varepsilon}_d = \frac{\varepsilon}{|\partial Q^+|} \sum_{i \in S^+} \sum_{j \in N_i} C_{ij}^{(2)}(1 - t_j^{net}). \tag{5.34}$$

Here $\{t_i^{net}, i \in I\}$ is the solution of the minimization problem (5.31) (or, that is the same, problem (5.32) and (5.33)). Then $\widehat{\varepsilon}_d$ is the minimum value of the quadratic form (5.31).

In continuum model, the total flux through the boundary ∂Q^+ is a sum of fluxes through the matrix and flux into the boundary-touching particles, see Fig. 5.5.

In a finite domain, there exist two types of necks: necks between particles and the necks between the near-boundary particles and boundary of the domain i, see Fig. 5.5. By using the notion of the pseudo-particles, we can treat the necks between particles and the boundary of the domain as necks between particles and the corresponding pseudo-particles. The leading part of fluxes is localized in the necks between the particles and between the particles and the boundary, see Fig. 5.5. In accordance with these notes, the flux through the matrix can be represented as a sum of fluxes between particles and the pseudo-particles.

We denoted S^+ the set of indices of particles and the pseudo-particles belonging to the boundary ∂Q^+. Now we separate particles and pseudo-particles belonging to S^+ into two groups. We denote S^+_{btp} the set of indices of the particles belonging to (touching) the boundary ∂Q^+, and denote S^+_{pp} the set of indices of the pseudo-particles belonging to ∂Q^+. Then the right-hand part of formula (5.34) can be written as a sum of two summands:

$$J_{btp} = \frac{\varepsilon}{|\partial Q^+|} \sum_{i \in S^+_{btp}} \sum_{j \in N_i} C^{(2)}_{ij}(1 - t^{net}_j) \tag{5.35}$$

and

$$J_{pp} = \frac{\varepsilon}{|\partial Q^+|} \sum_{i \in S^+_{pp}} \sum_{j \in N_i} C^{(2)}_{ij}(1 - t^{net}_j). \tag{5.36}$$

The quantity J_{btp} defined by (5.35) corresponds to the flux into the particles belonging to (touching) the boundary. It is the analog of flux to polar bodies of a system of bodies considered in Chapter 3. The quantity J_{pp} defined by (5.36) corresponds to the flux through the boundary of the matrix.

5.1.5. Formulation of the principle theorems

The principle theorems in the case under consideration are the same as in Chapters 3 and 4: about NL zones, about the asymptotic shielding, about network approximation and about approximation of potentials of particles. The proofs of the theorems presented in Chapter 3 and 4 must be modified to take into account that the problem is considered in finite domain. The notion of pseudo-particles makes the proofs of the theorems for finite domain similar to the proofs presented in Chapters 3 and 4. For this reason, we do not present these proofs in this book (the detailed proof can be found in [183, 185]) and present formulations of the theorems, only.

As above, we denote

$$\mathcal{J} = \{(i,j) : \delta_{ij}(\delta) \to 0 \text{ as } \delta \to 0\}$$

the pairs of the near-touching particles (including pseudo-particles) and

$$K = \bigcup_{(i,j) \in \mathcal{J}} K_{ij}$$

the system of channels between the near-touching particles (including pseudo-particles). We denote by D the system of the near touching particles (including pseudo-particles).

Theorem 5. *Theorem about NL zones.*
For the particles (pseudo-particles) and channels described above, the energy outside the channels

$$\int_{R^n \setminus (K \cup D)} |\nabla \varphi(\mathbf{x})|^2 d\mathbf{x} \leq C, \tag{5.37}$$

where $C < \infty$ does not depend on δ. Here $\varphi(\mathbf{x})$ is the solution of the problem (5.22)–(5.26).

Theorem 6. *Theorem about the asymptotic equivalence of capacities.*
If the particles (pseudo-particles) D_i and D_j are neighbors and the condition

$$C_{ij}^{(2)} \to \infty \quad as \ \delta \to 0 \tag{5.38}$$

holds, then for the channels described in Theorem 1, the capacities $C_{ij}^{(2)}$, $C^{S_{ij}}$ and $C^{R_{ij}}$ are asymptotically equivalent:

$$C_{ij}^{(2)} \sim C^{S_{ij}} \sim C^{R_{ij}}.$$

Theorem 7. *Network approximation theorem.*
Let capacities $C_{ij}^{(2)}$ of all neighbor particles (pseudo-particles) forming alive network(s) have the same order $f(\delta) \to \infty$ as $\delta \to 0$. Then effective conductivity $\widehat{\varepsilon} \to \infty$ as $\delta \to 0$.
The leading term of the effective conductivity $\widehat{\varepsilon}$ (5.16) corresponding to the continuum problem (5.3)–(5.7) is asymptotically equivalent, as $\delta \to 0$, to the effective conductivity

$$\widehat{\varepsilon}_d = \frac{\varepsilon}{|\partial Q^+|} \sum_{i \in S^+} \sum_{j \in N_i} C_{ij}^{(2)} (1 - t_j^{net})$$

computed by using the network model:

$$\widehat{\varepsilon} \sim \widehat{\varepsilon}_d \quad as \ \delta \to 0. \tag{5.39}$$

Here $\{t_i^{net}, \ i \in I\}$ is the solution of the finite-dimensional network problem (5.32) and (5.33) and $C_{ij}^{(2)}$ means the capacity.

Theorem 8. *Network approximation theorem for potentials of nodes.*
Let the particles (pseudo-particles) $\{D_i, i \in K\}$ satisfy the condition of uniformly dense packing in Π and capacities of any pair of the neighbor particles (pseudo-particles) satisfy the condition

$$C_{ij}^{(2)} \to \infty \quad as \ \delta \to 0 \ (i, j = 1, \dots N).$$

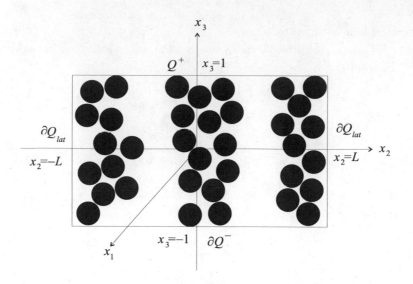

Figure 5.6. *System of particles, which has generated three alive networks.*

Then

$$|\mathbf{t}^{cont} - \mathbf{t}^{net}| \to 0 \ as \ \delta \to 0,$$

where $\mathbf{t}^{cont} = \{t_i^{cont}, \ i \in I\}$ *are the values of the potentials of the perfectly conducting particles determined from solution of the problem (5.22)–(5.26) and* $\mathbf{t}^{net} = \{t_i^{net}, \ i \in I\}$ *are the potentials of the nodes determined from solution of the network problem (5.32) and (5.33).*

The case of several alive networks

Considering composite, we deal with the system in which many particles are subjected to direct application of voltage. This kind of system can come apart into several networks, as displayed in Fig. 5.6. The chains of particles were observed in experiment, see, [380]. The theorems formulated above are valid for this case in the evident way.

5.2. Numerical analysis of transport properties of highly filled disordered composite material with network model

In this section, we present an application of the network model developed in Chapter 5 to numerical analysis of effective transport in high-contrast composite materials. The result of the numerical computations are taken from [41, 184].

5.2.1. Basic ideas of computation of transport properties of highly filled disordered composite material with network model

In Chapter 5, we express the leading term of capacity and effective conductivity $\widehat{\varepsilon}$ of high-contrast composite material through solution of the network problem (3.35), (3.36). The dimension of the network problem (3.35), (3.36) is significantly smaller than the dimension of a finite elements of finite differences approximations of the original problem (5.22)–(5.26). We demonstrate that the network approximation provides us with an effective tool for numerical analysis of high-contrast composite materials.

We consider models of composite material filled with mono- and polydispersed particles. We model particles by disks. A model of composite material is called monodisperse if the disks have the same radii. If the radii of the disks are varied, the composite material is called polydisperse.

Introducing the concepts of neighbors using the concept of characteristic distances

In Chapter 3, we introduced the concept of neighbors using the Voronoi method. Now we present the concept of neighbors based on the concept of characteristic distances [41], which is a useful alternative to the Voronoi method in numerical computations. The method of characteristic distance is in agreement with the fact of energy localization between closely spaced particles in the composite material noted in Chapter 2.

We introduce the following two characteristic distances between a given particle and all other particles:

(i) Distance between the nearest neighbors (closely spaced) particles. Denote it by δ^* (order of magnitude). Then

$$\delta^* < R$$

for densely packed particles.

(ii) The distance between a given particle and particles which are not close to it (non-neighboring). These particles form a second belt and the corresponding distances are greater than δ^*. For the sake of definiteness we choose this second order distance to be of order R.

The heuristic idea behind our approach is that we take into account the fluxes between the closely spaced particles only. To make the latter more precise we choose the cut-off distance δ^* in the interval $0.3R - 0.5R$. For this choice we obtain stable Delaunay graph even if the disks are not quite of identical sizes, but the variability in the sizes is not large (factor 2 or 3) [41]. It is important to not confuse cut-off distance δ^* with the distance δ between the particles. In numerical experiments from [41] $\delta << \delta^*$ (usually $\delta \approx 0.001 - 0.01R$).

We have also used another numerical criterion to support this choice of the cut-off distance δ^*. Namely, if δ^* is increased, then the effective conductivity practically does not change. The characteristic values of the fluxes corresponding to these distances are given by

$$q(\delta_{ij} = \delta^*) \approx \sqrt{\frac{R}{\delta^*}},$$

$$q(\delta_{ij} = R) \approx 1.$$

The first value is obtained from (3.167).

Then, we get

$$\frac{q(\delta_{ij} = \delta^*)}{q(\delta_{ij} = R)} \approx \sqrt{\frac{R}{\delta^*}}.$$

Thus, by results presented in Chapter 3, we have to neglect the fluxes of order 1 and keep the fluxes of order $\sqrt{\frac{R}{\delta^*}}$, where R is the characteristic radius of the particles and δ^* is the characteristic distance between the particles.

Organization of numerical analysis by using the discrete network approximation

We now define the discrete network, which corresponds to the original continuum model and formulate the discrete problem.

We create numerically a distribution of disks for a given radius (radii for polydisperse case) in the rectangle

$$\Pi = [-L, L] \times [-1, 1].$$

The center $\mathbf{x}$ of each disk is generated as a uniformly distributed in Π random variable. If a disk with the center $\mathbf{x}$ and radius R overlaps with any disk which is already present, then it is rejected, otherwise it is accepted and added to the list of the disks

$$L_N = \{\mathbf{x}_i, ; \, i = 1, ..., N\}. \tag{5.40}$$

In (5.40) we identify disks with those centers $\mathbf{x}_i$. If disks have various radii, the radii are included into the list of disks:

$$L_N = \{(\mathbf{x}_i, R_i); \, i = 1, ..., N\}. \tag{5.41}$$

We stop throwing in the disks when the prescribed volume fraction of the disks

$$V = \frac{1}{|\Pi|} \sum_{i=1}^{N} \pi R^2$$

is achieved. Here N is the current total number of the disks and $|\Pi|$ is the area of the rectangle Π.

We compute the distances δ_{ij} between the i-th and the j-th disks for the obtained distribution of disks. Then we determine the flux between the neighboring disks and neglect the fluxes between the non-neighboring disks. For this purpose we introduce the following set of numbers:

$$g_{ij} = \begin{cases} C_{ij}^{(2)} \text{ calculated with formula (3.168) if } \delta_{ij} \leq \delta^*, \\[2mm] 0 \text{ if } \delta_{ij} \geq \delta^*. \end{cases} \tag{5.42}$$

The quantity g_{ij} defined by (5.42) describes the flux between the i-th and the j-th disks. Thus we have defined a discrete network model,

$$\mathbf{G} = \{\mathbf{x}_i, g_{ij}; \; i, j = 1, ..., N\}, \tag{5.43}$$

which consists of points or sites $\mathbf{x}_i$ (centers of the disks) and edges with assigned numbers g_{ij}. The discrete model (5.43) does not explicitly contain the sizes of the disks and the distances between them, this information is incorporated via the set g_{ij}. Note that if the concentration is small, g_{ij} gives zero value of the effective conductivity whereas the conductivity of the matrix is 1. This is not a contradiction. It means that for high-contrast composite the value 1 becomes negligible as compared with the characteristic value of the effective conductivity.

The fluxes are concentrated in the channels between the disks. Any given pair of disks is connected by a unique channel. For the disks which are close to the boundaries $y = \pm 1$ the fluxes are concentrated in the channels between the disks and pseudo-disks.

The equations, which determine the values t_i of the potential on the sites $\mathbf{x}_i$ of the network (5.43) are

$$\sum_{j \in N_i} C_{ij}^{(2)}(t_i - t_j) = \sum_{j \in S^+} C_{ij}^{(2)} - \sum_{j \in S^-} C_{ij}^{(2)}. \tag{5.44}$$

The discrete formula for the effective conductivity is (see (5.34), in the case under consideration $|\partial Q^+| = 2L$):

$$\widehat{\varepsilon}_d = \frac{P^+}{2L}, \tag{5.45}$$

where

$$P^+ = \sum_{i \in S^+} \sum_{j \in N_i} g_{ij}(1 - t_j^{net}) \tag{5.46}$$

is the discrete formula for total flux through the boundary ∂Q^+.

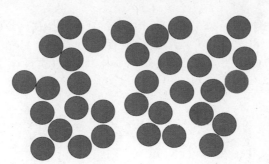

Figure 5.7. *The monodisperse composite material.*

5.2.2. Numerical simulation for monodisperse composite materials. The percolation phenomenon

A composite material filled with particles of the same sizes is called monodisperse, see Fig. 5.7.

We have implemented a numerical method to
1. Generate a configuration ω as a system of random disks.
2. Compute the coefficients $g_{ij}(\omega)$ (5.42) for this system of disks.
3. Solve the linear system (5.44).
4. Compute the total flux $P^+(\omega)$ using formula (5.46) for the configuration ω.

The control parameters are:
– the radius of the disks R,
– the total volume fraction V of the disks.

As a result of each run of all four codes we compute the value of the total flux denoted by $P^+(\omega)$, for a given random configuration ω.

In order to collect the statistical information, we run the program repeatedly S times and collect the data $\{P^+(\omega_s), s = 1, ..., S\}$. Then we compute

a) The expected value of the total flux

$$P = \mathbf{E}P^+(\omega).$$

b) The mean deviation of the total flux

$$DP = \mathbf{E}|P^+(\omega) - P|.$$

c) The maximal total flux over all collected data

$$m = \max_{\omega \in \Omega} P^+(\omega).$$

The expected value of the random function X is defined as

$$\mathbf{E}X(\omega) = \frac{1}{S}\sum_{s=1}^{S} X(\omega_s).$$

All the above quantities have been computed for given values of the quantities R and V which is why P, DP, m are functions of R and V.

5.2.3. Numerical results for monodisperse composite materials

The goal here is to test our numerical algorithms which are based on the network approximation, to make sure that we can consistently recover (in a very efficient way) known results, in particular, recover the percolation picture.

Description of the numerical procedure

In the numerical simulations, an integer lattice $\Pi_0 = 550 \times 400$ was used. That is

$$2L = \frac{550}{400} = 1.375.$$

This choice is made for convenience of the visual control on the computer screen. The centers of the disks have been generated in the form

$$\mathbf{x}_i = (random(550) + random(1), random(400) + random(1)),$$

where $random(n)$ is a pseudo-random number uniformly generated from the set $\{0, 1, ...n\}$ and $random(1)$ is a pseudo-random number uniformly generated from the interval $[0, 1]$. The disks near the boundary were retained if their centers were inside Π_0. The boundary values 1 and 0 were prescribed at the upper and lower edges of Π_0.

In the numerical procedure the disks of radii 35, 25, 20, 15, 10 (in undimensional coordinates) were used. In dimensional coordinates radii are $35/200$, $25/200$, $20/200$, $15/200$, $10/200$, since $[-1, 1]$ corresponds to $[0, 400]$). The average number of disks in the rectangle Π_0 was between 50 and 150.

The specific flux (capacity of a pair of bodies) is given by the following formulas:

$$C_{ij}^{(2)} = \pi \sqrt{\frac{R}{\delta_{ij}}}$$

for a pair disk–disk (formula (3.168) with $R_1 = R_2 = R$) and

$$C_{ij}^{(2)} = \pi \sqrt{\frac{2R}{\delta_{ij}}}$$

for a pair disk–pseudo-disk (formula (3.168) with $R_1 = R$ and $R_2 = \infty$).

Results of numerical analysis

Computer programs were developed for the simulation of random distributions of disks and computation of effective conductivity of composite material. The verification of the computer programs were carried out in the following way. All disks were of the same radius. The volume fraction V has been increased with a fixed step δV.

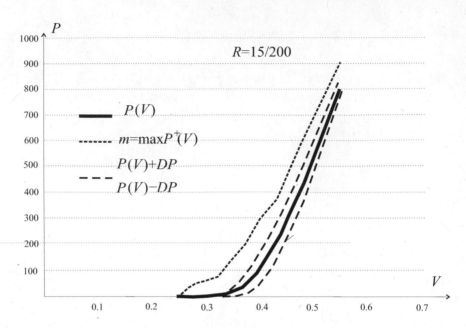

Figure 5.8. *Total flux as a function of volume fraction. The monodisperse composite material $R = 15/200$.*

For each value of V the total flux P has been computed using the above described procedure.

For each fixed V, the quantity $P^+(\omega_s)$ have been computed repeatedly for different random configurations (100 to 300 configurations for each value of V). After that, $P\ DP$ and m were computed. All these quantities depend on V.

Several sets of identical disks were used, that is, three different disk sizes (15/200, 20/200 and 25/200), to control the stability of our numerical computations.

Indeed, since the problem is scale invariant the choice of the radius should not change the effective conductivity. This effect was used for checking the numerical process. It was observed no difference in qualitative and quantitative dependence of $P^+(\omega_s)$ on the volume fraction V when the sizes were changed.

A typical plot of the function $P = P(V)$ (the total flux as function of the volume fraction of the disks) is shown in Fig. 5.8. The plot consists of three parts.

In the range

$$V \leq 0.2,$$

the total flux through the composite material layer is zero. More precisely, it is very small and it is very close to the OV-axis at the plot.

In the range

$$V = 0.2 - 0.35$$

the total flux is not positive for sure. Thus the plots of the function $P = P(V)$ reveal dependence which is typical in the percolation processes [315, 345]. The percolation threshold in all computations can be estimated as

$$V_0 \approx 0.3.$$

In the range

$$V \geq 0.35,$$

the total flux is positive (taking into account the mean deviation) and takes values as large as 500 to 1000 for V in the range 0.5 to 0.6.

Computations were carried out for several sizes of identical particles to make sure that there is no size effect, which in principle can appear as a numerical artifact. The computations demonstrate the absence of size effect [41, 184]. It means that numerical analysis based on network model captures basic features of the real physical phenomenon and provides realistic quantitative predictions.

We have used the graph of the maximal value $m(\omega_s)$ as an additional test for our code. Indeed, from the physical point of view it should behave (qualitatively) in the same way as the graph for the total flux, which is what one can see in Fig. 5.8.

Form of the function $P(V)$ (total flux as function of volume fraction of particles)

As was noted the plots of the function $P = P(V)$ reveal dependence which is typical in the percolation processes. Let us construct functions approximating the numerically obtained dependence $P = P(V)$. Note that we are interested in the dependence $P = P(V)$ between threshold value and maximal possible value of volume fraction of particles. The threshold is visually estimated as a value in the interval $[0.25, 0.35]$, see Fig. 5.8. This value is relatively far from the maximal possible value of volume fraction of particles, which is estimated as 0.6.

We construct approximating functions in the form of power function

$$P(V) = [a(V - V_0)]^b.$$

The parameters a, b and V_0 were determined in [184] from the numerically computed values of $P(V)$ using the method of least squares. The computations were carried out for the plot presented in Fig. 5.8 (other plots of the function $P(V)$ presented in [41] coincide with the plot presented in Fig. 5.8 with high accuracy). The parameters found for the power functions are the following:

$$a = 85,$$

the exponent

$$b = 2.17,$$

and the percolation threshold

$$V_0 = 0.3.$$

The function $P(V)$ has the form

$$P(V) = [85(V - 0.3)]^{2.17}. \tag{5.47}$$

Taking into account that the model was developed under condition that the filling particles are densely packed, we compute the parameters of the power function for

$$V \geq 0.4$$

(this is a high fullness beyond the threshold limit equal to 0.3 in the considered case, the maximal fullness is about 0.55).

In this case we obtain

$$a = 75,$$

and exponent

$$b = 2.25.$$

It is seen that the values of the parameters a and b change slightly. We recommend the last formula

$$P(V) = [75(V - 0.3)]^{2.27} \tag{5.48}$$

for volume fraction of the filling particles

$$V > 0.4.$$

Values of the function $P(V)$ computed using formulas (5.47) and (5.48) are presented in Table 5.1.

Note that the computations correspond to the difference of potential equal to 2 and the specimen length 2 and width

$$2L = 2 \times \frac{550}{400} = 2.75.$$

The total corresponding to (5.47) and (5.48) are

$$P(V) = [31(V - 0.3)]^{2.17} = 172(V - 0.3)^{2.17} \tag{5.49}$$

Table 5.1. *Values of the Function P(V); 2nd Row – Numerically Computed Data, 3rd Row – in Accordance with Formula (5.47) and 4th Row – in Accordance with Formula (5.48)*

V	0.300	0.325	0.350	0.375	0.400	0.425	0.450	0.475	0.500
$P(V)$	0	10	20	42	105	190	275	400	520
$P(V)$	0	5	25	63	117	190	282	392	525
$P(V)$	–	–	–	–	107	177	270	382	517

and
$$P(V) = [27(V - 0.3)]^{2.27} = 177(V - 0.3)^{2.27}, \tag{5.50}$$

respectively (we save only integer part of the multiplayer in (5.49) and (5.50)).
In accordance with (5.15), the effective conductivity

$$\widehat{\varepsilon} = \frac{P^+}{2L} = \frac{P^+}{2.75}.$$

The expected value

$$\mathbf{E}\widehat{\varepsilon} = \frac{P(V)}{2L}$$

of the effective conductivity (often this value is called effective conductivity for random media)

$$\mathbf{E}\widehat{\varepsilon}(V) = 63(V - 0.3)^{2.17}$$

and

$$\mathbf{E}\widehat{\varepsilon}(V) = 65(V - 0.3)^{2.27}.$$

5.2.4. The polydisperse highly filled composite material

In this section we apply the network model to the analysis of this problem for densely packed composite material and demonstrate that the network modeling is effective tool for the analysis of influence of polydispersity on the effective transport properties of composite materials.

Influence of polydispersity on the effective transport properties of composite material

There is considerable interest in the effect of polydispersity on the effective transport properties (permittivity and conductivity of composite materials, the effective viscosity of suspension of rigid inclusions in a fluid), see [16, 17, 43, 66, 76, 107, 133, 135, 145, 146, 151, 152, 156, 301, 302]. The main question here is whether the presence of polydispersity results in an increase or decrease of the effective conductivity. This question is of significant practical interest; when polymer/ceramic composites are prepared for capacitors or thermal insulting packages, should the ceramic powder be monodisperse or polydisperse to achieve desired (e.g., high) dielectric or thermal properties? How significant is the effect of polydispersity? Limited experimental data [76, 135, 151, 152, 156, 301, 306] are inconclusive and point to an increase [306] when the concentration of inclusions is not high.

Most theoretical works on this subject study polydispersity in the dilute limit case. In [356, 357, 359] the techniques of variational bounds [246] were used to estimate effective characteristic (effective conductivity or effective viscosity) $\widehat{\varepsilon}$ of polydispersed composite material in the low concentration regime in both two and three dimensions. They performed rigorous asymptotic analysis in V and evaluated the terms of order $O(V)$ and $O(V^2)$ in the expansion of $\widehat{\varepsilon}$ in powers of V, where V is the total volume fraction of the inclusions. Moreover, they performed partial analysis of the cubic term. The analysis of [359, 356, 357] shows that if the conductivity of inclusions is greater than the conductivity of matrix ($\varepsilon_i > \varepsilon_m$), then

$$\widehat{\varepsilon} > \widehat{\varepsilon}_{mono}, \tag{5.51}$$

where $\widehat{\varepsilon}_{mono}$ means the effective characteristic of monodispersed composite material with the same volume fraction of inclusions.

In [306] the problem of the effective conductivity was considered for spherical inclusions of two different sizes in three dimensions. Derivation of the effective conductivity was based on the Maxwell–Garnett formula. Note that the Maxwell–Garnett formula is an approximation in V for small V. This formula was derived for the dilute limit case when interactions between inclusions are small, see [246]. While it is known that for special geometric arrays the Maxwell–Garnett formula holds for fairly large concentrations (e.g., for periodic structure [32]) and other well separated geometries in general it only holds to the second order in V for macroscopically isotropic composites, see [246]. In particular, the Maxwell–Garnett approximation may not hold for high volume fractions in the presence of pronounced percolation effects, when the conductivity patterns dominate the behavior.

Considering not dilute (concentrated) composites, we meet additional problems, which is not solved completely for the present.

In the case of large concentration of filled particles, we can apply the method of network approximation to analysis of the polydispersion problem.

Flux between two not identical closely spaced disks

In this section, we consider the planar problem (two-dimensional domain filled with disks). Another widely used interpretation is a three-dimensional domain filled with a system of parallel cylinders.

In order to analyze polydispersed composite materials, we need to know the flux between two disks of different radii if the potential on each disk is a constant. Asymptotic formula for this flux was obtained in Section 3.7.3. The specific flux (capacity) $C_{ij}^{(2)}$ is given by (3.168) for a pair disk–disk and

$$C_{ij}^{(2)} = \pi \sqrt{\frac{2R_i}{\delta_{ij}}}$$

for a pair disk–pseudo-disk.

Numerical analysis of effective transport properties of polydisperse composite material

We call a composite material polydisperse, if it is filled with particles of different sizes, see Fig. 5.9. In our model it amounts to consideration of disks of different radii. The distribution of the sizes can be continuous or discrete. We consider the

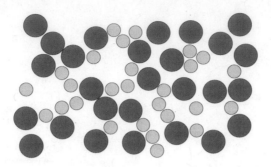

Figure 5.9. *The polydisperse composite material.*

discrete case that is we consider the disks of several fixed sizes. Namely, we consider the bimodal case, when the dispersion is formed of the disks of two fixed sizes. We describe a polydisperse composite material with the total volume fraction of disks V and relative volume fraction p_k of the disks of the k-th sort $(k = 1, 2)$. We assume that 1 is indexing of large disks and 2 is indexing of small disks. The volume fraction of the disks of k-th sort in the mixture is Vp_k.

We have

$$p_1 + p_2 = 1 \ (k = 1, 2)$$

and

$$Vp_1 + Vp_2 = V.$$

For the bimodal dispersion, we introduce a polydispersity parameter

$$p = p_2, \ 0 < p < 1,$$

which characterizes the relative volume fraction of small disks ($p = 0$ all disks are large, $p = 1$ all disks are small). Alternative polydispersity parameter

$$q = 1 - p = p_1$$

characterizes the relative volume fraction of large disks. For a fixed total volume fraction V, Vp is the total volume fraction of small disks and Vq is the total volume fraction of large disks. Our objective is to obtain the dependence of the effective conductivity $\widehat{\varepsilon}_{poly}$ on the polydispersity parameters.

We used the following two step procedure for generating the polydisperse (bimodal) random configurations of the disks.

First we used the algorithm described in Section 5.2.1 for the disks of larger size until we reached the volume fraction Vq. Next we used the same algorithm for the small disks with the volume fraction Vp, where V is the total volume fraction of disks (kept fixed). In the second step we take into account the large discs generated in the first step that is if a small disk overlaps with any other disk present (small or large) it is rejected. The cut-off distance for both steps is $\delta^* = 0.3R_2$.

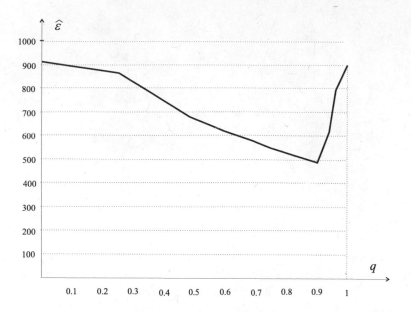

Figure 5.10. *The graphs of the effective conductivity $\widehat{\varepsilon}$ as a function of the relative volume fraction $q = p_1$ of large disks.*

We present result of the analysis of the influence of polydispersity on the transport properties of composite from [41]. The following mixture was analyzed in [41]. Radii of the disks:

$$R_1 = \frac{25}{200}, \quad R_2 = \frac{10}{200},$$

the total volume fraction of the disks

$$V = 0.55.$$

The relative volume fraction of the particles $p_1 = q$ and $p_2 = p$ $(p + q = 1)$ was varied from 0 to 1. In Fig. 5.10 the effective conductivity is presented as a function of p, which is the relative volume fraction of the disks of radius R_1.

This graph shows the same values of $\widehat{\varepsilon}$ for $p = 0$ and $p = 1$. It means there is no size effect, which agrees with the general theory, since the problem is scale invariant and it plunges down sharply in the interval $[0.5; 0.9]$. Thus we observe that for this mixture polydispersity can decrease the effective conductivity by approximately a factor of 2. This sharp drop occurs when the relative volume fraction p of the large particles is quite high. If we compare the minimum value

$$P(p = 0.9) = 484$$

with the corresponding value in a monodisperse composite material (it corresponds to volume fraction $0.9V \approx 0.5$):

$$P(0.5) = 680,$$

then we conclude that the small particles practically do not contribute the effective conductivity.

Chapter 6

EFFECTIVE TUNABILITY OF HIGH-CONTRAST COMPOSITES

This chapter addresses the rigorous treatment of the tunability effect (DC electric field driven variation of the permittivity) in a high-contrast composite of periodic structure (a nonlinear ferroelectric matrix with large dielectric constant filled with linear dielectric inclusions with normal dielectric constant). A combination of the homogenization theory and classical small parameter method makes possible carrying out the detailed analysis of the nonlinear problem for weak fields (the weak field theory attracts the attention of many researchers due to its potential applications in engineering, see, e.g., [395, 396]). We present recent results of numerical analysis of the nonlinear problem for strong fields [190]. The results presented in this chapter also give some information about the material design problem as applied to an engineering problem (about various aspects of material design problems see, e.g., [27, 29, 161]).

6.1. Nonlinear characteristics of composite materials

In general case, ferroelectric materials can appreciably change their dielectric permittivity ε under the application of a DC (direct current) electric field, or in other words, these materials can exhibit a *relative tunability* $T(E)$ defined as [171, 350]

$$T(E) = \frac{\varepsilon(0) - \varepsilon(E)}{\varepsilon(0)}, \tag{6.1}$$

where $\varepsilon(E)$ is the dielectric permittivity of the material corresponding to the field E applied to the material.

From the mathematical point of view, the relative tunability is a characteristic of nonlinearity of coefficients of boundary-value problem, see, e.g., equation (6.12) below.

Often the function $\varepsilon(E)$ is taken in the form

$$\varepsilon(E) = \varepsilon_0 - \mu E^2, \tag{6.2}$$

167

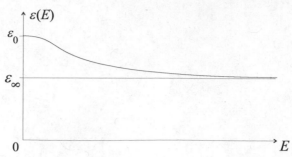

Figure 6.1. *Graph of the function $\varepsilon(E)$ (6.3).*

although the graph of the function

$$\varepsilon(E) = \varepsilon_\infty + \frac{\varepsilon_0 - \varepsilon_\infty}{1 + bE^2} \qquad (6.3)$$

fits with the experimental date. The constant ε_∞ means the dielectric permittivity "at the infinity" (practically, for large enough electric field) and ε_0 means the dielectric permittivity "in zero" (for very small electric field), see Fig. 6.1. For small field, we can use the quadratic (the second order) approximation (6.2) with no restriction.

If function $\varepsilon(E)$ is taken in the form (6.2) then formula (6.1) takes the form

$$T(E) = \frac{\mu}{\varepsilon_0} E^2, \qquad (6.4)$$

and (6.2) can be rewritten as

$$\varepsilon(E) = \varepsilon_0(1 - TE^2).$$

Relative tunability is the property that can be used for producing of electronic devices with parameters that are tunable under a DC control voltage. However, for various reasons related both to electronic devices and material science, only in the last decade intensive development efforts have been made (see, e.g., [79, 140, 327, 328, 331]). One of the developments in this field is the use of ferroelectric-dielectric composite materials.

A typical example is a $(Ba,Sr)TiO_3$ ceramic manufactured with addition of MgO or $MgTiO_3$ [12] (for other examples, see, e.g., [144, 158, 386]). Fig. 6.2 shows a scanning electron microscope image documenting the structure of the composite material [191]. The dielectric constants of these two phases are very different, see Table 6.1. The tunability values of these two phases are also significantly different; for a ferroelectric one can achieve $T_f = 0.2 - 0.3$ (here and afterward index "f" corresponds to ferroelectric and "d" corresponds to dielectric), whereas dielectrics (MgO, $MgTiO_3$) do not exhibit any measurable tuning at realistic values of the electric field. The volume fraction of the filled particles is not large (from 5 to 20 percent). In accordance with our classification, see Chapter 2, it is a high-contrast composite.

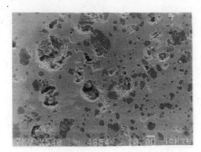

Figure 6.2. *Scanning electron microscope image (×540) of composite material: (Ba,Sr)TiO₃ matrix (the light phase) with MgO inclusions (the dark phase).*

"Anomalous" behavior of tunability of ferroelectric-dielectric composite

Taking into account the characteristics of a ferroelectric-dielectric composite, one expects that the dilution of ferroelectric with the dielectric leads to an appreciable reduction of the homogenized dielectric constant for the material. The reduction of the homogenized dielectric constant is observed in experiment [350]. By analogy, one might expect a reduction of the tunability of the composite material as the concentration of the linear dielectric component increases. At the same time the experiment shows that the addition of the linear, low permittivity dielectric does not affect the tunability of the material at least as far as the mixing rule predicts [191, 350]. We have even documented a pronounced increase of the tunability, see, e.g., [350].

The homogenized (called also effective or macroscopic) tunability is introduced as

$$\widehat{T}(E) = \frac{\widehat{\varepsilon}(0) - \widehat{\varepsilon}(E)}{\widehat{\varepsilon}(0)}, \tag{6.5}$$

where $\widehat{\varepsilon}(E)$ means the homogenized permittivity of composite material (recall that we use the hat symbol "^" to mark the homogenized characteristic corresponding to the local characteristic under consideration, see Chapter 1).

In [350], an averaging method developed in [203] was applied to evaluate the relative tunability for ferroelectric material filled with spherical dielectric inclusions. Under the condition that the volume fraction q of the inclusions is sufficiently small, the following formulas for computation of the homogenized permittivity $\widehat{\varepsilon}$ and ho-

Table 6.1. *Characteristics of Homogenized Materials (δ in the Combination $\tan\delta$ Means Loss Angle, not Characteristic Dimension of the Components).*

Material	Permittivity ε	Relative tunability T	Loss tangent $\tan\delta 10^{-4}$
(Ba,Sr)TiO₃	$1000 - 2000$	$0.2 - 0.3$	$50 - 100$
MgO	10	0	1

mogenized relative tunability $\widehat{T}$ of composite material were obtained:

$$\widehat{\varepsilon}(E) = \varepsilon_f(E)(1 - 1.5q), \tag{6.6}$$

$$\widehat{T}(E) = T_f(E)(1 + 0.2q). \tag{6.7}$$

where $\varepsilon_f(E)$ and $T_f(E)$ mean the permittivity and relative tunability of pure ferro-electric material.

Formula (6.6) for the homogenized permittivity agrees with the classic theory [348]. Formula (6.7) predicts increasing of the homogenized tunability. We estimate the value of possible increasing of tunability predicted by formula (6.7). Usually, formulas derived under condition $q \ll 1$ remain valid for the volume fraction of inclusions in the range 10 to 20%, but not more than 40 to 50% [315]. Then, the increasing of tunability predicted by formula (6.7) can be estimated as 2 to 4% maximum. These values are in the range of experimental error and it would be difficult to distinguish 2 to 4% increasing of tunability (if it exists) with stability or small decreasing of tunability.

We naturally arrive at the problem of the possible values of the homogenized tunability formulated as follows: *One has to estimate the interval of possible values of the homogenized tunability if characteristics of components of composite are known.* Little is known about the possible values problem for the homogenized tunability, although the problem of possible value of homogenized permittivity of composite dielectrics was intensively investigated, see, e.g., [34, 62, 63, 203, 259, 335, 373, 374] for the linear problem and [245, 299, 351, 352, 377] for the nonlinear problem. It is clear that the possible values problem for the homogenized tunability is important not only from the theoretical point of view. Solution of this problem would provide us with the information about characteristics, which are available (and which are not available, this is also important information) in the framework of composite material technology and give us orientation in the development of tunable materials for engineering applications.

At the present time there is no good reason to believe that the volume fraction q determines the homogenized relative tunability solely (or, at least, the volume fraction q is the leading characteristic determining the magnitude of the homogenized tunability of composite materials). At the same time, the investigators in the field of electroceramics do not pay proper attention to the influence of microgeometry and microtopology of components on the effective property of composite dielectrics, although this problem is discussed in recent literature, see, e.g., [271].

6.2. Homogenization procedure for nonlinear electrostatic problem

We start our analysis with the presentation of the homogenization procedure for nonlinear dielectric–ferroelectric composite (homogenization procedure was developed for the two-dimension problem in [177, 191] and for the three-dimension problem in [189]). Here, we present a version of the three-dimension homogenization procedure.

Local characteristics of composite material depend on the spatial variable $\mathbf{x}=(x_1, x_2, x_3)$. Then, in (6.2) ε_0 and μ are functions of the spatial variable $\mathbf{x}$, and (6.2) takes the form:

$$\varepsilon(\mathbf{x}, E) = \varepsilon_0(\mathbf{x}) - \mu(\mathbf{x})E^2, \tag{6.8}$$

where $E = |\mathbf{E}|$ means the magnitude of the electric field

$$\mathbf{E} = \nabla\varphi(\mathbf{x}), \tag{6.9}$$

$\varphi(\mathbf{x})$ denotes the electrostatic potential.

The local electric displacement $\mathbf{D}$ is related to $\mathbf{E}$ by the formula [340]

$$\mathbf{D} = \varepsilon(\mathbf{x}, E)\mathbf{E} = \varepsilon(\mathbf{x}, |\nabla\varphi(\mathbf{x})|)\nabla\varphi(\mathbf{x}) \tag{6.10}$$

and satisfies the equation [340]

$$\mathrm{div}\mathbf{D} = 0. \tag{6.11}$$

Substituting (6.9) into (6.11), we obtain

$$\mathrm{div}(\varepsilon(\mathbf{x}, |\nabla\varphi(\mathbf{x})|)\nabla\varphi(\mathbf{x})) = 0. \tag{6.12}$$

Theoretically, solving (6.12) with the appropriate boundary conditions, we can find the potential $\varphi(\mathbf{x})$. Due to inhomogeneity of material (thus, dependence of coefficient of the problem (6.12) on $\mathbf{x}$), it is a complicated problem, which cannot be accurately solved even using powerful modern computers.

Several authors have addressed computations of the effective properties for the composite materials similar to the one described here (see, e.g., [126, 162, 346]). We address specifically the case of the periodic composite illustrated in Fig. 6.3 in the weak nonlinearity limit, where the field-induced relative change of the dielectric constant is small ($T_f \ll 1$).

Homogenization procedure in the general case

We consider a composite possessing a periodic structure, see Fig. 6.3 (left), for which computation of the effective characteristics can be reduced to the analysis of a problem for one periodicity cell only. We assume that the periodicity cell Y is rectangular with the center at the origin of the coordinate axes and with the lengths of the sides equal to L and M. We assume that the periodicity cell contains one inclusion, which is symmetric with respect to the origin corresponding to the cell of the composite illustrated in Fig. 6.3 (right). These assumptions are made in order to simplify the computations to be presented below.

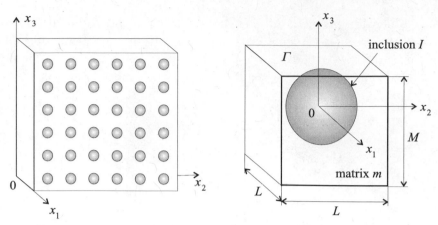

Figure 6.3. *Composite material of periodic structure (left) and its periodicity cell Y (right).*

We assume that the horizontal pair of sides $x_3 = \pm M/2$ of the periodicity cell Y are subjected to voltage $\pm EM/2$ (these values correspond to the application of overall electric field magnitude E to composite material) and on the vertical pair of sides the potential is periodic, see Fig. 6.3 (right). Because of the symmetry and equivalence of all periodic cells, we obtain the following conditions on the boundary of one cell:

$$\varphi(x_1, x_2, \pm M/2) = \pm EM/2, \qquad (6.13)$$

$$\frac{\partial \varphi}{\partial \mathbf{n}}(\pm L/2, x_2, x_3) = 0, \quad \frac{\partial \varphi}{\partial \mathbf{n}}(x_1, \pm L/2, x_3) = 0. \qquad (6.14)$$

The total value of the electric displacement $\mathbf{D} = \varepsilon(\mathbf{x}, |\nabla\varphi(\mathbf{x})|)\nabla\varphi(\mathbf{x})$ over the periodicity cell is the sum of the fluxes through all its sides. The total value of the electric displacement D_{total} over the upper side

$$\Gamma = \{-L/2 < x_1, x_2 < L/2, x_3 = M/2\} \qquad (6.15)$$

of the periodicity cell Y is

$$D_{total} = \int_{\Gamma} \varepsilon(\mathbf{x}, |\nabla\varphi(\mathbf{x})|)\frac{\partial \varphi}{\partial \mathbf{n}}(\mathbf{x})d\mathbf{x}.$$

Multiplying (6.12) by $\varphi(\mathbf{x})$ and integrating by parts, we obtain

$$-\int_{Y} \varepsilon(\mathbf{x}, |\nabla\varphi(\mathbf{x})|)|\nabla\varphi(\mathbf{x})|^2 d\mathbf{x} + \int_{\partial Y} \varepsilon(\mathbf{x}, |\nabla\varphi(\mathbf{x})|)\varphi(\mathbf{x})d\mathbf{x} = 0. \qquad (6.16)$$

The flux through the lateral sides of the cell is equal to zero (see boundary condition (6.14)), then

$$\int_{\partial Y \cap \{x_1=\pm L/2,\ x_2=\pm L/2\}} \varepsilon(\mathbf{x}, |\nabla\varphi(\mathbf{x})|)\varphi(\mathbf{x})d\mathbf{x} = 0. \tag{6.17}$$

Taking into account (6.17) and boundary condition (6.13), we obtain from (6.16) that

$$-\int_Y \varepsilon(\mathbf{x}, |\nabla\varphi(\mathbf{x})|)|\nabla\varphi(\mathbf{x})|^2 dx + \tag{6.18}$$

$$+ \int_{\partial Y \cap \{x_3=\pm M/2\}} \varepsilon(\mathbf{x}, |\nabla\varphi(\mathbf{x})|)\frac{\partial\varphi}{\partial\mathbf{n}}(\mathbf{x})[\pm EM/2]_p dx = 0.$$

Here $[]_p$ means "periodic" jump

$$[f(x_1, x_2, x_3)]_p = f(x_1, x_2, M/2) - f(x_1, x_2, -M/2),$$

i.e., the difference of values of the function $f(x_1, x_2, x_3)$ on the opposite sides $\partial Y \cap \{x_3 = \pm M/2\}$ of the periodicity cell Y.

The fluxes through the upper and lower sides of the cell Y are equal to

$$\int_{\partial Y \cap \{x_3=M/2\}} \varepsilon(\mathbf{x}, |\nabla\varphi(\mathbf{x})|)\frac{\partial\varphi}{\partial\mathbf{n}}(\mathbf{x})d\mathbf{x} = \tag{6.19}$$

$$= \int_{\partial Y \cap \{x_3=-M/2\}} \varepsilon(\mathbf{x}, |\nabla\varphi(\mathbf{x})|)\frac{\partial\varphi}{\partial\mathbf{n}}(\mathbf{x})d\mathbf{x}.$$

From (6.18) and (6.19) we obtain

$$\int_Y \varepsilon(\mathbf{x}, |\nabla\varphi(\mathbf{x})|)\nabla\varphi(\mathbf{x})dx = EM \int_{\partial Y \cap \{x_3=M/2\}} \varepsilon(\mathbf{x}, |\nabla\varphi(\mathbf{x})|)\frac{\partial\varphi}{\partial\mathbf{n}}(\mathbf{x})d\mathbf{x}. \tag{6.20}$$

From (6.18), (6.20) (we recall that the difference of potentials on the top and the bottom of the periodicity cell Y is equal to E) we obtain

$$D_{total} = \int_\Gamma \varepsilon(\mathbf{x}, |\nabla\varphi(\mathbf{x})|)\frac{\partial\varphi}{\partial\mathbf{n}}(\mathbf{x})dx = \tag{6.21}$$

$$= \frac{1}{EM} \int_Y \varepsilon(\mathbf{x}, |\nabla\varphi(\mathbf{x})|)|\nabla\varphi(\mathbf{x})|^2 dx.$$

We denote the specific flux

$$D = \frac{D_{total}}{L^2}$$

(flux divided by the area of the upper side Γ (6.15) of the periodicity cell, which is the square with the sides of the length L). The specific flux D divided by the overall electric field magnitude E is the homogenized permittivity $\widehat{\varepsilon}(E)$ of composite. By virtue of this note and (6.21) we arrive at the following formula:

$$\widehat{\varepsilon}(E) = \frac{D}{E} = \frac{1}{E^2 L^2 M} \int_Y \varepsilon(\mathbf{x}, |\nabla\varphi(\mathbf{x})|)|\nabla\varphi(\mathbf{x})|^2 d\mathbf{x}. \qquad (6.22)$$

Formula (6.22) corresponds to the general homogenization theory [30] if we denote by $|Y| = L^2 M$ the measure (the volume in the case under consideration) of the periodicity cell Y.

Case of small nonlinearity / small electric field

The case of small (also referred to as weak) nonlinearity is interesting from both the theoretical and engineering point of view [315, 395, 396]. In addition, as will be demonstrated below, this case is equivalent to the case of arbitrary nonlinearity under condition that electric field is small (this case also is referred to as a weak-field case).

We assume that the permittivity of the nonlinear material has the form

$$\varepsilon(\mathbf{x}, E) = \varepsilon_0(\mathbf{x}) + \lambda b(\mathbf{x}, E), \qquad (6.23)$$

where λ is a small parameter.

Analyzing small nonlinearity, we can combine the homogenization method with the classical method of small parameter [191]. Following to [191], we find the solution of (6.12), (6.14), (6.23) in the form of series in λ:

$$\varphi(\mathbf{x}) = \varphi_0(\mathbf{x}) + \lambda\varphi_1(\mathbf{x}) + \ldots = \sum_{n=0}^{\infty} \lambda^n \varphi_n(\mathbf{x}). \qquad (6.24)$$

Substituting (6.24) in (6.12), (6.14), (6.23), we have

$$\text{div}\,[\varepsilon_0(\mathbf{x}) + \lambda b(\mathbf{x}, |\nabla(\varphi_0 + \lambda\varphi_1 + \ldots)|)\nabla(\varphi_0 + \lambda\varphi_1 + \ldots)] = 0, \qquad (6.25)$$

$$(\varphi_0 + \lambda\varphi_1 + \ldots)(x_1, x_2, \pm M/2) = \pm E, \qquad (6.26)$$

$$\frac{\partial}{\partial \mathbf{n}}(\varphi_0 + \lambda\varphi_1 + \ldots)(\pm L/2, x_2, x_3) = 0, \qquad (6.27)$$

$$\frac{\partial}{\partial \mathbf{n}}(\varphi_0 + \lambda\varphi_1 + \ldots)(x_1, \pm L/2, x_3) = 0.$$

From (6.25)–(6.27) we obtain the equations of the 0-th order (terms corresponding to $\lambda^0 = 1$)

$$\text{div}[\varepsilon_0(\mathbf{x})\nabla\varphi_0] = 0, \qquad (6.28)$$

$$\varphi_0(x_1, x_2, \pm M/2) = \pm E/2, \qquad (6.29)$$

$$\frac{\partial\varphi_0}{\partial\mathbf{n}}(\pm L/2, x_2, x_3) = 0, \quad \frac{\partial\varphi_0}{\partial\mathbf{n}}(x_1, \pm L/2, x_3) = 0, \qquad (6.30)$$

and the equations of the 1st order (terms corresponding to λ)

$$\text{div}\left[\varepsilon_0(\mathbf{x})\nabla\varphi_1 + b(\mathbf{x}, |\nabla\varphi_0|)\nabla\varphi_0\right] = 0, \qquad (6.31)$$

$$\varphi_1(x_1, x_2, \pm M/2) = 0,$$

$$\frac{\partial\varphi_1}{\partial\mathbf{n}}(\pm L/2, x_2, x_3) = 0, \quad \frac{\partial\varphi_1}{\partial\mathbf{n}}(x_1, \pm L/2, x_3) = 0.$$

Substituting (6.24) in (6.22), we obtain

$$\widehat{\varepsilon}(E) = \frac{1}{E^2 M L^2} \int_Y (\varepsilon_0(\mathbf{x}) + \lambda b(\mathbf{x}, |\nabla(\varphi_0(\mathbf{x}) + \lambda\varphi_1(\mathbf{x}) + ...|)) \times \qquad (6.32)$$

$$\times |\nabla(\varphi_0(\mathbf{x}) + \lambda\varphi_1(\mathbf{x}) + ...)|^2 d\mathbf{x}.$$

Saving in (6.32) only the linear terms in λ, we obtain

$$\widehat{\varepsilon}(E) = \frac{1}{E^2 M L^2} \int_Y \varepsilon_0(\mathbf{x})|\nabla\varphi_0(\mathbf{x})|^2 d\mathbf{x} + \qquad (6.33)$$

$$+ \frac{\lambda}{E^2 M L^2} \int_Y [\varepsilon_0(\mathbf{x})\nabla\varphi_0\nabla\varphi_1(\mathbf{x}) + b(\mathbf{x}, |\nabla\varphi_0(\mathbf{x})|)\nabla\varphi_0(\mathbf{x})\nabla\varphi_0(\mathbf{x})]d\mathbf{x}.$$

Now we make some transformations. Multiplying (6.28) by $\varphi_1(\mathbf{x})$ and integrating over Y, we obtain

$$\int_Y \text{div}[\varepsilon_0(\mathbf{x})\nabla\varphi_0(\mathbf{x})]\varphi_1(\mathbf{x})d\mathbf{x} = 0. \qquad (6.34)$$

Integrating in (6.34) by parts, we have

$$-\int_Y \varepsilon_0(\mathbf{x})\nabla\varphi_0\nabla\varphi_1 d\mathbf{x} + \int_{\partial Y} \varepsilon_0(\mathbf{x})\frac{\partial\varphi_0}{\partial\mathbf{n}}(\mathbf{x})\varphi_1(\mathbf{x})d\mathbf{x} = 0. \qquad (6.35)$$

The boundary integral in (6.35) is the sum of integrals over the horizontal and vertical sides of the periodicity cell Y.

For the upper and lower sides of the periodicity cell, we have

$$\int_{\partial Y \cap \{y_3 = \pm M/2\}} \varepsilon_0(\mathbf{x}) \frac{\partial \varphi_0}{\partial \mathbf{n}} (\mathbf{x}) \varphi_1(\mathbf{x}) d\mathbf{x} = 0. \tag{6.36}$$

The equality (6.36) results from $\varphi_1(x_1, x_2, \pm L/2) = 0$ in accordance with (6.31).

For the lateral sides of the cell

$$\int_{\partial Y \cap \{x_1 = \pm L/2, \ x_2 = \pm L/2\}} \varepsilon_0(\mathbf{x}) \frac{\partial \varphi_0}{\partial \mathbf{n}} (\mathbf{x}) \varphi_1(\mathbf{x}) d\mathbf{x} = 0.$$

This equality results from

$$\frac{\partial \varphi_0}{\partial \mathbf{n}} (\pm L/2, x_2, x_3) = 0, \quad \frac{\partial \varphi_0}{\partial \mathbf{n}} (x_1, \pm L/2, x_3) = 0$$

in accordance with (6.30).

Then the boundary integral in (6.35) is equal to zero and from (6.35) it follows that

$$- \int_Y \varepsilon_0(\mathbf{x}) \nabla \varphi_0(\mathbf{x}) \nabla \varphi_1(\mathbf{x}) d\mathbf{x} = 0. \tag{6.37}$$

Note that we cannot obtain an analog of the equality (6.37) for an arbitrary function. We obtain (6.37) by using specific properties of the problem under consideration. Namely, we use the fact that the sides of the periodicity cell Y on which $\varphi_1(\mathbf{x}) = 0$ and $\frac{\partial \varphi_0}{\partial \mathbf{n}} (\mathbf{x}) = 0$ covers the boundary of the periodicity cell entirely.

Using (6.37), we can rewrite (6.33) as

$$\widehat{\varepsilon}(E) = \frac{1}{E^2 M L} \int_Y \varepsilon_0(\mathbf{x}) |\nabla \varphi_0(\mathbf{x})|^2 d\mathbf{x} + \frac{\lambda}{E M L^2} \int_Y b(\mathbf{x}, |\nabla \varphi_0(\mathbf{x})|) |\nabla \varphi_0(\mathbf{x})|^2 d\mathbf{x}.$$

Then

$$\widehat{\varepsilon}(E) = \widehat{\varepsilon}(0) + \lambda D(E), \tag{6.38}$$

where

$$\widehat{\varepsilon}(0) = \frac{1}{E^2 M L^2} \int_Y \varepsilon_0(\mathbf{x}) |\nabla \varphi_0(\mathbf{x})|^2 d\mathbf{x} = \tag{6.39}$$

$$= \frac{1}{M L^2} \int_Y \varepsilon_0(\mathbf{x}) \left| \frac{\nabla \varphi_0(\mathbf{x})}{E} \right|^2 d\mathbf{x},$$

and

$$D(E) = \frac{1}{E^2 M L^2} \int_Y b(\mathbf{x}, |\nabla \varphi_0(\mathbf{x})|) |\nabla \varphi_0(\mathbf{x})|^2 d\mathbf{x} = \qquad (6.40)$$

$$= \frac{1}{M L^2} \int_Y b(\mathbf{x}, |\nabla \varphi_0(\mathbf{x})|) \left| \frac{\nabla \varphi_0(\mathbf{x})}{E} \right|^2 d\mathbf{x}.$$

We denote $N(\mathbf{x})$ the solution of the linear cellular problem (6.28)–(6.30) corresponding to $E = 1$:

$$\operatorname{div}(\varepsilon_0(\mathbf{x}) \nabla N) = 0, \qquad (6.41)$$

$$N(x_1, x_2, \pm M/2) = \pm 1/2, \qquad (6.42)$$

$$\frac{\partial N}{\partial \mathbf{n}}(\pm L/2, x_2, x_3) = 0, \quad \frac{\partial N}{\partial \mathbf{n}}(x_1, \pm L/2, x_3) = 0. \qquad (6.43)$$

By virtue of the linearity of the problem (6.28)–(6.30), we have

$$\varphi_0(\mathbf{x}) = E N(\mathbf{x}), \qquad (6.44)$$

and we can write formulas (6.39) and (6.40) in the following form

$$\widehat{\varepsilon}(0) = \frac{1}{M L^2} \int_Y \varepsilon_0(\mathbf{x}) |\nabla N(\mathbf{x})|^2 d\mathbf{x}, \qquad (6.45)$$

and

$$D(E) = \frac{1}{M L^2} \int_Y b(\mathbf{x}, |E \nabla N(\mathbf{x})|) |\nabla N(\mathbf{x})|^2 d\mathbf{x}. \qquad (6.46)$$

Let us discuss the formulas obtained. The problem (6.41)–(6.43) is a special kind of the cellular problem (1.20). Formula (6.45) is an analog of the formula (1.23). It introduces the homogenized dielectric constant $\widehat{\varepsilon}(0)$ of a linear composite.

We consider the composite consisting of a nonlinear matrix filled with linear inclusions. In this case

$$b(\mathbf{x}, E) = \begin{cases} b_f(E) \neq 0 \text{ in matrix } m, \\ 0 \text{ in inclusion } I, \end{cases} \qquad (6.47)$$

$$\varepsilon_0(\mathbf{x}) = \begin{cases} \varepsilon_f \text{ in matrix } m, \\ \varepsilon_d \text{ in inclusion } I. \end{cases} \qquad (6.48)$$

Then formula (6.46) takes the form

$$D(E) = \frac{1}{ML^2} \int_m b_f(|E\nabla N(\mathbf{x})|)|\nabla N(\mathbf{x})|^2 d\mathbf{x},$$

so that we integrate over the matrix only. In this case $b_f(|E\nabla N|)$ do not depend on the spatial variables, but take values corresponding to the matrix.

Assume that $b(\mathbf{x}, 0) = 0$. Under this assumption the homogenized tunability $\widehat{T}(E)$ is computed in accordance with the formula

$$\widehat{T}(E) = \lambda \frac{D(E)}{\widehat{\varepsilon}(0)} = \lambda \frac{\int_m b_f(|E\nabla N(\mathbf{x})|)|\nabla N(\mathbf{x})|^2 d\mathbf{x}}{\int_Y \varepsilon_0(\mathbf{x})|\nabla N(\mathbf{x})|^2 d\mathbf{x}}. \tag{6.49}$$

The tunability of the material of the matrix is

$$T_f(E) = \lambda \frac{b_f(E)}{\varepsilon_f}. \tag{6.50}$$

When we dilute a matrix which has a large dielectric constant with particles which have low dielectric constant, the homogenized dielectric constant $\widehat{\varepsilon}(0)$ decreases. The ratio (6.49) can both increase and decrease against of $\lambda \frac{b_f(E)}{\varepsilon_f}$ even when $\widehat{\varepsilon}(0)$ (6.45) decreases.

Taking into account (6.47)–(6.50) we can write

$$\widehat{T}(E) - T_f(E) = \lambda \frac{D(E)}{\widehat{\varepsilon}(0)} - \lambda \frac{b_f(E)}{\varepsilon_f} = \tag{6.51}$$

$$= \lambda \frac{\int_Y b_f(|E\nabla N(\mathbf{x})|)|\nabla N(\mathbf{x})|^2 d\mathbf{x}}{\int_Y \varepsilon_0(\mathbf{x})|\nabla N(\mathbf{x})|^2 d\mathbf{x}} - \lambda \frac{b_f(E)}{\varepsilon_f} =$$

$$= \lambda \frac{\int_Y [\varepsilon_f b(\mathbf{x}, |E\nabla N(\mathbf{x})|) - b_f(E)\varepsilon_0(\mathbf{x})]|\nabla N(\mathbf{x})|^2 d\mathbf{x}}{\varepsilon_f \int_Y \varepsilon_0(\mathbf{x})|\nabla N(\mathbf{x})|^2 d\mathbf{x}} =$$

$$- \lambda \frac{\int_m [b_f(|E\nabla N(\mathbf{x})|) - b_f(E)]\nabla N(\mathbf{x})|^2 d\mathbf{x}}{\int_Y \varepsilon_0(\mathbf{x})|\nabla N(\mathbf{x})|^2 d\mathbf{x}}.$$

It is seen from (6.51) that the sign of the expression $\widehat{T}(E) - T_f(E)$ is determined by the term $b_f(|E\nabla N(\mathbf{x})|) - b_f(E)$, which depends on the distribution of the local electric field $N(\mathbf{x})$ over the periodicity cell.

The quadratic function $b(\mathbf{x}, E)$

In applications $\varepsilon_0(\mathbf{x})$ is often taken in the form of a quadratic function. In this case (see formula (6.8))

$$b(\mathbf{x}, E) = -\mu(\mathbf{x})E^2, \tag{6.52}$$

so that (6.46) becomes

$$D(E) = -BE^2. \tag{6.53}$$

Then

$$\widehat{\varepsilon}(E) = \widehat{\varepsilon}(0) - BE^2. \tag{6.54}$$

The coefficient B in (6.53) is given by the following formula:

$$B = \frac{1}{ML} \int_m \mu(\mathbf{x})|\nabla N(\mathbf{x})|^4 d\mathbf{x}. \tag{6.55}$$

The equations (6.53) and (6.54) mean that when the case of the local tunability is a quadratic function in E then the homogenized tunability is also a quadratic function in E. This conclusion cannot be expanded to nonlinearities of general form.

Small / weak electric field

Assume the potential applied to the composite material is small. Then the boundary condition (6.13) can be represented in the form

$$\varphi(x_1, x_2, \pm M/2) = \pm\xi EM/2,$$

where $\xi \ll 1$.

The potential of electric field in the composite material has the same order $\xi \ll 1$ and can be represented as

$$\varphi(\mathbf{x}) = \xi\varphi_1(\mathbf{x}).$$

For this case, formula (6.8) takes the form

$$\varepsilon(\mathbf{x}, |\nabla\varphi_1|) = \varepsilon(\mathbf{x}, 0) + b(\mathbf{x}, |\xi| \cdot |\nabla\varphi_1|) \tag{6.56}$$

and equation (6.12) takes the form (it is possible to divide the equation by $\xi \neq 0$)

$$\text{div}(\varepsilon(\mathbf{x}, |\xi| \cdot |\nabla\varphi_1|)\nabla\varphi_1) = 0. \tag{6.57}$$

Since $b(\mathbf{x}, 0) = 0$, the Taylor series for the function $b(\mathbf{x}, E)$ with respect to the variable E starts at the terms of the form $c + dE^2$. If $c = 0$ (this is our case) then (6.56) becomes

$$\varepsilon(\mathbf{x}, |\nabla\varphi_1|) = \varepsilon(\mathbf{x}, 0) + \xi^2 d(\mathbf{x})|\nabla\varphi_1|^2. \tag{6.58}$$

Comparing (6.57), (6.58) and (6.8), (6.12), we find that they coincide completely if we put $\varphi = \varphi_1$ and

$$\lambda = \xi^2,$$

$$b(\mathbf{x}, \nabla\varphi) = d(\mathbf{x})|\nabla\varphi|^2.$$

Thus, the formula (6.52) is valid for an arbitrary nonlinear composite if the potential of the electric field is small.

A formula similar to equations (6.53) and (6.55) was proposed in [126] for the general case. The method used in [126], however, cannot be considered as mathematically rigorous. Our analysis of this relation based on the homogenization method [30] guarantees the validity of (6.53) and (6.55) for small field or small nonlinearity limit only. The authors found no reasons to recommend (6.53) and (6.55) for the general case.

Macro anisotropic composite materials

Even when constitutive components of composite are isotropic, the structure of composite can be not isotropic (not invariant with respect to the interchange of coordinate axis) and the composite can demonstrate anisotropic homogenized characteristics (so called structural anisotropy [89, 169]). The homogenization procedure for macroanisotropic composite is the same as above. In Section 6.2 we consider the macroscopic electric field parallel to Ox_3-axis. Considering the macroscopic electric field parallel to Ox_1-axis, we arrive at the cellular problem

$$\operatorname{div}(a(\mathbf{x})\nabla N) = 0, \tag{6.59}$$

$$N(\pm L/2, x_2, x_3) = \pm 1/2, \tag{6.60}$$

$$\frac{\partial N}{\partial \mathbf{n}}(x_1, \pm L/2, x_3) = 0, \quad \frac{\partial N}{\partial \mathbf{n}}(x_1, x_2, \pm M/2) = 0. \tag{6.61}$$

Considering the macroscopic electric field parallel to Ox_2-axis, we arrive at the cellular problem

$$\operatorname{div}(a(\mathbf{x})\nabla N) = 0, \tag{6.62}$$

$$N(x_1, \pm L/2, x_3) = \pm 1/2, \tag{6.63}$$

$$\frac{\partial N}{\partial \mathbf{n}}(\pm L/2, x_2, x_3) = 0, \quad \frac{\partial N}{\partial \mathbf{n}}(x_1, x_2, \pm M/2) = 0. \tag{6.64}$$

The homogenized permittivity in the directions Ox_1 or Ox_2 is computed using formulas (6.38), (6.45) and (6.46) with the function $N(\mathbf{x})$ determined from solution of problem (6.59)–(6.61) or (6.62)–(6.64), correspondingly.

If the structure of composite material is invariant with respect to the interchange of coordinate axis, solution of problem (6.59)–(6.61) can be obtained from solution of problem (6.41)–(6.43) by interchanging variables x_1 and x_2.

Effective tunability of composite materials

The relative tunability of an electrically tunable material has been defined by equation (6.1). We will address the case of small tuning / weak electric field, so that the function $\varepsilon(E)$ can be approximated as $\varepsilon(E) = \varepsilon_f - \mu_f E^2$ (see formula (6.8)), and formula (6.1) takes the form (here and afterward we omit parameter λ, which has been used for mathematical rigor in the previous section)

$$T_f(E) = \frac{\mu_f}{\varepsilon_f} E^2. \tag{6.65}$$

In this case, the homogenized dielectric constant of the composite can be written in the form (see Section 6.2):

$$\widehat{\varepsilon}(E) = \widehat{\varepsilon}(0) - BE^2, \tag{6.66}$$

where

$$\widehat{\varepsilon}(0) = \frac{1}{|Y|} \int_Y \varepsilon_0(\mathbf{x})|\nabla N(\mathbf{x})|^2 d\mathbf{x} \tag{6.67}$$

and

$$B = \frac{1}{|Y|} \int_Y \mu(\mathbf{x})|\nabla N(\mathbf{x})|^4 d\mathbf{x}. \tag{6.68}$$

The homogenized (effective) tunability of the composite described by equation (6.66) is then

$$\widehat{T}(E) = \frac{B}{\widehat{\varepsilon}(0)} E^2. \tag{6.69}$$

Observing formulas (6.67)–(6.69) for the homogenized tunability of the composite, we find that the main difficulty in the calculation of the homogenized tunability is the computation of the function $N(\mathbf{x})$, which is solution of the cellular problem.

Effective tunability of high–contrast composite materials

In high-contrast composite, value of permittivity of inclusion I is small as compared with the value of permittivity of matrix $m = Y \setminus I$, that is large, see Table 6.1. The gradient of electric field $\nabla\varphi(\mathbf{x})$ in high-contrast composite material has specific form. Namely, it takes negligible small value in dielectric inclusions (especially as compared with its value in ferroelectric matrix), see Fig. 6.4 (left). As far as the electric displacement $\mathbf{D}$, it has the general form and takes similar (not large) values in dielectric inclusions and ferroelectric matrix, see Fig. 6.4 (right).

Then, we can neglect $\varepsilon_0(\mathbf{x})|\nabla N(\mathbf{x})|^2$ in the inclusions I. As a result, in the formula (6.67), the integral in denominator is carried out, in fact, over the matrix m only:

$$\int_Y \varepsilon_0(\mathbf{x})|\nabla N(\mathbf{x})|^2 dx \approx \int_m \varepsilon_0(\mathbf{x})|\nabla N(\mathbf{x})|^2 dx = \varepsilon_f \int_m |\nabla N(\mathbf{x})|^2 dx. \qquad (6.70)$$

Note that $\mu(\mathbf{x}) = 0$ in inclusion I, then

$$\int_Y \mu(\mathbf{x})|\nabla N(\mathbf{x})|^4 dx = \mu_f \int_m |\nabla N(\mathbf{x})|^4 dx.$$

By virtue of (6.70) and (6.70), we obtain from (6.67), (6.68), and (6.69) the following formula for the parameters $\widehat{\varepsilon}(0)$, $B(E)$ and $\widehat{T}(E)$ for high–contrast composite:

$$\widehat{\varepsilon}(0) = \varepsilon_f I_2, \ \ B(E) = \mu_f I_4, \qquad (6.71)$$

$$\widehat{T}(E) = \frac{\mu_f}{\varepsilon_f} J E^2, \qquad (6.72)$$

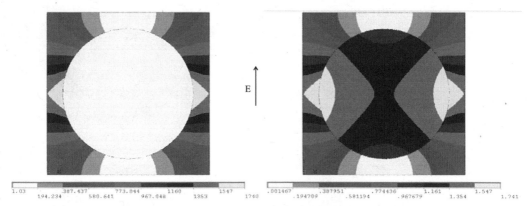

Figure 6.4. *The gradient of electric field* $|\nabla\varphi(\mathbf{x})|$ *(left) and electric displacement* $|\mathbf{D}(\mathbf{x})|$ *(right) in high-contrast composite.*

where we use the following notations:

$$I_2 = \frac{1}{|Y|} \int_m |\nabla N(\mathbf{x})|^2 d\mathbf{x} \approx \frac{1}{|Y|} \int_Y |\nabla N(\mathbf{x})|^2 d\mathbf{x}, \tag{6.73}$$

$$I_4 = \frac{1}{|Y|} \int_m |\nabla N(\mathbf{x})|^4 d\mathbf{x} \approx \frac{1}{|Y|} \int_Y |\nabla N(\mathbf{x})|^4 d\mathbf{x}, \tag{6.74}$$

$$J = \frac{I_4}{I_2}. \tag{6.75}$$

Note that the tunability of the pure ferroelectric is $\dfrac{\mu}{\varepsilon_f} E^2$. Then from equation (6.69) it follows that J is equal to the ratio of the homogenized tunability of composite to the tunability of the pure ferroelectric:

$$J = \frac{\widehat{T}(E)}{T_f(E)}. \tag{6.76}$$

Formula (6.76) allows us to evaluate the impact of dilution on the homogenized tunability $\widehat{T}(E)$.

6.2.1. Bounds on the effective tunability of a high-contrast composite

The estimating of effective characteristics is a popular method in the theory of composite materials and inhomogeneous structures, see, e.g., [225, 297, 315, 351, 376]. The idea of estimating of effective characteristics of composite materials turns to works of Voight [367] and Reuss [305].

For a high-contrast composite, the value of the tunability in accordance with formulas (6.72)–(6.75) can be expressed through the values of the integral functionals $I_2 = I_2(N)$ (6.73) and $I_4 = I_4(N)$ (6.74), where $N(\mathbf{x})$ is solution of the cellular problem (6.41)–(6.43). As has been noted, solution of the cellular problem (6.41)–(6.43) is a rather difficult task. But it is possible to estimate values of the functionals $I_2(N)$ (6.73) and $I_4(N)$ (6.74) without solving of the problem (6.41)–(6.43).

Low-sided estimate for tunability

We obtain estimates on the homogenized tunability using Cauchy-Bunyakovsky inequality [308]. We will use the special case of the mentioned equality, which has the following form:

$$\int_m f(\mathbf{x}) g(\mathbf{x}) d\mathbf{x} \leq \left(\int_m f(\mathbf{x})^2 d\mathbf{x} \right)^{1/2} \left(\int_m g(\mathbf{x})^2 d\mathbf{x} \right)^{1/2}.$$

We represent the integrand in the integral

$$I_2 = \int_m |\nabla N(\mathbf{x})|^2 d\mathbf{x}$$

in the following way:

$$f(\mathbf{x}) = |\nabla N(\mathbf{x})|^2, \; g(\mathbf{x}) = 1$$

and obtain

$$\int_m |\nabla N(\mathbf{x})|^2 dx \le \left(\int_m |\nabla N(\mathbf{x})|^4 dx\right)^{1/2} \left(\int_m dx\right)^{1/2} = \tag{6.77}$$

$$= \left(\int_m |\nabla N(\mathbf{x})|^4 dx\right)^{1/2} |m|^{1/2},$$

where $|m|$ means the measure of the domain occupied by the matrix ($|m|/|Y|$ is the volume ratio of the material of the matrix).

Then

$$\left(\int_m |\nabla N(\mathbf{x})|^2 dx\right)^2 \le \int_m |\nabla N(\mathbf{x})|^4 dx |m|. \tag{6.78}$$

Dividing (6.78) by $|Y|^2$, we obtain

$$\left(\frac{1}{|Y|}\int_Y |\nabla N(\mathbf{x})|^2 dx\right)^2 \le \frac{1}{|Y|}\int_Y |\nabla N(\mathbf{x})|^4 dx \frac{|m|}{|Y|}. \tag{6.79}$$

From (6.79) we have

$$I_2^2 \le I_4 \frac{|m|}{|Y|}$$

or

$$I_4 \ge I_2^2 \frac{|Y|}{|m|}.$$

Using the last inequality, we can estimate coefficient $\frac{\mu_f}{\varepsilon_f}J$ from (6.72) as follows:

$$\frac{\mu_f}{\varepsilon_f}J = \frac{B(E)}{\widehat{\varepsilon}(0)} \ge \frac{\mu_f}{\varepsilon_f}I_2\frac{|Y|}{|m|}. \tag{6.80}$$

Since

$$1 - \frac{|m|}{|Y|} = \frac{|I|}{|Y|} = q$$

is the volume fraction of dielectric inclusion, we can write

$$\frac{\mu_f}{\varepsilon_f}J = \frac{B}{\widehat{\varepsilon}(0)} \ge \frac{\mu_f}{\varepsilon_f}\cdot\frac{I_2}{1-q}.$$

Low-sided estimate for the homogenized tunability of dilute composite

For a dilute composite (a composite with a small volume fraction q of inclusions), the homogenized permittivity is given by the formula

$$\widehat{\varepsilon}(0) = \varepsilon_f(1 - 1.5q),$$

see formula (6.7). Thus,

$$I_2 = 1 - 1.5q.$$

From the last formula and (6.80) we have for dilute composite

$$\frac{\widehat{T}(E)}{E^2} = \frac{\mu_f}{\varepsilon_f}J = \frac{B}{\widehat{\varepsilon}(0)} \geq \frac{\mu_f}{\varepsilon_f} \cdot \frac{I_2}{1-q} = \frac{\mu_f}{\varepsilon_f} \cdot \frac{1-1.5q}{1-q} \approx \qquad (6.81)$$

$$\approx \frac{\mu_f}{\varepsilon_f}(1-1.5q)(1+q) \approx \frac{\mu_f}{\varepsilon_f}(1-0.5q).$$

It is seen from (6.81) that when we dilute ferroelectric matrix with dielectric inclusions, $\widehat{T}(E)$ decreases slowly than the homogenized tunability $\widehat{\varepsilon}(0)$ decreases.

6.2.2. Numerical computations of homogenized characteristics

We present some results of numerical computations from [191]. Table 6.2 [191] shows the values of the integrals I_2 (6.73) and I_4 (6.74) as functions of the volume fraction q of dielectric inclusion for the high-contrast composite ($\varepsilon_d = 0$). The quantity

$$I_2 = \frac{\widehat{\varepsilon}(0)}{\varepsilon_f}$$

the dielectric constant of the composite to the dielectric constant of the pure ferroelectric (the material of the matrix). The quantity

$$I_4 = \frac{B}{\mu_f}$$

is the ratio of the coefficient describing the quadratic non-linearity of the composite material to the corresponding coefficient for pure ferroelectric (see formula (6.66)). The quantity

$$J = \frac{I_4}{I_2} = \frac{\widehat{T}(E)}{T_f(E)}$$

is the ratio of the tunability of the composite to the tunability of the pure ferroelectric. Table 6.2 also gives the values of the parameter

$$J^* = \frac{B}{\widehat{\varepsilon}(0)}$$

calculated for a finite dielectric contrast between the matrix and inclusions ($\varepsilon_d = 10$ and $\varepsilon_f = 2000$). Comparing the calculated values of J and J^*, one finds them to be approximately equal.

Table 6.2. *Values of Parameters Controlling the Dielectric Properties of the Non-linear Periodic Composite Calculated as Functions of the Dielectric Volume Concentration q.*

q	I_2	I_4	$J = \dfrac{I_4}{I_2}$	J^*	$\dfrac{I_4}{I_2^2}$
0.03	0.94	0.98	1.04	1.04	1.11
0.05	0.91	0.96	1.06	1.06	1.17
0.09	0.84	0.92	1.10	1.09	1.31
0.13	0.78	0.87	1.12	1.12	1.45
0.20	0.67	0.79	1.17	1.16	1.75
0.28	0.56	0.68	1.22	1.21	2.18
0.35	0.48	0.60	1.25	1.25	2.60
0.39	0.40	0.57	1.30	1.27	2.94
0.46	0.36	0.50	1.38	1.35	3.79

6.2.3. Note on the decoupled approximation approach

In this section, we compare predictions of our analysis based on the homogenization theory with rather popular decoupled approximation approach in the effective medium theory [126, 315]. The decoupled approximation postulates that, for the spatial distribution of the electric field $\mathbf{E}(\mathbf{x})$ in the composite, there is a relation [346, 347]

$$\langle E^4(\mathbf{x}) \rangle = \langle E^2(\mathbf{x}) \rangle^2, \qquad (6.82)$$

where $\langle \rangle$ stands for the spatial average, for example, the averaging operator (1.22) and $E(\mathbf{x}) = |\mathbf{E}(\mathbf{x})|$.

Sometimes the more general relation (n is an integer number)

$$\langle E^{2n}(\mathbf{x}) \rangle = \langle E^2(\mathbf{x}) \rangle^n$$

is postulated [315, 347].

In terms of our variable, equation (6.82) states that

$$I_4 = I_2^2, \qquad (6.83)$$

where I_2 and I_4 are defined by (6.73) and (6.74).

From (6.79) we have

$$\frac{I_2^2}{I_4} \leq \frac{|m|}{|Y|}.$$

Taking into account that

$$\frac{|m|}{|Y|} = 1 - q < 1,$$

(q means the volume fraction of the dielectric inclusion), we have

$$I_2^2 \leq (1-q)I_4 < I_4$$

and conclude that the equality (6.83), which is the basic assumption of the decoupled approximation, can never be satisfied for the composite materials under consideration.

The results of our numerical computations presented in Section 6.2.2 enables evaluation of the validity of equation (6.83). It is seen from Table 6.2 that I_4 is much closer to I_2 rather than to I_2^2. Thus, our calculations give no support to the decoupled approximation in the effective medium approach.

6.3. Tunability of laminated composite

In this section, we compute homogenized dielectric characteristics of laminated composite material in the direction across the layers. This problem is interesting both theoretically (since the solution can be obtained in an explicit way and analyzed in detail) and practically (because manufacturing of laminated materials is a relatively simple technological process [169, 209, 267]).

We consider composite material formed of layers perpendicular to Ox-axis, see Fig. 6.5.

We assume the period L of the structure of the composite material to be equal to 1. Then the average value is computed as

$$\langle \bullet \rangle = \int\limits_0^1 \bullet\, dx.$$

Solution of the cellular problem (6.41)–(6.43) for laminated material can be obtained in explicit form. If layers are parallel to Ox-axis then the function $N(\mathbf{x})$

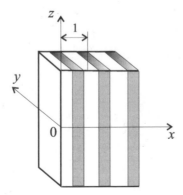

Figure 6.5. *Laminated composite material.*

is a function of unique variable x and cellular problem (6.41)–(6.43) takes the form

$$\frac{d}{dx}\left(\varepsilon_0(x)\frac{dN}{dx}(x)\right)=0,$$

$$N(\pm 1/2)=\pm 1/2.$$

Then

$$\frac{dN}{dx}(x)=\frac{C}{\varepsilon_0(x)}$$

and

$$N(x)=C\int_{-1/2}^{x}\frac{ds}{\varepsilon_0(s)}-1/2.$$

The constant C is determined from the condition

$$C\int_{-1/2}^{1/2}\frac{ds}{\varepsilon_0(s)}-1/2\doteq 1/2,$$

which may be written as

$$C\left\langle\frac{1}{\varepsilon_0(x)}\right\rangle=1.$$

Then

$$C=\frac{1}{\left\langle\dfrac{1}{\varepsilon_0(x)}\right\rangle}$$

and

$$\frac{dN}{dx}(x)=\frac{1}{\varepsilon_0(x)\left\langle\dfrac{1}{\varepsilon_0(x)}\right\rangle}. \qquad (6.84)$$

The tunability $\widehat{T}(E)$ is computed in accordance with formula (6.69), which can be rewritten in the form

$$\widehat{T}(E)=\frac{B}{\widehat{\varepsilon}(0)}E^2. \qquad (6.85)$$

Substituting (6.84) into (6.85), we obtain

$$\widehat{T}(E)=\frac{\displaystyle\int_0^1\frac{\mu(x)dx}{\left[\varepsilon_0(x)\left\langle\dfrac{1}{\varepsilon_0(x)}\right\rangle\right]^4}}{\displaystyle\int_0^1\frac{\varepsilon_0(x)dx}{\left[\varepsilon_0(x)\left\langle\dfrac{1}{\varepsilon_0(x)}\right\rangle\right]^2}}E^2=\frac{\displaystyle\int_0^1\frac{\mu(x)dx}{\left[\varepsilon_0(x)\left\langle\dfrac{1}{\varepsilon_0(x)}\right\rangle\right]^4}dx}{\displaystyle\int_0^1\frac{dx}{\varepsilon_0(x)\left\langle\dfrac{1}{\varepsilon_0(x)}\right\rangle^2}}E^2. \qquad (6.86)$$

The integral in the denominator of formula (6.86) is equal to

$$\int_0^1 \frac{dx}{\varepsilon_0(x) \left\langle \dfrac{dx}{\varepsilon_0(x)} \right\rangle^2} = \frac{1}{\left\langle \dfrac{1}{\varepsilon_0(x)} \right\rangle^2} \int_0^1 \frac{dx}{\varepsilon_0(x)} = \frac{1}{\left\langle \dfrac{1}{\varepsilon_0(x)} \right\rangle}. \qquad (6.87)$$

The quantity in the right-hand side of (6.87) equals the dielectric constant across the layers.

The integral in numerator of formula (6.86) is

$$\int_0^1 \frac{\mu(x)dx}{\left[\varepsilon_0(x) \left\langle \dfrac{1}{\varepsilon_0(x)} \right\rangle \right]^4} = \frac{1}{\left[\left\langle \dfrac{1}{\varepsilon_0(x)} \right\rangle \right]^4} \int_0^1 \frac{\mu(x)dx}{\varepsilon_0(x)^4}.$$

Then the fraction in (6.86) is equal to $\left[\left\langle \dfrac{1}{\varepsilon_0(x)} \right\rangle \right]^{-3} \displaystyle\int_0^1 \frac{\mu(x)dx}{\varepsilon_0(x)^4}$. Grouping terms in (6.86), we obtain

$$\widehat{T}(E) = \frac{E^2}{\left[\left\langle \dfrac{1}{\varepsilon_0(x)} \right\rangle \right]^3} \int_0^1 \frac{\mu(x)dx}{\varepsilon_0(x)^4} = \qquad (6.88)$$

$$= \frac{E^2}{\left[\left\langle \dfrac{1}{\varepsilon_0(x)} \right\rangle \right]^3} \int_0^1 \frac{\tau(x)dx}{\varepsilon_0(x)^3} = \frac{E^2}{\left[\left\langle \dfrac{1}{\varepsilon_0(x)} \right\rangle \right]^3} \left\langle \frac{\tau(x)}{\varepsilon_0(x)^3} \right\rangle,$$

where we use notation

$$\tau(x) = \frac{\mu(x)}{\varepsilon_0(x)}. \qquad (6.89)$$

We denote

$$\alpha(x) = \frac{1}{\varepsilon_0(x)}.$$

Using this notation, we can rewrite the right-hand part of (6.88) in terms of $\alpha(x)$ and $\tau(x)$ as follows:

$$\widehat{T}(E) = E^2 \left[\frac{1}{\langle \alpha(x) \rangle} \right]^3 \langle \tau(x)(\alpha(x))^3 \rangle = E^2 \frac{\langle \tau(x)(\alpha(x))^3 \rangle}{\langle \alpha(x) \rangle^3}. \qquad (6.90)$$

6.3.1. Tunability of laminated composite in terms of electric displacement

We have computed the homogenized tunability as a function of overall electric field E. The dielectric constant can be represented as a function of electric field E or as a function of electric displacement $|\mathbf{D}| = \varepsilon(E)E$. In the case of quadratic nonlinearity (6.2) and small tuning (or small electric field E), we can write

$$\widehat{\varepsilon}(E) = \varepsilon_0(1 + T(E)) = \varepsilon_0 \left(1 + \frac{\mu}{\varepsilon_0} E^2 \right) \approx \tag{6.91}$$

$$\approx \varepsilon_0 + \mu \frac{|\mathbf{D}|^2}{\varepsilon_0^2} = \varepsilon_0 \left(1 + \frac{\mu}{\varepsilon_0^3} |\mathbf{D}|^2 \right) = \varepsilon_0(1 + t(|\mathbf{D}|)) = \widehat{\varepsilon}(|\mathbf{D}|),$$

where

$$T(E) = \frac{\mu}{\varepsilon_0} E^2$$

is the relative tunability *in terms of electric field* and

$$t(|\mathbf{D}|) = \frac{\mu}{\varepsilon_0^3} |\mathbf{D}|$$

is the relative tunability *in terms of electric displacement*. These are two different functions [187]. Note that the original definition of tunability (6.1) was formulated in terms of electric field E.

Writing (6.91), we use the approximate equality

$$|\mathbf{D}| = (\varepsilon_0 + \mu E^2)E \approx \varepsilon_0 E,$$

which is valid for small electric field E.

The overall electric displacement is related to the overall electric field by the equality

$$|\mathbf{D}| = E \frac{1}{\langle \alpha(x) \rangle}.$$

Substituting this equality to (6.90), we can rewrite the nonlinear term $\widehat{T}(E)$ in the formula for homogenized dielectric constant as follows

$$E^2 \left\langle \tau(x)(\alpha(x))^3 \right\rangle = |\mathbf{D}|^2 \frac{\left\langle \tau(x)(\alpha(x))^3 \right\rangle}{\langle \alpha(x) \rangle^2}. \tag{6.92}$$

We denote

$$t(x) = \frac{\mu(x)}{\varepsilon_0^3(x)}.$$

Using (6.89), we can rewrite this equality as

$$t(x) = \frac{\tau(x)}{\varepsilon_0^2(x)},$$

and then write

$$\tau(x) = t(x)\varepsilon_0^2(x) = \frac{t(x)}{\alpha^2(x)}.$$

Then the right-hand side of equality (6.92) can be rewritten in the form

$$|\mathbf{D}|^2 \frac{\langle t(x)\alpha(x)\rangle}{\langle \alpha(x)\rangle}. \qquad (6.93)$$

The expression in (6.90) is $\widehat{t}(|\mathbf{D}|)$. Then homogenized tunability in terms of electric displacement is

$$\widehat{t}(|\mathbf{D}|) = \frac{\langle t(x)\alpha(x)\rangle}{\langle \alpha(x)\rangle}|\mathbf{D}|^2. \qquad (6.94)$$

The functions $t(x)$ and $\alpha(x)$ in (6.94) are non-negative. Then, it follows from (6.94) that

$$\min\{t_i(|\mathbf{D}|)\} \le \widehat{t} \le \max\{t_i(|\mathbf{D}|)\}, \qquad (6.95)$$

where $t_i(|\mathbf{D}|)$ means the tunability of the i–th component of the composite in the term of the electric displacement.

6.3.2. Analysis of possible values of effective tunability using convex combinations technique

There is no simple estimate on the quantity $\widehat{T}$ (6.90). It is seen from (6.90) that value of $\widehat{T}$ is determined by the functionals $\langle \alpha(x)\rangle$ and $\langle \tau(x)(\alpha(x))^3\rangle$. We consider the case when n homogeneous materials are available to manufacture a composite of laminated structure. In the case under consideration, the possible values of the pair $(\langle \alpha(x)\rangle, \langle \tau(x)(\alpha(x))^3\rangle)$ form the set (see Appendix B)

$$\Lambda = \mathrm{conv}\{(\alpha_i, \tau_i\alpha_i^3) : i = 1, ..., n\}, \qquad (6.96)$$

where "conv" means the convex hull [307].

The method described in Appendix B can be used to design a laminated composite with the necessary $\widehat{T}$. We introduce the coordinate system Oxy and denote $x = \alpha$, $y = \tau\alpha^3$. In these notations

$$\Lambda = \mathrm{conv}\{\mathbf{x}_i : i = 1, ..., n\}, \qquad (6.97)$$

where

$$\{\mathbf{x}_i = (x_i, y_i) : x_i = \alpha_i, \ y_i = \tau_i\alpha_i^3; \ i = 1, ..., n.\}$$

Correspondingly,

$$\langle x \rangle = \langle \alpha \rangle, \ \langle y \rangle = \langle \tau\alpha^3 \rangle.$$

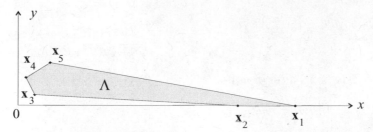

Figure 6.6. *The convex hull (five materials are available: materials 1 and 2 are dielectrics; materials 3, 4 and 5 are ferroelectrics).*

The polyhedron Λ in the coordinates Oxy is shown in Fig. 6.6.

We note that the points (x_i, y_i) corresponding to dielectrics belongs to Ox axis because $y_i = t_i \alpha_i^3 = 0$ for dielectric materials.

In accordance with Appendix B, a composite with prescribed characteristics (x, y) can be manufactured if and only if the condition

$$(x, y) \in \Lambda$$

is satisfied.

In accordance with the formula (6.90), the effective tunability can take any values determined by the formula

$$\widehat{T} = \frac{y}{x^3} \tag{6.98}$$

under condition $(x, y) \in \Lambda$.

We can rewrite (6.98) as

$$y = \widehat{T} x^3 \tag{6.99}$$

The equality (6.99) determines cubic parabola $y = cx^3$.

We formulate the following problem: one has to determine maximum coefficient c, such that the cubic parabola $y = cx^3$ touches the polyhedron Λ.

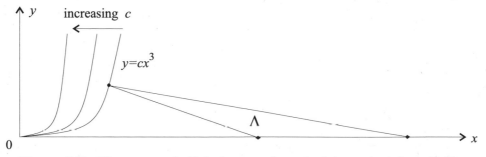

Figure 6.7. *The convex hull Λ for one ferroelectric material available.*

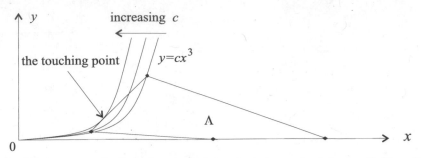

Figure 6.8. *The convex hull Λ for two ferroelectric materials available. Two right parabolas correspond to pure ferroelectric material. The left parabola corresponds to a two-component composite.*

It is evident that the solution of this problem determines the composite possessing the maximum effective tunability. The problem can be analyzed in detail if we take into account that the polyhedron Λ is a convex domain and the graph of the function $y = kx^3$ is a convex curve.

If one tunable material is available to produce the composite, maximum corresponds to this material, see Fig. 6.7. If two tunable materials are available, one can design a composite which has effective tunability $\widehat{T}$ greater than the tunabilities of those components, see Fig. 6.8.

6.3.3. Two-component laminated composite

We consider the two-component laminated composite for which

$$\varepsilon_0(x) = \varepsilon_1, \quad \mu(x) = \mu_1 \text{ in the 1st material,}$$

$$\varepsilon_0(x) = \varepsilon_2, \quad \mu(x) = \mu_2 \text{ in the 2nd material,}$$

and denote λ_i the volume fraction of the i-th material ($i = 1, 2$). In this case

$$\langle F(x) \rangle = F_1 \lambda_1 + F_2 \lambda_2 = F_1 X + F_2(1 - X),$$

where $X = \lambda_1 \in [0,1]$.

In Fig. 6.9, plots of the function

$$\widehat{T}(1) = \frac{\widehat{T}(E)}{E^2},$$

see (6.90), and the function

$$\widehat{t}(1) = \frac{\widehat{t}(|\mathbf{D}|)}{|\mathbf{D}|^2},$$

see (6.94), as functions of the variable $X = \lambda_1$ are displayed. It is seen from Fig. 6.9 (left) that the tunability of composite material $\widehat{T}(E)$ (6.90) as a function of electric

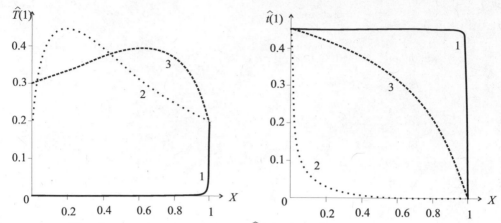

Figure 6.9. *The homogenized tunability $\widehat{T}(1)$ (6.90) as a function of electric field (a) and the homogenized tunability $\widehat{t}(1)$ (6.94) as a function of electric displacement (b) in dependence on volume fraction of components. $(\varepsilon_1,\varepsilon_2,K_1,K_2)=$ (0,01; 1; 0.2; 0) - line 1; (3; 1; 0.2; 0.2) - line 2; (1; 2; 0.2; 0.3) - line 3.*

field (which corresponds to the original definition of tunability (6.1)) can take values greater than tunability of its components, T_1 and T_2.

The tunability of composite $\widehat{t}(|\mathbf{D}|)$ (6.94) as a function of electric displacement lies between the limits pointed in (6.95). The estimate (6.95) corresponds to computation of the homogenized tunability of laminated composite materials, carried out in [350] (the computation of overall tunability of laminated composite in [350] corresponds to $\widehat{t}(|\mathbf{D}|)$, but not to $\widehat{T}(E)$).

6.4. Tunability amplification factor of composite

In this section, we introduce the notion of tunability amplification factor (TAF), which can be used to compare the homogenized tunability of composite with the tunability of materials forming the composite.

Quadratic function $\varepsilon(E)$

First, we consider a material with the quadratic function with respect to variable E, see formula (6.2). In accordance with the results presented in the previous sections, the homogenized permittivity $\widehat{\varepsilon}(E)$ of composite material in the direction Oy_3 is given by the formula (6.66) where $\widehat{\varepsilon}(0)$ and B are given by formulas (6.67) and (6.68), correspondingly.

We consider the ferroelectric-dielectric composite. Since $\mu(\mathbf{x}) = 0$ in dielectric inclusion, integration in (6.95) is carried out over the matrix only and (6.95) can be written as

$$B = \frac{\mu_f}{|Y|} \int_m |\nabla N(\mathbf{x})|^4 d\mathbf{x}, \tag{6.100}$$

("f" is indexing the ferroelectric).

The homogenized relative tunability (the relative tunability of composite material) $\widehat{T}(E)$ corresponding to (6.92) is

$$\widehat{T}(E) = \frac{B}{\widehat{\varepsilon}(0)} E^2.$$

By virtue of (6.100) and (6.67), we can write this formula in the form

$$\widehat{T}(E) = \frac{B}{\widehat{\varepsilon}(0)} = \frac{\mu_f \displaystyle\int_m |\nabla N(\mathbf{x})|^4 d\mathbf{x}}{\displaystyle\int_Y \varepsilon_0(\mathbf{x})|\nabla N(\mathbf{x})|^2 d\mathbf{x}} E^2.$$

We denote

$$T_f(E) = \frac{\mu_f}{\varepsilon_f} E^2$$

the relative tunability of pure ferroelectric material. In order to compare the homogenized tunability of composite material and tunability of pure ferroelectric material, we introduce the ratio

$$K = \frac{\widehat{T}(E)}{T_f(E)}, \tag{6.101}$$

named *the tunability amplification factor (TAF)* of composite material.

In the considered case

$$K = \frac{\widehat{T}(E)}{T_f(E)} = \frac{\dfrac{B}{\widehat{\varepsilon}(0)} E^2}{\dfrac{\mu_f}{\varepsilon_f} E^2} = \frac{\varepsilon_f \displaystyle\int_m |\nabla N(\mathbf{x})|^4 d\mathbf{x}}{\displaystyle\int_Y \varepsilon_0(\mathbf{x})|\nabla N(\mathbf{x})|^2 d\mathbf{x}}. \tag{6.102}$$

It is seen that the tunability amplification factor K (6.102) for the quadratic function $\varepsilon(E)$ does not depend on the magnitude of electric field E. This implies that the tunability amplification factor is determined by material characteristics and geometry of constitutive components of composite material. Note that the tunability amplification factor K (6.102) does not depend on the tunability of ferroelectric (tunable) component of composite. It is determined by dielectric constants of the components of composite only.

Function $\varepsilon(E)$ of general form

In general, the composite can contain n tunable components with tunabilities $\{T_1, ..., T_n\}$. We compare the homogenized tunability of the composite with the maximum tunability of those components

$$T_{max} = \max_{i=1,...,n} T_i(E).$$

and introduce the tunability amplification factor by the formula

$$K = \frac{\widehat{T}(E)}{T_f(E)} = \frac{\displaystyle\int_m \mu(\mathbf{x})|\nabla N(\mathbf{x})|^4 d\mathbf{x}}{T_{max}\displaystyle\int_Y \varepsilon_0(\mathbf{x})|\nabla N(\mathbf{x})|^2 d\mathbf{x}}. \tag{6.103}$$

It is seen that the main difficulty in the computation of the homogenized tunability is the computation of solution $N(\mathbf{x})$ of the cellular problem (6.41)–(6.43). The exact solution of this problem is known for few special cases only (for example, for laminated materials, see Section 6.3 above). Generally, this problem can be analyzed in two ways. The first way is the use of a purely mathematical approach, which usually provides us only with bounds on the homogenized characteristics. The advantage of the bounds is that they are valid for a wide class of composite materials. The bounds on the homogenized tunability were presented in Section 6.2.1. Another method is a numerical analysis of the problem. This method provides us with detailed information about a specific problem. Some results of numerical analysis were presented in Section 6.2.2 and additional results will be presented in the Section 6.5 below.

The tunability amplification factor (6.101) is a ratio of tunabilities of composite and its tunable component. Another characteristic, relative tunability amplification factor

$$K_1 = K - 1 = \frac{\widehat{T}(E) - T_f(E)}{T_f(E)},$$

describes the relative change of tunability amplification factor. It is a visible characteristic of increasing of tunability.

6.5. Numerical design of composites possessing high tunability amplification factor

In the previous sections (see also [191, 331, 329]), we demonstrated that ferroelectric–dielectric composite materials with the homogenized tunability amplification factor

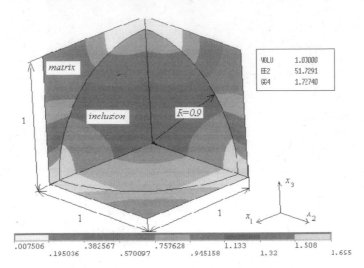

Figure 6.10. *Distribution of electric field $E = |\nabla\varphi(\mathbf{x})|$ in the cell containing spherical inclusion and results of computations (boxed): VOLU=$|Y|$, EE2=$\widehat{\varepsilon}(0)$, GG4=B.*

slightly greater than 1 (namely, $1.04 - 1.38$) exist. The numerical prototypes (designs integrated with the finite element method tools) of microstructures demonstrating tunability amplification factor K significantly greater than 1 were announced first in [188]. Below we present some designs from [177, 189].

6.5.1. Ferroelectric–dielectric composite materials

Typical values of characteristics of materials under consideration are presented in Table 6.1.

We analyze the values of the homogenized characteristics of composite as compared with the corresponding characteristics of pure materials (usually, pure ferroelectric), i.e., the relative change of characteristics. The relative change of characteristic is determined by relative values of material characteristics of composite material components. Consequently, we can use relative characteristics of materials in the computations. For example, the values of tunability 1 and 10 in our numerical computations may correspond to actual values of tunability 0.05 and 0.5 or 0.01 and 0.1.

The spherical inclusion

We consider a composite material consisting of a ferroelectric nonlinear matrix with a periodic system of dielectric linear spherical inclusions. The periodicity cell of the composite material is a cube with a sphere placed in its center. Due to the symmetry of the problem, we can solve the cellular problem for $\frac{1}{4}$ of the periodicity cell displayed in Fig. 6.10. The size of $\frac{1}{4}$ of the periodicity cell is noted $1\times1\times1$, the radius of the dielectric inclusion is noted $R = 0.9$. The volume fraction of dielectric is $q = 0.63$. In numerical computation, the permittivity of the ferroelectric matrix and dielectric inclusion are $\varepsilon_f = 100$ and $\varepsilon_d = 1$, correspondingly. The relative tunability is $T_f(E) = 0.01E^2$ for ferroelectric and $T_d = 0$ for dielectric.

Results of our numerical computation are presented in Fig. 6.10. The picture shows the distribution of the electric field $E(\mathbf{x}) = |\nabla N(\mathbf{x})|$ (the overall electric field is applied in the direction of Ox_3-axis and has the form $E = (0, 0, 1)$). The numbers in the box are results of numerical computations carried out using ANSYS software.

Numerically computed coefficients (6.67) and (6.68) are the following:

$$\widehat{\varepsilon}(0) = \frac{1}{|Y|} \int_Y \varepsilon_0(\mathbf{x})|\nabla N(\mathbf{x})|^2 dx = 52, \; B = \frac{1}{|Y|} \int_Y \mu(\mathbf{x})|\nabla N(\mathbf{x})|^4 dx = 1.7.$$

The first coefficient is equal to the homogenized permittivity of the composite material. The homogenized relative tunability of composite material is

$$\widehat{T}(E) = \frac{B}{\widehat{\varepsilon}(0)} E^2 = \frac{1.7}{52} \approx 0.033E^2.$$

The tunability amplification factor is

$$K = \frac{\widehat{T}(E)}{T_f(E)} \approx \frac{0.033}{0.01} = 3.3 = 330\% \; (K_1 = 2.3 = 230\%).$$

Due to the symmetry of the periodicity cell, see Fig. 6.10, the composite is macroisotropic and it can be described with the scalar homogenized permittivity $\widehat{\varepsilon}(0)$.

The composite material under consideration has relative tunability three times greater than the relative tunability of ferroelectric and permittivity two times smaller than the permittivity of ferroelectric. The design can be estimated as simple as compared with the designs used in modern technologies of composite materials [169].

The slot tunability amplifier

Now, we present designs significantly (more than 30 times) increasing the homogenized tunability and describe the microscopic mechanisms of the increasing phenomenon of the homogenized tunability in composite materials.

In the considered case, composite is macro anisotropic. It brings no problem in numerical analysis except that it is necessary to pay attention to direction of electric field and appropriate indexing of the homogenized characteristics, see Section 6.2.

In our numerical computations, the permittivity of ferroelectric and dielectric components are $\varepsilon_f = 1000$ and $\varepsilon_d = 1$, correspondingly. The tunabilities are $T_f = 0.001E^2$ and $T_d - 0$, correspondingly.

The periodicity cell is right-angle 2×10 occupied by two ferroelectric right-angle inclusions, see Fig. 6.11. The length of the inclusions is 0.5 and the thickness of

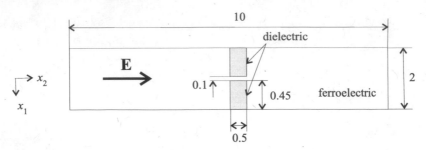

Figure 6.11. *Drawing of the periodicity cell with inclusion.*

the inclusions is 0.95. Thus, a slot of thickness 0.1 between the left and right inclusions exists, see Fig. 6.11. The volume fraction of dielectric is $q = 0.0475$. When periodically repeated in vertical direction, the periodicity cell generates a two-dimensional sandwich structure: two ferroelectric layers with a perforated dielectric layer between them. Direction of the overall electric field $\mathbf{E}$ is shown in Fig. 6.11 (it is parallel to Ox_2-axis and has the form $\mathbf{E} = (0, 1)$). The electric field flows through the slots in the dielectric layer. As a result, a large gradient of electric potential arises in the slots occupied by the ferroelectric layers, see Fig. 6.12.

Numerically computed coefficients (6.67) and (6.68) are the following:

$$\int_Y \varepsilon_0(\mathbf{x})|\nabla N(\mathbf{x})|^2 d\mathbf{x} = 8920, \quad \int_Y \mu(\mathbf{x})|\nabla N(\mathbf{x})|^4 d\mathbf{x} = 318.$$

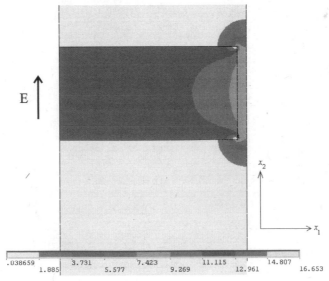

Figure 6.12. *Numerically computed distribution of electric field $E(\mathbf{x}) = |\nabla \varphi(\mathbf{x})|$ in the neighbor of slot.*

The homogenized permittivity $\widehat{\varepsilon}_2(0)$ of the composite material in the direction Ox_2 is equal to the first integral divided by the volume of the periodicity cell $|Y| = 20$:

$$\widehat{\varepsilon}_2(0) = \frac{8920}{20} = 446$$

and

$$B = \frac{318}{20} = 15.9.$$

The homogenized relative tunability of the composite material is

$$\widehat{T}(E) = \frac{B}{\widehat{\varepsilon}_2(0)} E^2 = \frac{15.9}{446} E^2 \approx 0.036 E^2.$$

The tunability amplification factor is

$$K \approx \frac{0.036}{0.001} = 36 = 3600\% \ (K_1 = 35 = 3500\%).$$

Our numerical analysis demonstrated that the homogenized characteristics of the composite structure under consideration strongly depend on the variation of geometrical parameters of the structure (the length and the width of the slot). In order to find the structure possessing a tunability amplification factor as high as 35, the authors solved problems for various geometrical parameters of the slot (in fact, made iterations of shape design procedure). Note that the random choice of the size of the slot gave a value of the tunability amplification factor K slightly greater than 1 or less than 1 (i.e., slow increasing or decreasing of the homogenized tunability of composite as compared with tunability of pure ferroelectric). This means that random search is not an effective method to design highly tunable structures.

The hole tunability amplifier

The three-dimensional analog of a slot is a hole, see Fig. 6.13. The tunability amplification factor of three-dimensional structure with a hole was investigated. In

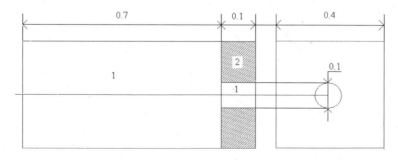

Figure 6.13. *Drawing of the periodicity cell (1 – ferroelectric, 2 – dielectric).*

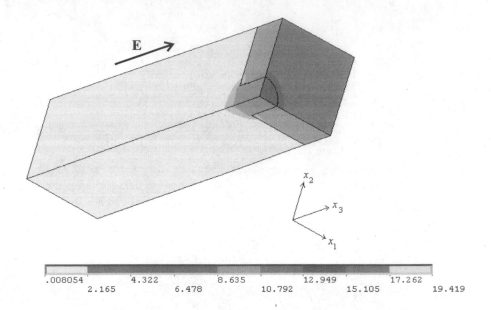

Figure 6.14. *Distribution of $|\nabla N(\mathbf{x})|^2$ in the periodicity cell.*

numerical computation, we take $\varepsilon_f = 1000$ and $\varepsilon_d = 1$, correspondingly. The tunability of ferroelectric is $T_f = 0.001E^2$ and tunability of dielectric is $T_d = 0$. The dimensions of the elements forming the periodicity cell are shown in Fig. 6.14. The overall electric field is parallel to Ox_3–axis and has the form $\mathbf{E} = (0,0,1)$.

The domains indexed by 1 are occupied by ferroelectric. The domain indexed by 2 is occupied by dielectric. The domain occupied by dielectric has a form of layer with the cylindrical hole. When periodically repeated, this periodicity cell generates a three-dimensional sandwich structure: two layers of ferroelectric with a perforated dielectric layer between them, which is a three-dimensional analog of the two-dimensional structure considered in the previous section. The volume fraction of dielectric for the cell shown in Fig. 6.13 is $q \approx 0.09$.

Now, the electric field flows through the hole in the dielectric layer and large gradient of electric potential arises in the hole occupied by ferroelectric, see Fig. 6.14.

Numerically computed values of the integrals are the following:

$$\int_Y \varepsilon_0(\mathbf{x})|\nabla N(\mathbf{x})|^2 d\mathbf{x} = 48, \quad \int_Y \mu(\mathbf{x})|\nabla N(\mathbf{x})|^4 d\mathbf{x} = 1.12.$$

The homogenized permittivity $\widehat{\varepsilon}_3(0)$ in the direction Ox_3 and homogenized tunability B of the composite material are equal to the integrals divided by $|Y| = 0.1446$:

$$\widehat{\varepsilon}_3(0) = \frac{48}{0.144} \approx 333$$

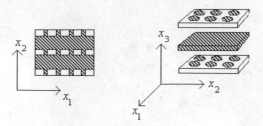

Figure 6.15. *Two- and three-dimensional laminated composite materials with perforated layers.*

and

$$B = \frac{1.12}{0.144} \approx 7.78.$$

The relative tunability of composite material is

$$\widehat{T}(E) = \frac{7.78}{333} E^2 \approx 0.023 E^2.$$

The corresponding tunability amplification factor is

$$K \approx \frac{0.023}{0.001} \approx 23 = 2300\% \ (K_1 = 22 = 2200\%).$$

6.5.2. Isotropic composite materials

The last two designs correspond to the sandwich (laminated) composite materials with perforated middle layer, see Fig. 6.15.

The laminated composite materials are anisotropic. Using the designs obtained, one can design *isotropic* material with the tunability amplification factor close to the tunability amplification factor of the corresponding laminated composite material.

We consider the cellular structures shown in Fig. 6.16. They can be described as two- and three-dimensional boxes with perforated walls (the walls are assumed to be

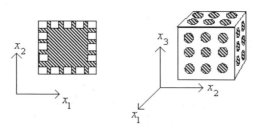

Figure 6.16. *Cellular cells (microstructure) boxes with perforated walls (the hatched domains are occupied by ferroelectric material).*

parallel to the coordinate planes). If an electric field is parallel to a coordinate axis, the walls parallel to this axis influence the field slowly. As far as the wall perpendicular to the direction of the field, it works similarly to the laminated composite material described above. Then the structures shown in Fig. 6.16 have the tunability amplification factor approximately equal to that of the sandwich structure. Due to the symmetry of the boxes, the homogenized materials with the periodicity cells shown in Fig. 6.16 are isotropic.

6.5.3. Ferroelectric–ferroelectric composite material

The permittivity of ferroelectric varies in the range 500 to 2000. This variation of permittivity allows development of slow-contrast ferroelectric–ferroelectric composite materials, which can demonstrate the property of concentration of electric field and possess the effective tunability greater than the tunabilities of those components.

   We consider a ferroelectric matrix with a periodic system of ferroelectric spherical inclusions (an analog of ferroelectric-dielectric composite material considered in Section 6.5.1). The dimensions of $\dfrac{1}{4}$ of the periodicity cell is $1\times1\times1$, the radius of inclusion is 0.9, see Fig. 6.17. The volume fraction of ferroelectric 2 is $q = 0.63$.

   In our numerical computation, we take the permittivity of the components of the composite material $\varepsilon_{f1} = 1$, $\varepsilon_{f2} = 5$ and tunability of the components $T_{f1} = E^2$, $T_{f2} = 0.2E^2$ (1 and 2 are indices of the ferroelectric materials, the parameter μ is equal to 1 for both materials). As above, we use here the relative values of characteristics of components.

   Numerically computed coefficients (6.67) and (6.68) are the following:

$$\widehat{\varepsilon}(0) = \frac{1}{|Y|} \int\limits_{Y} \varepsilon_0(\mathbf{x})|\nabla N(\mathbf{x})|^2 d\mathbf{x} = 1.9, \; B = \frac{1}{|Y|} \int\limits_{Y} \mu(\mathbf{x})|\nabla N(\mathbf{x})|^4 d\mathbf{x} = 5.1.$$

   The effective relative tunability of composite material is

$$\widehat{T}(E) = \frac{5.1}{1.9} \approx 2.6E^2.$$

   The tunability amplification factor (computed with respect to the maximum value of the tunability of components of composite material) is

$$K = \frac{2.6}{1} = 2.6 = 260\% \; (K_1 = 1.6 = 160\%).$$

   Note that the tunability amplification factor of the ferroelectric-ferroelectric composite material under consideration is not very different from the tunability amplification factor of the ferroelectric-dielectric composite material considered in section 5.1 ($K_1 = 160\%$ and $K_1 = 230\%$, correspondingly) while the difference in the contrast of components is significant ($\dfrac{\varepsilon_{f2}}{\varepsilon_{f1}} = 5$ and $\dfrac{\varepsilon_f}{\varepsilon_d} = 100$, correspondingly).

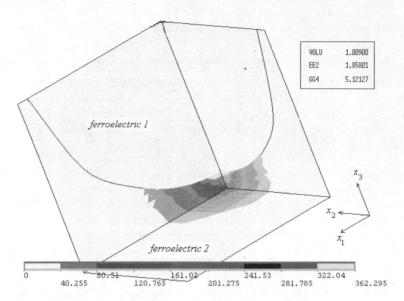

Figure 6.17. *Distribution of $|\nabla N(\mathbf{x})|^4$ in the periodicity cell of composite ferroelectric matrix – ferroelectric spherical inclusion and results of computations (boxed):* $VOLU = |Y|$, $EE2 = A_3$, $GG4 = B$.

In the case under consideration, the increase of tunability is a result of the phenomenon of the concentration of electric field in the form of an energy channel, see Chapters 2 and 3 (compare with Fig. 2.6). The energy channel is clearly seen in Fig. 6.17. We conclude that the relatively low contrast of component properties can lead to a marked increase of the effective tunability.

6.6. The problem of maximum value for the homogenized tunability amplification factor

In [350] the conjecture that the homogenized tunability of ferroelectric–dielectric composite can take value greater than the tunability of its ferroelectric component was formulated. In terms of the tunability amplification factor it means that the tunability amplification factor can take value greater than 1. Later, composite material with the homogenized tunability amplification factor slightly greater than 1 was designed [191]. In the previous sections, we have demonstrated that the homogenized tunability amplification factor can take values as large as 30 for practically realizable structures. The following question arises: *is the value of the tunability amplification factor K bounded above or K can take arbitrary large values?*

To answer this question, we consider planar model of composite material of periodic structure, see Fig. 2.1, with nonlinear matrix ($\mu_m \neq 0$ in matrix) filled with linear inclusions with high dielectric constant

$$\mu_I(\mathbf{x}) = 0, \ \varepsilon_I(\mathbf{x}) \gg \varepsilon_m \tag{6.104}$$

and estimate the homogenized tunability amplification factor K from below for this composite material. As above, index I corresponds to inclusion, index m corresponds to matrix.

The inequality (6.104) means that the inclusions have high transport property as compared with the transport property of matrix. Materials of such kind were analyzed numerically in Chapter 2.

From the pictures presented in Fig. 2.12 it follows that in the disk the value of electric field $|\nabla N(\mathbf{x})|$ and the density of local energy are negligibly small as compared with the corresponding values in the matrix. Note that the flux $\varepsilon(\mathbf{x})|\nabla N(\mathbf{x})|$ is not small in the disks, see Fig. 2.12.

We neglect the energy stored in the disk and write approximate equality

$$\int_Y \varepsilon(\mathbf{x})|\nabla N(\mathbf{x})|^2 dx \approx \varepsilon_m \int_m |\nabla N(\mathbf{x})|^2 dx. \tag{6.105}$$

Due to $\mu(\mathbf{x}) = 0$ in the inclusions, we have

$$\int_Y \mu(\mathbf{x})|\nabla N(\mathbf{x})|^4 dx = \mu_m \int_m |\nabla N(\mathbf{x})|^4 dx, \tag{6.106}$$

where μ_f is value of $\mu(\mathbf{x})$ in the matrix. It is seen that the problem of possible value of tunability amplification factor is determined by possible values of functionals I_2 and I_4 and it is densely related to the truncated moments problem [173].

We use inequality (6.77) and write

$$\int_m |\nabla N(\mathbf{x})|^2 dx \leq \left(\int_m |\nabla N(\mathbf{x})|^4 dx \right)^{1/2} |m|^{1/2}, \tag{6.107}$$

where $|m|$ means the area of the domain occupied by matrix.

By virtue of (6.67)–(6.69) and (6.105)–(6.107), we can write the following estimate with respect to the tunability amplification factor:

$$\widehat{T}(E) = \frac{\int_Y \mu(\mathbf{x})|\nabla N(\mathbf{x})|^4 dx}{\int_Y \varepsilon_0(\mathbf{x})|\nabla N(\mathbf{x})|^2 dx} E^2 = \frac{\mu_m}{\varepsilon_m} \cdot \frac{\int_m |\nabla N(\mathbf{x})|^4 dx}{\int_m |\nabla N(\mathbf{x})|^2 dx} E^2 \geq \tag{6.108}$$

$$\geq \frac{\mu_m}{\varepsilon_m} \cdot \frac{\int_m |\nabla N(\mathbf{x})|^2 dx}{|m|} E^2 = T_m(E) \cdot \frac{\int_m |\nabla N(\mathbf{x})|^2 dx}{|m|},$$

where

$$T_m(E) = \frac{\mu_m}{\varepsilon_m} E^2$$

is tunability of the material of the matrix.

Using (6.108), we obtain the following low-sided estimate with respect to the amplification factor of the composite:

$$K = \frac{\widehat{T}(E)}{T_m(E)} \geq \frac{\int_m |\nabla N(\mathbf{x})|^2 d\mathbf{x}}{|m|}. \tag{6.109}$$

Now, we obtain low estimate for the right-hand side of (6.109). The area $|m|$ of the domain occupied by the matrix is determined by the size of the disks. For densely packed disks it is about 0.4 to 0.5 of total area of the domain occupied by composite.

The integral $\int_m |\nabla N(\mathbf{x})|^2 d\mathbf{x}$ is equal to the energy accumulated in the matrix divided by ε_m.

As it was demonstrated in Chapters 2 and 3 for high-contrast highly filled composites, almost all energy is accumulated in the matrix (namely, in the energy channels between the neighbor particles, see Fig. 2.12) and it can be approximated by the energy of the network model, which has the order of $\pi\sqrt{\dfrac{R}{\delta}}$, where R is the radius of the disk (inclusion) and δ is the distance between neighbor disks, see Chapter 3. Then the tunability amplification factor of densely packed high–contrast composite material has the order of $\dfrac{\pi}{\varepsilon_m}\sqrt{\dfrac{R}{\delta}}$ and increases *infinitely* as δ becomes small. Note that this is a conclusion of theoretical nature. For particle-filled composite materials produced by using traditional technologies

$$0.001 < \frac{\delta}{R} < 0.01$$

and the tunability amplification factor is estimated as

$$\frac{30}{\varepsilon_m} < K < \frac{100}{\varepsilon_m}.$$

Since ε_m is a large number, K is not large for particle-filled composite materials.

6.7. What determines the effective characteristics of composites?

In Table 6.3 we display some characteristics of the composite materials considered above. It is seen from Table 6.3 that there is no correlation between the volume fraction of components and the tunability amplification factor for composite materials. In addition, there is no correlation between the volume fraction of components

Table 6.3. *The Homogenized Characteristics of Ferroelectric–Dielectric Composites for Various Values of the Volume Fraction q of Dielectric Component ($\dfrac{\widehat{\varepsilon}(0)}{\varepsilon_f}$ – the Homogenized Permittivity as Compared with the Permittivity of Ferroelectric Component).*

The design	q	$\dfrac{\widehat{\varepsilon}(0)}{\varepsilon_f}$	K	$K_1\%$
Fig. 6.12	0.05	0.446	36	3500
Table 6.2	0.05	0.91	1.06	6
Fig. 6.14	0.09	0.012	23	2200
Formula (6.7)	0.09	0.85	1.018	1.8
Table 6.2	0.09	0.84	1.1	10
Fig. 6.10	0.63	0.517	3.3	230

and the relative homogenized permittivity $\dfrac{\widehat{\varepsilon}(0)}{\varepsilon_f}$ for these composite materials (here $\varepsilon_f = \varepsilon_f(0)$ denotes the permittivity of ferroelectric matrix near zero). We recall that $\widehat{\varepsilon}(0)$ is equal to the homogenized permittivity of linear material.

We present one more example of the absence of correlation between the volume fractions of components and homogenized permittivity and tunability amplification factor of composites. In Section 6.6, an example of material with a large tunability amplification factor was presented. The tunability amplification factor of composite with densely packed circular inclusions was estimated as $\dfrac{\pi}{\varepsilon_f}\sqrt{\dfrac{R}{\delta}}$, where δ is the characteristic distance between the inclusions. Small variation of δ leads to small variation of the volume fraction q. At the same time small variation of δ can lead to significant increasing of the quantity $\dfrac{\pi}{\varepsilon_f}\sqrt{\dfrac{R}{\delta}}$ (thus, homogenized permittivity and the tunability amplification factor).

As a result of our analysis of the homogenization characteristics, our answer to the question in the title of this section is as follows: *"The local topology and geometry of components of composites, not those volume fractions, are the main factors which control the effective properties of composite material."* This conclusion is valid for both linear and nonlinear composite materials.

Correlation between the volume fraction of components and the homogenized characteristics of composite material may take place for materials with a specific type of microtopology and microgeometry (for example, correlation between the volume fraction of components and the homogenized permittivity for takes place for particles filled composite [92, 128, 227, 348], see also Table 6.1).

6.8. The difference between design problems of tunable composites in the cases of weak and strong fields

We have considered some aspects of design problems for weak and strong electric fields. The relationship between the constitutive low (6.4) valid for arbitrary value of electric field and approximation (6.2) is illustrated by Fig. 6.18. The graph of the function (6.2) must be the tangential parabola for the graph of the function (6.4)). While value of the local electric field $E = |\mathbf{E}|$ remains in the interval

$$0 \leq E \leq E_{small},$$

near zero, we can use the approximation (6.2). But for $E > E_{small}$, the approximation (6.2) cannot be accepted (for example, because for large E, the approximation (6.2) takes nonrealistic negative value). Note that value E_{small} can be determined with no problem for any specific curve ε–E. It can be done for curve ε–E obtained from theoretical analysis as well obtained from experiment.

Note that exact solutions obtained for strong field for laminated materials in [177] demonstrates coincidence of the homogenized characteristics computed using constitutive lows (6.2) and (6.4) for weak field (of course, if graph of the function (6.2) is the tangential parabola for the graph of the function (6.4)).

The exact solutions corresponding to strong fields were obtained for laminated composite materials only (this is normal situation, because solution of strongly nonlinear problem is difficult task). The solutions demonstrate difference between design problem for weak and strong fields. To explain this difference in the clearest way, we introduce the notion of *control resource* of nonlinear material. We call control recourse of nonlinear material the interval of possible values of a characteristic of the material when the field takes all possible values (usually, from zero to infinity).

For example, if dielectric constant of material is described by the equation (see

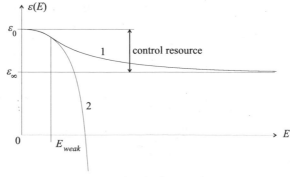

Figure 6.18. *Graph of the function (6.4) (line 1) and graph of the corresponding tangential parabola, function (6.2) (line 2).*

equation (6.3))

$$\varepsilon(E) = \varepsilon_\infty + \frac{\varepsilon_0 - \varepsilon_\infty}{1 + bE^2},$$

the control resource is the interval $[\varepsilon_0, \varepsilon_\infty]$, see Fig. 6.18.

The control resource is a material characteristic (it is the property of material and does not depend on the fields applied to the material).

Now, we make the following note. In the case of weak average field we can amplify and concentrate local fields in specific domain(s). It is not a simple problem, it is especially difficult to understand in which domain(s) it is necessary to concentrate the field. Here we do not discuss the methods of solution of this problem. We now accept that the problem can be solved in one way or another (examples of solutions were presented in the previous sections). Then, all control resources of components of composite material are available for the designer in the case of a weak field. In the case of a strong field, the control resources are the same. But the amplification and concentration of local field do not bring substantial changes in the behavior of composite (the local field is usually large if overall field is large, in other words, we can significantly amplify but cannot significantly weaken the local field as compared with the overall field). As a result, we conclude that the homogenized (overall) characteristics are structural design sensitive for weak fields and they become less sensitive for strong fields.

If we analyze tunability amplification factor, it means that we can design structures possessing large TAF for the case corresponding to a weak field. At the same time, it is more difficult (if possible) to design structures possessing large TAF for the case corresponding to a strong field. To illustrate this conclusion, we present the following example. Let $\varepsilon_0 = 1500$, $\varepsilon_\infty = 1000$ and dielectric constant of dielectric inclusion $\varepsilon_d = 5$. For near zero overall field the homogenized dielectric constant $\widehat{\varepsilon}(0)$ of the composite is equal to the homogenized dielectric constant of linear composite with the dielectric constant of the matrix $\varepsilon_0 = 1500$ and dielectric constant of dielectric inclusion $\varepsilon_d = 5$. It follows, e.g., from formula (6.66). For strong overall electric field the homogenized dielectric constant $\widehat{\varepsilon}(\infty)$ of the composite is equal to the homogenized dielectric constant of linear composite with the dielectric constant of the matrix $\varepsilon_0 = 1000$ and dielectric constant of dielectric inclusion $\varepsilon_d = 5$. Then for strong overall electric field the homogenized tunability value is about

$$\widehat{T}_{max} = \frac{\widehat{\varepsilon}(0) - \widehat{\varepsilon}(\infty)}{\widehat{\varepsilon}(0)}.$$

An example

We present the result of the numerical computation of $\widehat{T}_{max}$ for the slot tunability amplifier displayed in Fig. 6.12 . We consider nonlinear material with the dielectric constant $\varepsilon(E)$ given by (6.4) for

$$\varepsilon_0 = 1000, \ \varepsilon_\infty = 700.$$

Table 6.4. *The Homogenized Permittivity of Composite with Linear Components.*

dielectric constant of matrix	$\widehat{\varepsilon}$
1000	222.9
700	157.1

The dielectric constant of the inclusion is assumed to be equal to 1.

The homogenized permittivity of composite in zero $\widehat{\varepsilon}(0)$ can be computed by solution of linear homogenization problem with dielectric constant of matrix equal to 1000. The homogenized permittivity of composite "at the infinity" $\widehat{\varepsilon}(\infty)$ can be computed by solution of linear homogenization problem with dielectric constant of matrix equal to 700. In Table 6.4 the results of numerical computation of the homogenized permittivities are presented.

The tunabilities for pure ferroelectric and composite are presented in Table 6.5. It is seen from Table 6.4 that tunability of composite corresponding to strong electric field is practically equal to the tunability of pure ferroelectric.

It is a question: Which overall electric field E may be accepted as strong field?

For pure ferroelectric, we accept a field as strong if $bE^2 > 100$. In this case the relative difference between ε_∞ and $\varepsilon(E)$ is less than 1%.

For composite material, amplification of overall electric field E generates nonuniform electric field with potential $\varphi(\mathbf{x})$. The overall electric field E may be accepted as strong if the local field $\nabla\varphi(\mathbf{x})$ can be accepted as strong for any $\mathbf{x}$. It may be done if the local electric field $\nabla\varphi(\mathbf{x})$ satisfies the condition $b|\nabla\varphi(\mathbf{x})|^2 > 100$, see notations in (6.4).

In the case under consideration $|\nabla\varphi(\mathbf{x})| > 0.038659E$ (in accordance with numerical computations, 0.038659 is the minimal value of the electric field in Fig. 6.12). Thus, we may accept the overall electric field E as strong if

$$b(0.038659E)^2 > 100 \qquad (6.110)$$

or

$$E > \sqrt{\frac{100}{b}} \cdot \frac{1}{0.038659}.$$

We note that $\sqrt{\frac{100}{b}}$ was accepted as a typical value of strong electric field in pure ferroelectric,

$$\frac{1}{0.038659} \approx 25.9.$$

Table 6.5. *Tunabilities.*

for pure ferroelectric	for composite
0.3	0.2952

The value E determined by (6.110) is approximately 26 times greater than the value of electric field, which we have accepted as strong for pure ferroelectric.

6.9. Numerical analysis of tunability of composite in strong fields

In this section, we present the solution of the homogenization problem when the function $\varepsilon(E)$ has the form (6.3). We arrive at a strongly nonlinear cellular problem, which can be solved by using the numerical method developed in [190].

6.9.1. Numerical method for analysis of the problem

For the sake of simplicity, we consider a planar problem. The periodicity cell of composite under consideration is shown in Fig. 6.19.

We assume that the composite is subjected to the electric field with the overall (average) strength E in the direction Ox_2 and denote $\varphi(\mathbf{x})$ the local potential of electric field.

In the case under consideration by virtue of the periodicity the problem of electrostatics in the periodic array is reduced to the following problem on the periodicity cell:

$$\operatorname{div}(\varepsilon(\mathbf{x}, |\nabla\varphi(\mathbf{x})|)\nabla\varphi(\mathbf{x})) = 0. \tag{6.111}$$

By virtue of the symmetry of the periodicity cell, the following boundary conditions take place:

$$\varphi(x_1, -0.5) = -E/2, \quad \varphi(x_1, 0.5) = E/2, \tag{6.112}$$

$$\frac{\partial\varphi}{\partial\mathbf{n}}(-0.5, x_2) = \frac{\partial\varphi}{\partial\mathbf{n}}(0.5, x_2) = 0. \tag{6.113}$$

For the approximation (6.3), the local dielectric constant $\varepsilon(\mathbf{x}, |\nabla\varphi(\mathbf{x})|)$ is given

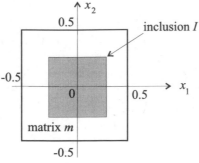

Figure 6.19. *Periodicity cell Y.*

by the formula

$$\varepsilon(\mathbf{x}, |\nabla\varphi(\mathbf{x})|) = \begin{cases} 1 \text{ if } \mathbf{x} \in I, \\ \varepsilon_\infty + \dfrac{\varepsilon_0 - \varepsilon_\infty}{1 + b|\nabla\varphi|^2} \text{ if } \mathbf{x} \in m = Y\backslash I. \end{cases} \quad (6.114)$$

We recall that multiplying the function $\varepsilon(\mathbf{x}, |\nabla\varphi(\mathbf{x})|)$ by any nonzero number, does not change the solution of the problem (6.111) and (6.112). Thus, the assumption that the dielectric constant of the matrix is equal to unity does not restrict the generality of our consideration.

The problem (6.111)–(6.114) is classified as strongly nonlinear problem [122] because its coefficient depends on the gradient of unknown function. Such kind problem cannot be solved using standard software. We describe a numerical method, which can be used to solve the problem under consideration.

First, we demonstrate that there exists an integral functional $\int_Y q(|\nabla\varphi(\mathbf{x})|)d\mathbf{x}$, whose minimizing function is the solution to the problem (6.111)–(6.114).

We multiply both sides of the equation (6.111) by a trial function $-\delta\varphi(\mathbf{x})$ and integrate by part over the domain Y. As a result, we obtain

$$-\int_Y \delta\varphi(\mathbf{x})\mathrm{div}(\varepsilon(|\nabla\varphi(\mathbf{x})|)\nabla\varphi(\mathbf{x}))d\mathbf{x} = \int_Y \varepsilon(|\nabla\varphi(\mathbf{x})|)\nabla\delta\varphi(\mathbf{x})\nabla\varphi(\mathbf{x})d\mathbf{x}. \quad (6.115)$$

We note that the integral over the boundary of the periodicity cell Y

$$\int_{\partial Y} \varepsilon(|\nabla\varphi(\mathbf{x})|)\delta\varphi\frac{\partial\varphi}{\partial\mathbf{n}}(\mathbf{x})d\mathbf{x} = 0$$

because on the left and right boundaries of the domain $\frac{\partial\varphi}{\partial\mathbf{n}}(\mathbf{x}) = 0$ in accordance with (6.113) and on the top and bottom boundaries $\delta\varphi(\mathbf{x}) = 0$ in accordance with (6.112).

The variation of the functional $\int_Y q(|\nabla\varphi(\mathbf{x})|)d\mathbf{x}$ is

$$\delta\int_Y q(|\nabla\varphi(\mathbf{x})|)d\mathbf{x} = \int_Y q'(|\nabla\varphi(\mathbf{x})|)\frac{\nabla\varphi(\mathbf{x})\nabla\delta\varphi}{|\nabla\varphi(\mathbf{x})|}d\mathbf{x}. \quad (6.116)$$

The integrands in the right-hand sides of (6.115) and (6.116) are equal in the domain m occupied by matrix if the following equation is satisfied:

$$\varepsilon(|\nabla\varphi(\mathbf{x})|) = q'(|\nabla\varphi(\mathbf{x})|)\frac{1}{|\nabla\varphi(\mathbf{x})|}. \quad (6.117)$$

Since all the functions forming the equation (6.117) depend on the same variable $|\nabla\varphi(\mathbf{x})|$ only, we denote $z = |\nabla\varphi(\mathbf{x})|$ and obtain from (6.117) the following ordinary differential equation:

$$q'(z) = \varepsilon(z)z. \quad (6.118)$$

For the function (6.3), the equation (6.118) takes the form

$$q'(z) = \left(\varepsilon_\infty + \frac{\varepsilon_0 - \varepsilon_\infty}{1 + b\,|z|^2} \right) z. \tag{6.119}$$

From the physical meaning of the function $q(z)$ it follows that

$$q(0) = 0. \tag{6.120}$$

Solving the problem (6.119), (6.120), we find the function $q(z)$ in the matrix m occupied by nonlinear material (ferroelectric):

$$q(z) = \frac{\varepsilon_\infty}{2} z^2 + \frac{\varepsilon_0 - \varepsilon_\infty}{2b} \ln(1 + bz^2).$$

In the inclusion I occupied by linear material (dielectric) $q(z) = z^2/2$. Thus, the required functional has the form

$$\mathcal{F}(\varphi) = \frac{1}{2} \int_I |\nabla\varphi(\mathbf{x})|^2 d\mathbf{x} +$$

$$+ \int_m \left(\frac{\varepsilon_\infty}{2} |\nabla\varphi(\mathbf{x})|^2 + \frac{\varepsilon_0 - \varepsilon_\infty}{2b} \ln(1 + b|\nabla\varphi(\mathbf{x})|)^2 \right) d\mathbf{x}.$$

Solution of the problem

$$\mathcal{F}(\varphi) \to \min \tag{6.121}$$

under the conditions (6.112) is equivalent to solution of the boundary value problem (6.111)–(6.113).

Now, we can construct a discrete approximation of the minimization problem (6.121). It can be done in various ways. We use rightangle uniform mesh with the distance between the neighbor nodes equal to h. The approximation for gradient is taken in the form

$$|\nabla\varphi|^2 \approx \frac{(\varphi_{i,j+1} - \varphi_{i,j})^2}{h^2} + \frac{(\varphi_{i+1,j} - \varphi_{i,j})^2}{h^2},$$

where $\varphi_{i,j}$ denotes value of the function $\varphi(\mathbf{x})$ in the node with the coordinates $x_1 = ih$, $x_2 = jh$ $(i, j = 1, \ldots, N = [1/h])$.

After that, we can write the following approximation for the functional $\mathcal{F}(\varphi)$:

$$\mathcal{F}(\varphi) = \int_I \frac{1}{2} |\nabla\varphi(\mathbf{x})|^2 \, d\mathbf{x} + \tag{6.122}$$

$$+ \int_m \frac{1}{2k} \left((\varepsilon_0 - \varepsilon_\infty) \ln(1 + k\,|\nabla\varphi(\mathbf{x})|^2) + \varepsilon_\infty\,|\nabla\varphi(\mathbf{x})|^2 \right) d\mathbf{x} \approx$$

$$\approx \frac{\varepsilon_0 - \varepsilon_\infty}{2k} \sum_{ih,\,jh\in m} \ln\left(1 + \frac{(\varphi_{i,j+1} - \varphi_{i,j})^2}{h^2} + \frac{(\varphi_{i+1,j} - \varphi_{i,j})^2}{h^2} \right) h^2 +$$

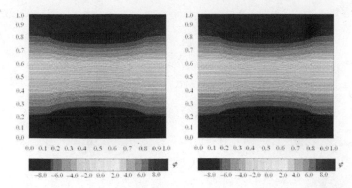

Figure 6.20. *Distribution of local potential over the periodicity cell: solution of non-linear problem (left) and solution of the corresponding linear problem (right) for E = 20.*

$$+\frac{\varepsilon_\infty}{2} \sum_{ih,\,jh\in m} \left(\frac{(\varphi_{i,j+1} - \varphi_{i,j})^2}{h^2} + \frac{(\varphi_{i+1,j} - \varphi_{i,j})^2}{h^2}\right) h^2 +$$

$$+\frac{1}{2} \sum_{ih,\,jh\in I} \left(\frac{(\varphi_{i,j+1} - \varphi_{i,j})^2}{h^2} + \frac{(\varphi_{i+1,j} - \varphi_{i,j})^2}{h^2}\right) h^2 = F(\varphi_{ij})$$

We arrive at the finite dimensional minimization problem

$$F(\varphi_{ij}) \to \min \tag{6.123}$$

with conditions

$$\varphi_{i1} = -E/2,\ \varphi_{iN} = E/2\ (i = 1, ..., N). \tag{6.124}$$

The problem (6.123) and (6.124) was solved by using the method of gradient descent (see for details [190]). To test the program, we use the property of the problem (6.111)–(6.112) described in Section 6.8. In accordance with this property, for strong field, solution of the nonlinear problem (6.111)–(6.112) coincides with solution of the corresponding linear problem, namely, the problem with dielectric constant given by the formula (compare with (6.114))

$$\varepsilon(\mathbf{x}) = \begin{cases} 1 \text{ if } \mathbf{x} \in I, \\ \varepsilon_\infty \text{ if } \mathbf{x} \in m = Y\backslash I \end{cases}. \tag{6.125}$$

The nonlinear problem was solved for the following values of the parameters:

$$\varepsilon_0 = 150, \varepsilon_\infty = 100, b = 0.5$$

For $b = 0.5$ and $E = 20$, the quantity $bE^2 = 200$. Thus the electric field $E = 20$ can be accepted as strong (see formula (6.3) and Section 6.8) for the problem under consideration.

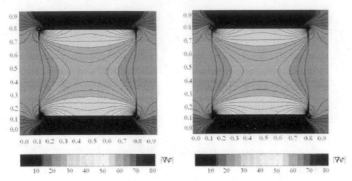

Figure 6.21. *Distribution of local driving force over the periodicity cell: solution of nonlinear problem (left) and solution of the corresponding linear problem (right) for E = 20.*

Figs. 6.20 and 6.21 present results of numerical computations. In the figures, the left figures correspond to the solution of the nonlinear problem (6.123) and (6.124) (which is equivalent to the original boundary value problem (6.111)–(6.113)) for $E = 20$. The right figures correspond to the solution of the linear problem with the coefficient (6.125). Solution of the linear problem with the coefficient (6.125) was obtained by using ANSYS finite element software. It is seen from Figs 6.20 and 6.21 that solution of the nonlinear problem for $E = 20$ and solution of the corresponding linear problem are in agreement.

6.9.2. Numerical analysis of effective tunability

Now, we present some results of numerical computations of effective tunability of ferroelectric–dielectric composite material from [190]. The problem was solved for the parameters indicated above ($\varepsilon_0 = 150, \varepsilon_\infty = 100, b = 0.5$).

The numerical computations were made for overall electric field $E = 0$, 1, 2, 4, 8, 10, 14. We note that working voltage takes values of tens (up to hundred) kilovolt/sm. for real materials [350]. In our computations the relative values of constants and overall electric field are used (see comment in the beginning of Section 6.5). It does not restrict the generality of consideration (for example, one can assume that in the problem considered the unity of voltage is equal to 1 kilovolt/sm.).

Effective permittivity of composite

The plots of the function $\widehat{\varepsilon}(E)$ are displayed in Figs. 6.22 and 6.23. The dots in the figures correspond to the numerically computed values of $\widehat{\varepsilon}(E)$.

Computations were done for the size of the periodicity cell 1×1 and sizes of the dielectric inclusion (Fig. 6.22) and 0.7×0.7 (Fig. 6.23). The volume fractions of the dielectric inclusion are 0.16 and 0.49, correspondingly.

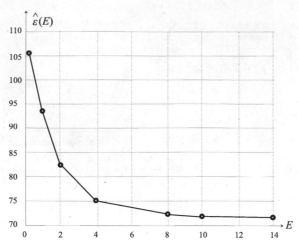

Figure 6.22. *Effective permittivity of composite in dependence on the overall electric field (volume fraction of inclusion 0.16).*

It is seen that the effective permittivity of composite with volume fraction of inclusions equal to 0.16 is approximately two times greater than the effective permittivity of composite with volume fraction of inclusions equal to 0.49. This fact correlates with the known property of effective permittivity of linear composites.

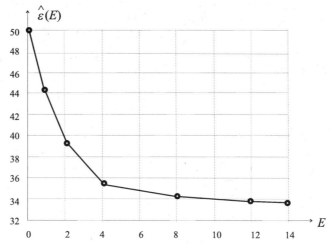

Figure 6.23. *Effective permittivity of composite material in dependence on the overall electric field (volume fraction of inclusion 0.49).*

Effective tunability of composite material

Effective tunability $\widehat{T}(E)$ of composite material is determined by the formula (6.5) and it is easily calculated if effective permittivity $\widehat{\varepsilon}(E)$ of composite material is known. In Fig. 6.24, the plot of the effective tunability $\widehat{T}(E)$ is displayed for composite material with the periodicity cell 1×1 and inclusion 0.7×0.7. In Fig. 6.24 also the plot of the effective tunability is displayed for pure ferroelectric. The dots correspond to numerically computed values, lines correspond to interpolation. It is seen that the plots of tunability of composite material and tunability of pure ferroelectric are very close.

Computation for composite material with the periodicity cell 1×1 and inclusion 0.4×0.4 gave values of the effective tunability very close to the values of the effective tunability computed for composite with 0.7×0.7 inclusion [190].

We can conclude that in the case of strong fields, the effective tunability also demonstrates stability when we dilute ferroelectric matrix with dielectric inclusions. At the same time, the phenomenon of amplification of tunability does not take place for strong fields. The amplification of tunability is possible only for small fields.

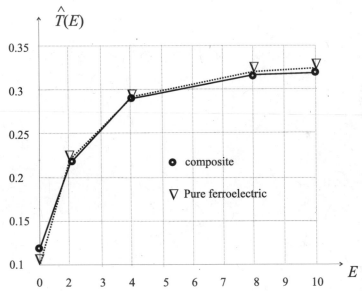

Figure 6.24. *Effective tunability of composite material and pure ferroelectric.*

Chapter 7

EFFECTIVE LOSS OF HIGH-CONTRAST COMPOSITES

In this chapter, we discuss the problem of effective loss of composite materials. These problems attract attention due to the potential utility of ferroelectric materials in high frequency tunable devices, see, e.g., [331, 394]. Although the problem is known for a long time, it has not been investigated completely, see, e.g., [12, 330, 350]. Probably, it is impossible solve this problem in explicit form in the general case. However, in the special cases it can be solved.

7.1. Effective loss of particle-filled composite

We consider two types of high-contrast composite materials: particle-filled and laminated and demonstrate that behavior of effective loss is different for these materials. First, we analyze the effective loss of high-contrast particle-filled composite.

7.1.1. Two-sided bounds on the effective loss tangent of composite material

We consider the case of small nonlinearities or small fields. In this case, we use the following formula for the imaginary part $\widehat{\varepsilon}''$ of the homogenized permittivity of composite material:

$$\widehat{\varepsilon}'' = \frac{1}{|Y|} \int_Y \varepsilon''(\mathbf{x})|\nabla N(\mathbf{x})|^2 d\mathbf{x}, \qquad (7.1)$$

where $\varepsilon''(\mathbf{x})$ means the imaginary part of permittivity of components of composite and $N(\mathbf{x})$ is the potential of local electric field. We remind that the real part $\widehat{\varepsilon}(0)$ of the homogenized permittivity of composite is assumed to be given by the formula (6.67), that is $\widehat{\varepsilon} = \widehat{\varepsilon}(0)$.

Using the notion of loss tangent for the components of composite materials

$$\tan \delta(\mathbf{x}) = \frac{\varepsilon''(\mathbf{x})}{\varepsilon(\mathbf{x})},$$

where $\varepsilon(\mathbf{x})$ means the real part of permittivity of components of composite. Then, we can rewrite (7.1) as

$$\widehat{\varepsilon}'' = \frac{1}{|Y|} \int\limits_{Y} \tan\delta(\mathbf{x})\varepsilon(\mathbf{x})|\nabla N(\mathbf{x})|^2 d\mathbf{x}. \tag{7.2}$$

We see from this equation that formula (7.1) is based on the following two assumptions:

1) the local loss is proportional to the local energy,
2) the effective (total) loss is the sum of the local losses.

These assumptions are usually accepted in the theory of heterogeneous materials, see, e.g., [350].

We introduce the effective loss tangent of composite as

$$\tan\widehat{\delta} = \frac{\widehat{\varepsilon}''}{\widehat{\varepsilon}}.$$

Substituting $\widehat{\varepsilon} = \widehat{\varepsilon}(0)$ in accordance with (6.67) and $\widehat{\varepsilon}''$ in accordance with (7.2), we obtain

$$\tan\widehat{\delta} = \frac{\displaystyle\int\limits_{Y} \tan\delta(\mathbf{x})\varepsilon(\mathbf{x})|\nabla N(\mathbf{x})|^2 d\mathbf{x}}{\displaystyle\int\limits_{Y} \varepsilon(\mathbf{x})|\nabla N(\mathbf{x})|^2 d\mathbf{x}}. \tag{7.3}$$

We see the term $\varepsilon(\mathbf{x})|\nabla N(\mathbf{x})|^2$ in the integrals in numerator and denominator of formula (7.3). In addition, all terms forming the integrands in (7.3) are positive. Using these notes and a well-known estimate for integral [308], we obtain the following two-sided bound on the homogenized loss tangent of composite:

$$\min_{\mathbf{x}\in Y} \tan\delta(\mathbf{x}) \leq \tan\widehat{\delta} \leq \max_{\mathbf{x}\in Y} \tan\delta(\mathbf{x}).$$

For two component ferroelectric–dielectric composite material, it takes the form

$$\tan\delta_d \leq \tan\widehat{\delta} \leq \tan\delta_f \tag{7.4}$$

because $\tan\delta_d \leq \tan\delta_f$ ($\tan\delta_d$ and $\tan\delta_f$ are loss tangents of dielectric and ferroelectric components of the composite, correspondingly).

7.1.2. Effective loss tangent of high-contrast composites

Estimate (7.4) means that dilution of ferroelectric material with dielectric inclusions cannot increase the effective loss tangent of composite material. It may sound attractive for one interested in development of low loss materials. Nevertheless, dilution of

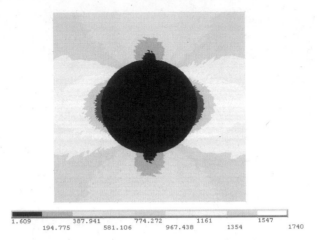

| 1.609 | | 387.941 | | 774.272 | | 1161 | | 1547 | |
| | 194.775 | | 581.106 | | 967.438 | | 1354 | | 1740 |

Figure 7.1. *Energy distribution in high-contrast composite (the white area in the figure corresponds to value close to zero). Permittivity of inclusion $\varepsilon_I = 1$, permittivity of matrix $\varepsilon_m = 1000$.*

ferroelectric material with dielectric inclusions does not decrease homogenized loss in *any case*. We demonstrate that dilution of ferroelectric with the dielectric does not decrease the homogenized loss if components are highly contrasted.

It is seen from formula (7.3) that dielectric loss is determined by the distribution of the local energy $\frac{1}{2}\varepsilon(\mathbf{x})|\nabla N(\mathbf{x})|^2$ over the periodicity cell of composite. It was noted in Section 6.2 that for the high-contrast composite under consideration distribution of the local energy has specific form. Fig. 7.1 demonstrates the distribution of the local energy high-contrast composite (result of numerical computation is presented). It is seen that the local energy in the dielectric inclusion is negligible as compared with the magnitude of the local energy in the ferroelectric matrix. Then for the composite ferroelectric matrix–dielectric inclusions we have that

$$\int_Y \varepsilon(\mathbf{x})|\nabla N(\mathbf{x})|^2 d\mathbf{x} \approx \varepsilon_f \int_m |\nabla N(\mathbf{x})|^2 d\mathbf{x}$$

and (7.3) takes the form

$$\tan\widehat{\delta} = \frac{\tan\delta_f \varepsilon_f \int_m |\nabla N(\mathbf{x})|^2 d\mathbf{x}}{\int_m \varepsilon_0(\mathbf{x})|\nabla N(\mathbf{x})|^2 d\mathbf{x}} = \frac{\tan\delta_f \varepsilon_f \int_m |\nabla N(\mathbf{x})|^2 d\mathbf{x}}{\varepsilon_f \int_m |\nabla N(\mathbf{x})|^2 d\mathbf{x}} = \tan\delta_f \qquad (7.5)$$

(recall that we assume that the matrix is ferroelectric with loss tangent $\tan\delta_f$ and permittivity ε_f).

Formula (7.5) means that the homogenized loss tangent of high-contrast ferroelectric matrix–dielectric inclusions composite material is equal to the loss tangent of the ferroelectric component. This formula in conjunction with Fig. 2.12 indicates the nature of stability of loss tangent in ferroelectric filled with dielectric inclusions. When we dilute the ferroelectric with the dielectric inclusions, the local energy $\frac{1}{2}\varepsilon(\mathbf{x})|\nabla N(\mathbf{x})|^2$ (which determines the total loss of composite) is forced out from dielectric inclusions into ferroelectric matrix and the homogenized loss of composite is determined, in fact, by the loss property of the matrix.

In accordance with our computations, the imaginary part $\widehat{\varepsilon}''$ of the homogenized permittivity is

$$\widehat{\varepsilon}'' = \widehat{\varepsilon}\tan\widehat{\delta} = \widehat{\varepsilon}\tan\delta_f. \qquad (7.6)$$

It means that the imaginary part $\widehat{\varepsilon}''$ of the homogenized permittivity is proportional to the real part $\widehat{\varepsilon} = \widehat{\varepsilon}(0)$ of homogenized permittivity. Since

$$\widehat{\varepsilon} < \varepsilon_f,$$

the imaginary parts $\widehat{\varepsilon}''$ of the homogenized permittivity of composite is smaller than the imaginary part ε_f of the ferroelectric matrix.

7.2. Effective loss of laminated composite material

Now, we analyze the effective loss of high-contrast laminated composite material.

Solution of the cellular problem (6.41)–(6.43) for laminated composite material (see Section 6.3 and Fig. 6.5) has the following form (x is the coordinate across the layers):

$$\frac{dN}{dx}(x) = \frac{1}{\varepsilon(x)\left\langle\dfrac{1}{\varepsilon(x)}\right\rangle}.$$

Then

$$\varepsilon(x)\left(\frac{dN}{dx}(x)\right)^2 = \frac{1}{\varepsilon(x)\left\langle\dfrac{1}{\varepsilon(x)}\right\rangle^2}. \qquad (7.7)$$

Substituting (7.7) into (7.3), we obtain

$$\tan\widehat{\delta} = \frac{\left\langle\dfrac{\tan\delta(x)}{\varepsilon(x)}\right\rangle}{\left\langle\dfrac{1}{\varepsilon(x)}\right\rangle}. \qquad (7.8)$$

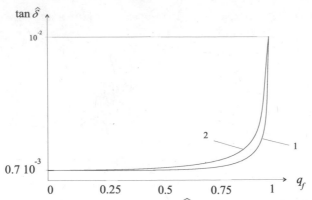

Figure 7.2. *The homogenized loss tangent* $\tan\widehat{\delta}$ *as function of volume fraction* q_f *of tunable component; composite 1 – line 1, composite 2 – line 2.*

We denote $\beta = \dfrac{1}{\varepsilon(x)}$ and rewrite (7.8) as

$$\tan\widehat{\delta} = \frac{\langle \tan\delta(x)\beta(x)\rangle}{\langle\beta(x)\rangle}. \tag{7.9}$$

Formula (7.9) leads us to two–sided estimate (7.4).

We consider two-component ferroelectric–dielectric laminated composite material with the following characteristics of the layers:

$$\varepsilon(x) = \varepsilon_f, \quad \tan\delta(x) = \tan\delta_f \neq 0 \text{ in ferroelectric layers,}$$

$$\varepsilon(x) = \varepsilon_d, \quad \tan\delta(x) = 0.7\cdot 10^{-3} \text{ in dielectric layers,}$$

and denote by q_f the volume fraction of ferroelectric.

In the case under consideration formula (7.9) takes the form

$$\tan\widehat{\delta} = \frac{\tan\delta_f + \tan\delta_d \dfrac{\varepsilon_f}{\varepsilon_d}\cdot\left(\dfrac{1}{q_f}-1\right)}{1 + \dfrac{\varepsilon_f}{\varepsilon_d}\cdot\left(\dfrac{1}{q_f}-1\right)}. \tag{7.10}$$

In Fig. 7.2, plots of the homogenized loss tangent $\tan\widehat{\delta}$ (7.10) as functions of volume fraction q_f of tunable component are shown. The plots are presented for two composites with the following characteristics of layers:

Composite 1: dielectric $\tan\delta = 0.7\cdot 10^{-3}$, $\varepsilon_d = 10$, ferroelectric $\tan\delta = 10^{-2}$, $\varepsilon_f = 1000$;

Composite 2: dielectric $\tan\delta = 0.7\cdot 10^{-3}$, $\varepsilon_d = 5$; ferroelectric $\tan\delta = 10^{-2}$, $\varepsilon_f = 2000$.

It is seen that the homogenized loss tunability of laminated composite material decreases fast when q_f decreases. For $q_f = 0.75$ (relative thickness of dielectric layer is 0.25) the homogenized loss decreases ten times for the composite 2. For the composite 1, the homogenized loss decreases ten times for $X \approx 0.9$ (relative thickness of dielectric layer is about 0.1).

The difference in behavior of effective loss of particle-filled composite and laminated composite is clear seen. This difference is related to the difference of topologies of the composite materials considered above.

Chapter 8

TRANSPORT AND ELASTIC PROPERTIES OF THIN LAYERS

This chapter is devoted to asymptotic behavior of transport and elastic properties of thin layers, which cover or join solid bodies. If the thin layer possesses transport or elastic characteristics compared to the characteristics of the solid bodies, it does not affect global transport / elastic properties of the structure. In this case, thin layers affect the characteristics of local nature, for example, the strength of the structure. In the case where the characteristics of the layers strongly differ from the characteristics of the bodies, one usually arrives at a limit problem with interface condition on the surface corresponding to the limit position of the thin layer. The rigorous asymptotic analysis of such problems started in [317, 337, 338], (see also bibliographical comment in [318]), which were followed by [42, 71, 87, 204, 229, 263, 289, 290].

We present asymptotic analyses for two problems of practical interest. The first is a problem of transport properties of thin covers. In this problem, we arrive at the third type of boundary condition for the limit problem instead of the original first type. The second is problems of elastic properties of a thin layer between two elastic bodies (this is interpreted as the problem of glued bodies). In this problem, we arrive at interface boundary conditions. The methods presented below can be applied to a wide class of problems. For example, one can modify the methods presented for analysis of elastic bodies with thin soft covers or write an electrostatic problem for a body with a thin crack. Note that the trial functions developed for analysis of thin joints (see [180] and Section 8.2 below) were predecessors of the trial functions used in [41] for developing of network models for high-contrast dispersed composites.

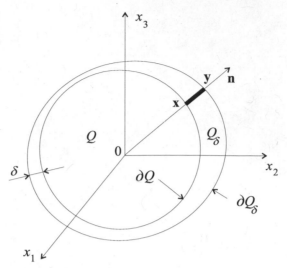

Figure 8.1. *A body with a thin cover.*

8.1. Asymptotic of first boundary-value problem for elliptic equation in a region with a thin cover

In this section, we consider the asymptotic behavior of the boundary-value problem in a domain with a thin cover, which has a low transport characteristic (conductivity, permittivity, etc.). Problems of such kind arise in the description of processes in bodies, which have various kinds of coatings, for example, heat-insulating, and also in the description of the diffusion process through thin membranes.

The specifics of the problem under consideration is that the first type of boundary condition for the original problem is replaced by the third type boundary condition for the limit problem. The mentioned effect demonstrates how the third type of boundary condition may arise in the mathematical way (usually, the third type of boundary condition is introduced in an empirical manner, see, e.g., [24, 326]).

8.1.1. Formulation of the problem

We consider a bounded domain in R^n of the type shown in Fig. 8.1 consisting of a domain Q surrounded by a thin covering domain Q_δ (δ characterizes the thickness of the domain Q_δ). For simplicity (which does not restrict the generality of our consideration) we assume that Q is a sphere of radius r_0 with the center in the origin of the coordinate system. We assume that the domains have boundaries ∂Q and ∂Q_δ (∂Q_δ means the exterior boundary of Q_δ, see Fig. 8.1) as smooth as necessary for the following computations. The exterior normal to ∂Q is denoted by $\mathbf{n(x)}$.

Now we formulate assumptions on the domains occupied by the bodies and the transport characteristics of the bodies. We assume that for all $\mathbf{x} \in \partial Q$ the length of the segment $[\mathbf{x}, \mathbf{y}]$ of the normal $\mathbf{n}(\mathbf{x})$ with the endpoints $\mathbf{x} \in \partial Q$ and $\mathbf{y} \in \partial Q_\delta$ is equal to $\delta \Phi(\mathbf{x})$. We assume that

$$0 < C_1 \le \Phi(\mathbf{x}) \text{ for any } \mathbf{x} \in \partial Q$$

and

$$\Phi \in C^1(\partial Q), \|\Phi\|_{C^1(\partial Q)} \le C_2 < \infty,$$

where the constants C_1 and C_2 do not depend on δ.

We denote the tensor of transport characteristics of the material of the body Q by $a_{ij}(\mathbf{x})$, the tensor of transport characteristics of the material of the cover Q_δ by $\delta b_{ij}(\mathbf{x})$ and introduce the tensor

$$A_{ij}^\delta(\mathbf{x}) = \begin{cases} a_{ij}(\mathbf{x}) \text{ in } Q, \\ \delta b_{ij}(\mathbf{x}) \text{ in } Q_\delta, \end{cases} \tag{8.1}$$

which describes the local properties of the inhomogeneous system body–cover.

We assume that

$$a_{ij}(\mathbf{x}) = a_{ji}(\mathbf{x}), b_{ij}(\mathbf{x}) = b_{ji}(\mathbf{x}) \in C^1(R^n),$$

$$\|a_{ij}\|_{C^1(R^n)} \le C_3, \ \|b_{ij}\|_{C^1(R^n)} \le C_3 < \infty$$

and

$$0 < c_1|\mathbf{z}|^2 \le a_{ij}(\mathbf{x})z_i z_j \le c_2|\mathbf{z}|^2,$$

$$0 < c_1|\mathbf{z}|^2 \le b_{ij}(\mathbf{x})z_i z_j \le c_2|\mathbf{z}|^2$$

for any $\mathbf{x} \in Q \bigcup Q_\delta$, $\mathbf{z} \in R^n$, where $C_3 < \infty$, and $0 < c_1, c_2 < \infty$ do not depend on the parameter δ.

In the domain $Q \bigcup Q_\delta$ we seek the solution of the differential equation

$$\frac{\partial}{\partial x_i}\left(A_{ij}^\delta(\mathbf{x})\frac{\partial \varphi^\delta}{\partial x_j}\right) = f(\mathbf{x}), \tag{8.2}$$

with the *first* type of boundary condition:

$$\varphi^\delta(\mathbf{x}) = T(\mathbf{x}) \text{ on } \partial Q_\delta, \tag{8.3}$$

and condition of ideal conjugation on the interface surface ∂Q:

$$[\varphi^\delta]_{\partial Q} = 0, \ \left[A_{ij}^\delta(\mathbf{x})\frac{\partial \varphi^\delta}{\partial x_j}n_i\right]_{\partial Q} = 0. \tag{8.4}$$

Here $[\bullet]_{\partial Q}$ means jump of a function on the surface ∂Q (difference of values of a function on opposite sides of the surface ∂Q). We assume that $T(\mathbf{x}) \in C^1(Q \bigcup Q_\delta)$.

Under the conditions formulated above, problem (8.2)–(8.4) can be formulated in the form of the following minimization problem (see, e.g., [113, 122]):

$$I_\delta(\varphi) + \langle f, \varphi \rangle \to \min, \ \varphi(\mathbf{x}) \in T(\mathbf{x}) + H_0^1(Q \bigcup Q_\delta), \tag{8.5}$$

where

$$I_\delta(\varphi) = \frac{1}{2} \int_Q A_{ij}^\delta(\mathbf{x}) \frac{\partial \varphi}{\partial x_i}(\mathbf{x}) \frac{\partial \varphi}{\partial x_j}(\mathbf{x}) d\mathbf{x} =$$

$$= \frac{1}{2} \int_Q a_{ij}(\mathbf{x}) \frac{\partial \varphi}{\partial x_i}(\mathbf{x}) \frac{\partial \varphi}{\partial x_j}(\mathbf{x}) d\mathbf{x} + \frac{\delta}{2} \int_{Q_\delta} b_{ij}(\mathbf{x}) \frac{\partial \varphi}{\partial x_i}(\mathbf{x}) \frac{\partial \varphi}{\partial x_j}(\mathbf{x}) d\mathbf{x}$$

and

$$\langle f, \varphi \rangle = \int_{Q \cup Q_\delta} f(\mathbf{x}) \varphi(\mathbf{x}) d\mathbf{x}.$$

Here $H_0^1(Q \bigcup Q_\delta)$ means the set of functions belonging to $H^1(Q \bigcup Q_\delta)$, which vanish on ∂Q_δ.

Our aim is analysis of the problem (8.2)–(8.4) or, that is the same, problem (8.5) under the condition $\delta \to 0$.

8.1.2. Estimates for solution of the problem (8.2)–(8.4)

We shall need estimates of functions belonging to $H_0^1(Q \bigcup Q_\delta)$ in terms of their derivatives similar to standard ones [212, 256, 318], but containing explicit expressions of the constants in the terms of δ.

Lemma 11. *For any function* $\varphi(\mathbf{x}) \in T(\mathbf{x}) + H_0^1(Q \bigcup Q_\delta)$ *the following inequality*

$$\int_\Gamma |\varphi(\mathbf{x})|^2 d\mathbf{x} \leq C \left(1 + \delta^{1/2} \|\nabla \varphi\|_{L_2(Q_\delta)} \right)^2 \tag{8.6}$$

holds. Here $\Gamma \subset Q_\delta$ *is a measurable set, lying on the sphere*

$$S(r) = \{ \mathbf{x} : |\mathbf{x}| = r \}, \ r \in [r_0, r_0 + \delta \Phi_{max}].$$

It is denoted

$$\Phi_{max} = \max_{\mathbf{x} \in \partial Q} \Phi(\mathbf{x}).$$

Proof. We consider first the case when $\varphi(\mathbf{x}) \in C^1(Q \bigcup Q_\delta)$, $T(\mathbf{x}) \in C^1(Q \bigcup Q_\delta)$ and $\varphi(\mathbf{x}) = T(\mathbf{x})$ on ∂Q_δ and prove the estimate (8.6) for this case.

The point $\mathbf{x} \in Q_\delta$ can be specified using the spherical coordinates (ρ, Θ), where $\rho = |\mathbf{x}|$ and $\Theta = (\theta_1, ..., \theta_{n-1})$ is a system of orthogonal coordinates on the sphere $\rho = const$ (these are coordinates similar to polar angles). Moreover, we can select the coordinates (ρ, Θ) in such way that this system be orthonormal (it means that

(ρ, Θ) is a modification of spherical coordinate system but not exactly a spherical coordinate system). In this case, for an arbitrary $r \geq r_0$, we have the equality

$$\varphi(r, \Theta) = T(\Theta) + \int_{r_0 + \delta\Phi(\Theta)}^{r} \frac{\partial\varphi}{\partial\rho}(\rho, \Theta) d\rho. \qquad (8.7)$$

Hereafter

$$T(\Theta) = T(r_0 + \delta\Phi(\Theta), \Theta) = T(\mathbf{x})$$

means value of T on the surface ∂Q_δ, see (8.3). We assume that $(r_0 + \delta\Phi(\Theta), \Theta)$ are coordinates of the point $\mathbf{x}$ in the coordinate system (ρ, Θ).

We square both sides of the equality (8.7). Using Cauchy-Bunyakovsky inequality [308], we obtain

$$|\varphi(r, \Theta)|^2 = \|T\|^2_{C(\partial Q_\delta)} + 2\|T\|_{C(\partial Q_\delta)} \int_{r_0 + \delta\Phi(\Theta)}^{r} \left| \frac{\partial\varphi}{\partial\rho}(\rho, \Theta) \right| d\rho + \qquad (8.8)$$

$$+ \left(\int_{r_0 + \delta\Phi(\Theta)}^{r} \frac{\partial\varphi}{\partial\rho}(\rho, \Theta) d\rho \right)^2 \leq$$

$$\leq \|T\|^2_{C(\partial Q_\delta)} + 2\|T\|_{C(\partial Q_\delta)} \int_{r_0 + \delta\Phi(\Theta)}^{r} \left| \frac{\partial\varphi}{\partial\rho}(\rho, \Theta) \right| d\rho +$$

$$+ \left(\int_{r_0 + \delta\Phi(\Theta)}^{r} d\rho \right) \left(\int_{r_0 + \delta\Phi(\Theta)}^{r} \left| \frac{\partial\varphi}{\partial\rho}(\rho, \Theta) \right|^2 d\rho \right) =$$

$$= \|T\|^2_{C(\partial Q_\delta)} + 2\|T\|_{C(\partial Q_\delta)} \int_{r_0 + \delta\Phi(\Theta)}^{r} \left| \frac{\partial\varphi}{\partial\rho}(\rho, \Theta) \right| d\rho +$$

$$+ \delta\Phi(\Theta) \int_{r_0 + \delta\Phi(\Theta)}^{r} \left| \frac{\partial\varphi}{\partial\rho}(\rho, \Theta) \right|^2 d\rho.$$

Integrating both sides of the inequality (8.8) over the set Γ, we have

$$\int_{\Gamma} |\varphi(r, \Theta)|^2 d\Theta \leq \|T\|^2_{C(\partial Q_\delta)} |\Gamma| + \qquad (8.9)$$

$$+ 2\|T\|_{C(\partial Q_\delta)} \int_{\Gamma} d\Theta \int_{r_0 + \delta\Phi(\Theta)}^{r} \left| \frac{\partial\varphi}{\partial\rho}(\rho, \Theta) \right| d\rho +$$

$$+ \delta\Phi_{max} \int_{\Gamma} d\Theta \int_{r_0 + \delta\Phi(\Theta)}^{r} \left| \frac{\partial\varphi}{\partial\rho}(\rho, \Theta) \right|^2 d\rho \leq$$

$$\leq ||T||^2_{C(\partial Q_\delta)}|\Gamma| + \tilde{C}||T||_{C(\partial Q_\delta)}\delta^{1/2}\Phi_{max}^{1/2}||\nabla\varphi||_{L_2(Q_\delta)} +$$

$$+\delta\Phi_{max}||\nabla\varphi||^2_{L_2(Q_\delta)} \leq$$

$$\leq ||T||^2_{C(\partial Q_\delta)}|\Gamma| + \tilde{C}||T||_{C(\partial Q_\delta)}\delta^{1/2}\Phi_{max}^{1/2}||\nabla\varphi||_{L_2(Q_\delta)} +$$

$$+\tilde{C}\delta\Phi_{max}||\nabla\varphi||^2_{L_2(Q_\delta)},$$

where $\tilde{C} < \infty$ and $|\Gamma|$ means the measure of Γ.

The right-hand side of (8.9) can be estimated from above by the quantity

$$C(1 + 2\delta^{1/2}||\nabla\varphi||_{L_2(Q_\delta)} + \delta||\nabla\varphi||^2_{L_2(Q_\delta)})$$

with the constant $C < \infty$. Then the inequality (8.9) follows from the inequality (8.6).

Deriving (8.9), we have used the inequality

$$\int_\Gamma d\Theta \int_{r_0+\delta\Phi(\Theta)}^r \left|\frac{\partial\varphi}{\partial\rho}(\rho,\Theta)\right| d\rho \leq \int_{Q_\delta} |\nabla\varphi(\mathbf{x})|d\mathbf{x} \leq$$

$$\leq |Q_\delta| \cdot ||\nabla\varphi||_{L_2(Q_\delta)} \leq 2\pi\delta\Phi_{max}||\nabla\varphi||_{L_2(Q_\delta)}.$$

Remark 1. Note that by virtue of the orthnormality of the initial Cartesian coordinate system and the system of coordinates (ρ,Θ), we have

$$|\nabla_\mathbf{x}\varphi(\mathbf{x})| = |\nabla_{(\rho,\Theta)}\varphi(\rho,\Theta)|,$$

where $\nabla_\mathbf{x}$ and $\nabla_{(\rho,\Theta)}$ are the gradients in the corresponding coordinate systems. Hence

$$||\nabla_\mathbf{x}\varphi||_{L_2(Q_\delta)} = ||\nabla_{(\rho,\Theta)}\varphi||_{L_2(Q_\delta)}.$$

Then, we shall use the notation $||\nabla\varphi||_{L_2(Q_\delta)}$ without indicating the system of coordinates (with the exception for specially stipulated cases).

Now we justify the inequality (8.6) for the function

$$\varphi(\mathbf{x}) \in T(\mathbf{x}) + H_0^1(Q \bigcup Q_\delta).$$

We consider an approximating sequence $\{\varphi_n(\mathbf{x})\} \subset T(\mathbf{x}) + C^1(Q \bigcup Q_\delta)$, such that

$$\varphi_n(\mathbf{x}) \to \varphi(\mathbf{x}) \text{ in } H_0^1(Q \bigcup Q_\delta)$$

as $n \to \infty$.

For these sequence $\{\varphi_n(\mathbf{x})\}$ the inequality (8.6) is valid. By virtue of the embedding theorem [212]

$$\varphi_n(\mathbf{x}) \to \varphi(\mathbf{x}) \text{ in } L_2(\Gamma)$$

as $n \to \infty$.

Now, if we write the inequality (8.6) for $\varphi_n(\mathbf{x})$ and then pass to limit as $n \to \infty$, we obtain the inequality (8.6) for the function $\varphi(\mathbf{x})$. **Proof is completed**.

Lemma 12. *For any function* $\varphi(\mathbf{x}) \in T(\mathbf{x}) + H_0^1(Q \bigcup Q_\delta)$ *the following inequality*

$$\|\varphi\|_{L_2(Q_\delta)} \leq \delta^{1/2} C \left(1 + \delta^{1/2} \|\nabla \varphi\|_{L_2(Q_\delta)} \right) \tag{8.10}$$

holds. Here $C < \infty$ *does not depend on* δ.

Proof. We denote

$$\Gamma(r) = \{\mathbf{x} : |\mathbf{x}| = r\} \cap Q_\delta = \{\mathbf{x} = (r, \Theta) : r = r_0 + \delta \Phi(\Theta)\}. \tag{8.11}$$

The notation $\mathbf{x} = (\rho, \Theta)$ means that the coordinates (ρ, Θ) determine the point $\mathbf{x}$. The equation $r = r_0 + \delta \Phi(\Theta)$ in (8.11) follows from the definition of the domain Q_δ. Since $\Phi(\Theta)$ is smooth function, the set $\Gamma(r)$ is measurable [53, 160] and the inequality (8.6) can be applied to it. We have

$$\int_{Q_\delta} |\varphi(\mathbf{x})|^2 d\mathbf{x} = \int_{r_0}^{r_0 + \delta \Phi(\Theta)} d\rho \int_{\Gamma(\rho)} |\varphi(\rho, \Theta)|^2 d\Theta \leq \tag{8.12}$$

$$\leq C \int_{r_0}^{r_0 + \delta \Phi(\Theta)} d\rho \left(1 + \delta^{1/2} \|\nabla \varphi\|_{L_2(Q_\delta)} \right)^2 \leq$$

$$\leq C \delta \Phi_{max} \left(1 + \delta^{1/2} \|\nabla \varphi\|_{L_2(Q_\delta)} \right)^2.$$

Taking into account algebraic inequality

$$\sqrt{1 + x^2} \leq 1 + x \text{ for } x \geq 0, \tag{8.13}$$

we obtain from (8.12) the required inequality (8.10). **Proof is completed**.

Lemma 13. *For any function* $\varphi(\mathbf{x}) \in T(\mathbf{x}) + H_0^1(Q \bigcup Q_\delta)$ *the following inequality*

$$\|\varphi\|_{L_2(Q)} \leq C \left(1 + \|\nabla \varphi\|_{L_2(Q)} + \delta^{1/2} \|\nabla \varphi\|_{L_2(Q_\delta)} \right) \tag{8.14}$$

holds. Here $C < \infty$ *does not depend on* δ.

Proof. Using the Poincare inequality [256], we write

$$\|\varphi\|_{L_2(Q)} \leq C_1 \|\nabla \varphi\|_{L_2(Q)} + C_1 \|\varphi\|_{L_2(\partial Q)}. \tag{8.15}$$

From the inequality (8.6) written for $\Gamma = \partial Q$ and the inequality (8.13) there follows that

$$\|\varphi\|_{L_2(\partial Q)} \leq C\left(1 + \delta^{1/2}\|\nabla\varphi\|_{L_2(Q_\delta)}\right). \tag{8.16}$$

From the inequalities (8.15) and (8.16) there follows the required inequality (8.14). **Proof is completed**.

Using Lemmas 11–13, we can derive estimates with respect to the solution $\varphi^\delta(\mathbf{x})$ of the problem (8.2)–(8.4).

Lemma 14. *The following estimates hold*

$$\|\nabla\varphi^\delta\|_{L_2(Q)}, \ \delta^{1/2}\|\nabla\varphi^\delta\|_{L_2(Q_\delta)}, \ \|\varphi^\delta\|_{L_2(Q\bigcup Q_\delta)} \leq C, \tag{8.17}$$

where $C < \infty$ does not depend on δ.

Proof. We take arbitrary function $\psi(\mathbf{x}) \in T(\mathbf{x}) + H_0^1(Q\bigcup Q_\delta)$. Since $\varphi^\delta(\mathbf{x})$ is the solution of the minimization problem (8.5) then for any δ

$$I_\delta(\varphi^\delta) \leq I_\delta(\psi) \leq C,$$

where $C < \infty$ does not depend on δ by virtue of the conditions on the coefficients $a_{ij}(\mathbf{x})$ and $b_{ij}(\mathbf{x})$. From here, taking into account that the matrices $a_{ij}(\mathbf{x})$ and $b_{ij}(\mathbf{x})$ are positively defined with the constants $c_1 > 0$ independent on δ, we have

$$c_1\|\nabla\varphi^\delta\|^2_{L_2(Q)} + \delta c_1\|\nabla\varphi^\delta\|^2_{L_2(Q_\delta)} \leq \tag{8.18}$$

$$\leq \left|\int_Q a_{ij}(\mathbf{x})\frac{\partial\varphi^\delta}{\partial x_i}(\mathbf{x})\frac{\partial\varphi^\delta}{\partial x_j}(\mathbf{x})d\mathbf{x} + \delta\int_{Q_\delta} b_{ij}(\mathbf{x})\frac{\partial\varphi^\delta}{\partial x_i}(\mathbf{x})\frac{\partial\varphi^\delta}{\partial x_j}(\mathbf{x})d\mathbf{x}\right| \leq$$

$$\leq \left|\int_{Q\bigcup Q_\delta} f(\mathbf{x})\varphi^\delta(\mathbf{x})d\mathbf{x}\right| + C \leq \|f\|_{L_2(Q\bigcup Q_\delta)}\|\varphi^\delta\|_{L_2(Q\bigcup Q_\delta)} + C.$$

By using the inequalities (8.10) and (8.14), we conclude that there exists a number $\delta_0 > 0$ such that for all $\delta \in (0, \delta_0]$

$$\|\varphi^\delta\|_{L_2(Q\bigcup Q_\delta)} \leq \|\varphi^\delta\|_{L_2(Q)} + \|\varphi^\delta\|_{L_2(Q_\delta)} \leq \tag{8.19}$$

$$\leq C_1\left(1 + \delta^{1/2}\|\nabla\varphi^\delta\|_{L_2(Q)} + \delta^{1/2}(1 + \delta^{1/2}\|\nabla\varphi^\delta\|_{L_2(Q_\delta)})\right) \leq$$

$$\leq C_1(1 + \delta_0^{1/2})\left(1 + \|\nabla\varphi^\delta\|_{L_2(Q)} + \delta^{1/2}\|\nabla\varphi^\delta\|_{L_2(Q_\delta)}\right).$$

Substituting (8.19) into (8.18), we obtain the inequality

$$||\nabla\varphi^\delta||^2_{L_2(Q)} + \delta||\nabla\varphi^\delta||^2_{L_2(Q_\delta)} \leq \qquad (8.20)$$

$$\leq C_1(1 + \delta_0^{1/2})c_1^{-1}||f||_{L_2(R^n)}\left(1 + ||\nabla\varphi^\delta||_{L_2(Q)} + \delta^{1/2}||\nabla\varphi^\delta||_{L_2(Q_\delta)}\right).$$

The set of solutions of algebraic inequality

$$x^2 + y^2 \leq A(x + y) + B$$

is a bounded set in R^2 if A, $B > 0$ (this is a circular disk of the radius $\sqrt{B + A^2/2}$ with the center in the point $(A/2, A/2)$). By using this note, we derive from inequality (8.19) that the quantities $||\nabla\varphi^\delta||_{L_2(Q)}$ and $\delta^{1/2}||\nabla\varphi^\delta||_{L_2(Q_\delta)}$ are restricted uniformly for $\delta \in (0, \delta_0]$. **Proof is completed.**

Lemma 15. *The sequence $\{\varphi^\delta(\mathbf{x})\}$ of solutions of the problem (8.2)–(8.4) is weakly compact in $H^1(Q)$, Then there exists a subsequence $\{\varphi^\eta(\mathbf{x})\} \subset \{\varphi^\delta(\mathbf{x})\}$ such that*

$$\varphi^\eta(\mathbf{x}) \to \varphi^*(\mathbf{x}) \text{ weakly in } H^1(Q) \text{ (strong in } L_2(Q)) \text{ as } \eta \to 0.$$

The proof follows from Lemma 14 and the weak compactness of a bounded sequence in a Hilbert space [392].

Now we write the variational form [212] of the problem (8.2)–(8.4) or, that is the same as problem (8.5). It has the form

$$\varphi(\mathbf{x}) \in T(\mathbf{x}) + H_0^1(Q \bigcup Q_\delta), \qquad (8.21)$$

$$\int_Q a_{ij}(\mathbf{x})\frac{\partial\varphi^\delta}{\partial x_i}(\mathbf{x})\frac{\partial\phi}{\partial x_j}(\mathbf{x})d\mathbf{x} + \int_{Q_\delta} b_{ij}(\mathbf{x})\frac{\partial\varphi^\delta}{\partial x_i}(\mathbf{x})\frac{\partial\phi}{\partial x_j}(\mathbf{x})d\mathbf{x} + \qquad (8.22)$$

$$+ \int_{Q \bigcup Q_\delta} f(\mathbf{x})\phi(\mathbf{x})d\mathbf{x} = 0$$

for any $\phi(\mathbf{x}) \in H_0^1(Q \bigcup Q_\delta)$. The vanishing of the trial function $\phi(\mathbf{x})$ on ∂Q_δ is essential [212] with the condition (8.21).

The equivalence of the problem (8.2)–(8.4), the problem (8.5) and the problem (8.21) and (8.22) is known, for example, from [212].

8.1.3. Construction of special trial function

Now we introduce trial functions of special form, which takes into account specific of the problem under consideration. We consider the following trial function:

$$\phi(r, \Theta) = \begin{cases} \psi(r, \Theta) \in C^\infty(Q) \text{ in } Q, \\ \\ \psi(r_0, \Theta)\left(1 - \dfrac{r - r_0}{\delta\Phi(\Theta)}\right) \text{ in } Q_\delta, \end{cases} \qquad (8.23)$$

where (r, Θ) are spherical coordinates and $\psi(r, \Theta)$ is arbitrary function belonging to $C^{\infty}(Q)$.

The formula (8.23) guarantees that

$$\phi(r, \Theta) = 0 \text{ for } r = r_0 + \delta\Phi(\Theta),$$

i.e., on the boundary ∂Q_δ. Thus, the function $\phi(r, \Theta) \subset H_0^1(Q \bigcup Q_\delta)$.

Lemma 16. *For the trial function $\phi(r, \Theta)$ (8.23) the following relations*

$$|\phi(r, \Theta)|, \left|\frac{\partial\phi(r, \Theta)}{\partial\theta_\alpha}\right| \le C\|\psi\|_{C^1(Q)} \, (\alpha = 1, ..., n-1) \text{ in } Q \cup Q_\delta, \qquad (8.24)$$

$$\frac{\partial\phi}{\partial r}(r, \Theta) = -\frac{\partial\psi(r_0, \Theta)}{\delta\Phi(\Theta)} \text{ in } Q_\delta \qquad (8.25)$$

hold.

The verification of the inequalities (8.24) and (8.25) is carried out by the differentiating of the formulas (8.23) and taking into account that $|r - r_0| \le \delta\Phi(\Theta)$.

Lemma 17. *For the subsequence $\{\varphi^\eta(\mathbf{x})\}$ and trial function (8.23) with a fixed $\psi(r, \Theta)$, the following limits take place*

$$\int_Q a_{ij}(\mathbf{x}) \frac{\partial\varphi^\eta}{\partial x_j}(\mathbf{x}) \frac{\partial\psi}{\partial x_i}(\mathbf{x}) d\mathbf{x} \to \int_Q a_{ij}(\mathbf{x}) \frac{\partial\varphi^*}{\partial x_j}(\mathbf{x}) \frac{\partial\psi}{\partial x_i}(\mathbf{x}) d\mathbf{x}, \qquad (8.26)$$

$$\int_{Q \bigcup Q_\eta} f(\mathbf{x})\psi(\mathbf{x}) d\mathbf{x} \to \int_Q f(\mathbf{x})\psi(\mathbf{x}) d\mathbf{x}$$

as $\eta \to 0$. Here $\psi(\mathbf{x})$ means $\psi(r, \Theta)$ in the Cartesian coordinates $\mathbf{x} = (x_1, ..., x_n)$.

Proof. The first limit holds by virtue of Lemma 15. In order to prove the second limit it is sufficient (taking into account Lemma 15) to note that by virtue of the inequalities (8.10) and (8.17) we have

$$\left|\int_{Q_\eta} f(\mathbf{x})\psi(\mathbf{x}) d\mathbf{x}\right| \le \|f\|_{L_2(Q_\eta)} \|\psi\|_{L_2(Q_\eta)}$$

In order to estimate $\|\psi\|_{L_2(Q_\eta)}$, we write

$$\int_{Q_\eta} \psi^2(\mathbf{x}) d\mathbf{x} \le \int_{S(r_0)} \psi^2(r_0, \Theta) d\Theta \int_{r_0} (r_0 + \eta\Phi(\Theta)) \left(1 - \frac{r - r_0}{\delta\Phi(\Theta)}\right) dr \le$$

$$\le \eta \int_{S(r_0)} \psi^2(r_0, \Theta) d\Theta \Phi(\Theta)) dr,$$

where $S(r_0) = \{\mathbf{x} : |\mathbf{x}| = r_0\}$. The last expression tends to zero as $\eta \to 0$. **Proof is completed**.

Remark 2. We make a note on the change of variables in the integrals. We use change of variables

$$\mathbf{x} \leftrightarrow (r, \Theta),$$

where $\mathbf{x}$ are coordinates in a standard Cartesian coordinate system and (r, Θ) are modified spherical coordinates – spherical coordinates with orthonormal corresponding frame. We denote

$$\{\mathcal{B}_{ij}(r, \Theta)\} = \left\{\frac{\partial \mathbf{x}}{\partial(r, \Theta)}\right\}$$

Jacobi matrix for the function $\mathbf{x} = \mathbf{x}(r, \Theta)$ and denote

$$B_{ij}(r, \Theta) = \mathcal{B}_{ik}(r, \Theta)\mathcal{B}_{jl}(r, \Theta)b_{kl}(r, \Theta).$$

By virtue of the orthonormality of the corresponding frame of both coordinate systems, we have

$$\det\{\mathcal{B}_{ij}(r, \Theta)\} = 1.$$

By using Remark 2, we write

$$\int_{Q_\eta} b_{ij}(\mathbf{x})\frac{\partial \varphi^\eta}{\partial x_j}(\mathbf{x})\frac{\partial \phi}{\partial x_j}(\mathbf{x})d\mathbf{x} = \qquad (8.27)$$

$$= \int_{Q_\eta} b_{ij}(r, \Theta)\mathcal{B}_{ik}(r, \Theta)\mathcal{B}_{jl}(r, \Theta)\frac{\partial \varphi^\eta}{\partial \theta_k}(r, \Theta)\frac{\partial \phi}{\partial \theta_l}(r, \Theta)drd\Theta =$$

$$= \int_{Q_\eta} B_{kl}(r, \Theta)\frac{\partial \varphi^\eta}{\partial \theta_k}(r, \Theta)\frac{\partial \phi}{\partial \theta_l}(r, \Theta)drd\Theta,$$

where in the last integral the indices k and l number the coordinates $(r, \theta_1, ..., \theta_{n-1})$.

Since the function $\mathbf{x} = \mathbf{x}(r, \Theta)$ is inversible, Jacobi matrix $\{\mathcal{B}_{ij}(r, \Theta)\}$ is nondegenerated. Then the coefficients $B_{ij}(r, \Theta)$ preserve the properties of the coefficients $b_{ij}(\mathbf{x})$ indicated in the beginning of this section (in general, with different constants).

8.1.4. The convergence theorem and the limit problem

The variables $(r, \theta_1, ..., \theta_{n-1})$ are naturally separated into two groups: the variable r and the variables $\theta_1, ..., \theta_{n-1}$. We will use index r to refer to the variable r and the indices α, β to refer to the variables $\theta_1, ..., \theta_{n-1}$. Using this notations, we rewrite the last integral in (8.27) as

$$\int_{Q_\eta} B_{rr} \frac{\partial \varphi^\eta}{\partial r}(r, \Theta) \frac{\partial \phi}{\partial r}(r, \Theta) dr d\Theta + \tag{8.28}$$

$$+ \int_{Q_\eta} B_{r\beta} \frac{\partial \varphi^\eta}{\partial r}(r, \Theta) \frac{\partial \phi}{\partial \theta_\beta}(r, \Theta) dr d\Theta +$$

$$+ \int_{Q_\eta} B_{\alpha r} \frac{\partial \varphi^\eta}{\partial \theta_\alpha}(r, \Theta) \frac{\partial \phi}{\partial r}(r, \Theta) dr d\Theta +$$

$$+ \int_{Q_\eta} B_{\alpha\beta} \frac{\partial \varphi^\eta}{\partial \theta_\alpha}(r, \Theta) \frac{\partial \phi}{\partial \theta_\beta}(r, \Theta) dr d\Theta.$$

We consider the limit values of the summands in (8.28) as $\eta \to 0$ separately.

Lemma 18. *For the subsequence $\{\varphi^\eta(\mathbf{x})\}$ and trial function $\phi(r, \Theta)$ (8.23) with a fixed $\psi(r, \Theta)$, the following limits take place as $\eta \to 0$*

$$\textbf{(A)} \quad \int_{Q_\eta} B_{r\alpha}(r, \Theta) \frac{\partial \varphi^\eta}{\partial r}(r, \Theta) \frac{\partial \phi}{\partial \theta_\alpha}(r, \Theta) dr d\Theta \to 0, \tag{8.29}$$

$$\textbf{(B)} \quad \int_{Q_\eta} B_{\alpha r}(r, \Theta) \frac{\partial \varphi^\eta}{\partial \theta_\alpha}(r, \Theta) \frac{\partial \phi}{\partial r}(r, \Theta) dr d\Theta \to 0,$$

$$\textbf{(C)} \quad \int_{Q_\eta} B_{\alpha\beta}(r, \Theta) \frac{\partial \varphi^\eta}{\partial \theta_\alpha}(r, \Theta) \frac{\partial \phi}{\partial \theta_\beta}(r, \Theta) dr d\Theta \to 0,$$

where the index r corresponds to the variable r and the indices α and β correspond to variables θ_α and θ_β.

Proof. The expression in (8.27) consists of the sum of the terms indicated in the items A to C. We consider these cases separately.

Case A. We have

$$\left| \eta \int_{Q_\eta} B_{r\alpha}(r, \Theta) \frac{\partial \varphi^\eta}{\partial r}(r, \Theta) \frac{\partial \phi}{\partial \theta_\alpha}(r, \Theta) dr d\Theta \right| \le \tag{8.30}$$

$$\le \eta \int_{Q_\eta} \left| B_{r\alpha}(r, \Theta) \frac{\partial \varphi^\eta}{\partial r}(r, \Theta) \right| \cdot \left| \frac{\partial \phi}{\partial \theta_\alpha}(r, \Theta) \right| dr d\Theta.$$

By using the equality (8.23) and the inequality $\|B_{r\alpha}\|_{C(Q_\eta)} \le C < \infty$, see Remark 2, we obtain that the right-hand side of (8.30) does not exceed

$$\eta \int_{Q_\eta} C \|\psi\|_{C^1(Q)} \left| \frac{\partial \varphi^\eta}{\partial r}(r, \Theta) \right| dr d\Theta \le \eta C \|\psi\|_{C^1(Q)} \eta^{1/2} \|\nabla \varphi^\eta\|_{L_2(Q_\eta)}. \tag{8.31}$$

Deriving (8.31), we use the Cauchy-Bunyakovsky inequality and orthonormality of the coordinate systems used, see Remark 1.

Case B. In view of inequality (8.25) we have

$$\eta \int_{Q_\eta} B_{\alpha r}(r,\Theta) \frac{\partial \varphi^\eta}{\partial \theta_\alpha}(r,\Theta) \frac{\partial \phi}{\partial r}(r,\Theta) dr d\Theta = \tag{8.32}$$

$$= -\eta \int_{Q_\eta} B_{r\alpha}(r,\Theta) \frac{\partial \varphi^\eta}{\partial \theta_\alpha} \frac{\psi(r_0,\Theta)}{\eta \Phi(\Theta)} dr d\Theta.$$

We consider the last integral in (8.32) and perform the change of variable

$$\rho = r_0 + \frac{r - r_0}{\Phi(\Theta)},$$

which transforms the domain Q_η into the spherical layer

$$P = \{(\rho,\Theta) : r_0 \leq \rho \leq r_0 + \eta\}.$$

Under this substitution

$$\frac{\partial}{\partial r} = \frac{1}{\Phi(\Theta)} \frac{\partial}{\partial \rho}$$

and

$$\det \left\{ \frac{\partial(r,\Theta)}{\partial(\rho,\Theta)} \right\} = \Phi(\Theta).$$

After that the integral in the right-hand side of (8.32) takes the form

$$\int_P B_{\alpha r}(\rho,\Theta) \psi(r_0,\Theta) \frac{\partial \varphi^\eta}{\partial \theta_\alpha} \Phi(\Theta) d\rho d\Theta. \tag{8.33}$$

We represent (8.33) as iterated integral on $P = \partial Q \times [r_0, r_0 + \eta]$ and in the inner integral we integrate by parts. After this we obtain

$$\int_P B_{\alpha r}(\rho,\Theta) \psi(r_0,\Theta) \frac{\partial \varphi^\eta}{\partial \theta_\alpha}(r,\Theta) d\rho d\Theta = \tag{8.34}$$

$$= \int_{r_0}^{r_0+\eta} d\rho \int_{S(\rho)} B_{\alpha r}(\rho,\Theta) \psi(r_0,\Theta) \frac{\partial \varphi^\eta}{\partial \theta_\alpha}(r,\Theta) d\Theta =$$

$$= -\int_{r_0}^{r_0+\eta} d\rho \int_{S(\rho)} \frac{\partial}{\partial \theta_\alpha} \left[B_{\alpha r}(\rho,\Theta) \psi(r_0,\Theta) \right] \varphi^\eta(r,\Theta) d\Theta,$$

where

$$S(R) = \{(\rho,\Theta) : \rho = R\}.$$

The term outside integral is missing by virtue of the periodicity of the functions in (8.34) with respect to the variables $\Theta \in S(\rho)$ for a fixed r.

By virtue of the conditions on the functions $b_{ij}(\mathbf{x})$, $\Phi(\mathbf{x})$ and Remark 2, we have

$$\left| \frac{\partial}{\partial t_\alpha} \Big[B_{\alpha r}(\rho, \Theta) \psi(r_0, \Theta) \Big] \right| \leq C \|\psi\|_{C^1(Q)}$$

for any $(\rho, \Theta) \in P$.

In view of this inequality, the absolute value of the integral in the right-hand side of (8.34) does not exceed the quantity

$$\int_{r_0}^{r_0+\eta} d\rho \int_{S(\rho)} C \|\psi\|_{C^1(Q)} |\varphi^\eta(\rho, \Theta)| d\Theta \leq \qquad (8.35)$$

$$\leq \frac{C}{C_1} \int_{r_0}^{r_0+\eta} d\rho \int_{S(\rho)} |\varphi^\eta(\rho, \Theta)| d\Theta \leq$$

$$\leq \frac{C}{C_1} \int_{Q_\eta} |\varphi^\eta(\rho, \Theta)| d\rho d\Theta \leq \frac{C}{C_1} \eta^{1/2} \|\varphi^\eta\|_{L_2(Q_\eta)}.$$

In order to write the last inequality in (8.35) we have used the Cauchy-Bunyakovsky inequality.

By virtue of the inequality (8.17), the right-hand side of (8.35) tends to zero as $\eta \to 0$. Then the right-hand side of (8.32) tends to zero.

Case C. According to the conditions on the functions $\{b_{ij}\}$, Remark 2, and inequality (8.24), we have the inequality

$$\left| B_{\alpha\beta}(\rho, \Theta) \frac{\partial \phi}{\partial \theta_\beta}(\rho, \Theta) \right| \leq \|\psi\|_{C^1(Q)},$$

where $C < \infty$ does not depend on the variables $(\rho, \Theta) \in Q_\eta$. As a result,

$$\left| \eta \int_{Q_\eta} B_{\alpha\beta}(\rho, \Theta) \frac{\partial \varphi^\eta}{\partial \theta_\alpha}(\rho, \Theta) \frac{\partial \phi}{\partial \theta_\beta}(\rho, \Theta) d\rho d\Theta \right| \leq \qquad (8.36)$$

$$\leq \eta \int_{Q_\eta} \left| B_{\alpha\beta}(\rho, \Theta) \frac{\partial \phi}{\partial \theta_\beta}(\rho, \Theta) \right| \cdot \left| \frac{\partial \varphi^\eta}{\partial \theta_\alpha}(\rho, \Theta) \right| d\rho d\Theta \leq$$

$$\leq \eta C \eta^{1/2} \|\nabla \varphi^\eta\|_{L_2(Q_\eta)}.$$

In order to write the last inequality in (8.35) we have used the Cauchy-Bunyakovsky inequality and the orthogonality of the frames of the coordinate systems $\mathbf{x}$ and (r, Θ).

By virtue of the inequality (8.24), the expression in the right-hand side of (8.36) tends to zero. **Proof is completed.**

Now, we compute the limit value of the remaining integral in (8.22).

Lemma 19. *For the subsequence* $\{\varphi^{\eta}(\mathbf{x})\}$ *and trial function* $\phi(r, \Theta)$ *(8.23) with a fixed* $\psi(r, \Theta)$, *the following limit takes place:*

$$\eta \int_{Q_{\eta}} B_{rr}(r, \Theta) \frac{\partial \varphi^{\eta}}{\partial r}(r, \Theta) \frac{\partial \phi}{\partial r}(r, \Theta) dr d\Theta \rightarrow \qquad (8.37)$$

$$\rightarrow -\int_{\partial Q} \frac{B_{rr}(r_0, \Theta)}{\Phi(\Theta)} (T(\Theta) - \varphi^*(r_0, \Theta)) \psi(r_0, \Theta) d\Theta$$

as $\eta \rightarrow 0$.

Proof. Using the equality (8.25), we have

$$\eta \int_{Q_{\eta}} B_{rr}(r, \Theta) \frac{\partial \varphi^{\eta}}{\partial r}(r, \Theta) \frac{\partial \phi}{\partial r}(r, \Theta) dr d\Theta = \qquad (8.38)$$

$$= -\eta \int_{Q_{\eta}} B_{rr}(r, \Theta) \frac{\partial \varphi^{\eta}}{\partial r}(r, \Theta) \frac{\psi(r_0, \Theta)}{\eta \Phi(\Theta)} dr d\Theta.$$

We approximate the sequence $\{\varphi^{\eta}(\mathbf{x})\} \subset H^1(Q \bigcup Q_{\eta})$ by the sequence $\{\zeta_{\eta}(\mathbf{x})\} \subset C^{\infty}(Q \bigcup Q_{\eta})$, such that

$$\|\zeta_{\eta} - \varphi^{\eta}\|_{H^1(Q \bigcup Q_{\eta})} \rightarrow 0 \text{ as } \eta \rightarrow 0 \qquad (8.39)$$

and equality

$$\zeta_{\eta}(\mathbf{x}) = T(\mathbf{x}) \text{ on } \partial Q_{\eta} \qquad (8.40)$$

holds. The sequence $\{\psi_{\eta}(\mathbf{x})\}$ can be constructed by using the standard diagonalization procedure [160].

Representing the domain Q_{η} in the form

$$Q_{\eta} = \partial Q \times [r_0, r_0 + \eta \Phi(\Theta)],$$

we consider the integral

$$-\int_{Q_{\eta}} B_{rr}(r, \Theta) \frac{\partial \zeta_{\eta}}{\partial r}(r, \Theta) \frac{\psi(r_0, \Theta)}{\Phi(\Theta)} dr d\Theta = \qquad (8.41)$$

$$= -\int_{\partial Q} \frac{\psi(r_0, \Theta)}{\Phi(\Theta)} d\Theta \int_{r_0}^{r_0 + \eta \Phi(\Theta)} B_{rr}(r, \Theta) \frac{\partial \zeta_{\eta}}{\partial r}(r, \Theta) dr.$$

Integrating by parts in the inner integral in the right-hand side of (8.41), we have

$$\int_{r_0}^{r_0 + \eta \Phi(\Theta)} B_{rr}(r, \Theta) \frac{\partial \zeta_{\eta}}{\partial r}(r, \Theta) dr =$$

$$= B_{rr}(r, \Theta) \zeta_{\eta}(r, \Theta) \Big|_{r_0}^{r_0 + \eta \Phi(\Theta)} - \int_{r_0}^{r_0 + \eta \Phi(\Theta)} \frac{\partial B_{rr}}{\partial r}(r, \Theta) \zeta_{\eta}(r, \Theta) dr. \qquad (8.42)$$

Then the right-hand side of (8.41) is equal to

$$-\int_{\partial Q} \frac{\psi(r_0, \Theta)}{\Phi(\Theta)} \left[B_{rr}(r_0 + \eta\Phi(\Theta), \Theta) T(\Theta) - B_{rr}(r_0, \Theta) \zeta_\eta(r_0, \Theta)) \right] d\Theta + \quad (8.43)$$

$$+ \int_{\partial Q} \frac{\psi(r_0, \Theta)}{\Phi(\Theta)} d\Theta \int_{r_0}^{r_0 + \eta\Psi(\Theta)} \frac{\partial B_{rr}}{\partial r}(r, \Theta) \zeta_\eta(r, \Theta) dr.$$

In the computations of (8.43) we have taken into account condition (8.40), by virtue of which we have

$$\zeta_\eta(r_0 + \eta\Phi(\Theta)) = T(\mathbf{x}).$$

We consider the integrals in (8.43). We start with the second one. By virtue of the condition on the functions $b_{ij}(\mathbf{x})$, $\Phi(\mathbf{x})$ and Remark 1, we have

$$\left| \frac{\psi(r_0, \Theta)}{\Phi(\Theta)} \cdot \frac{\partial B_{rr}}{\partial r}(r, \Theta) \right| \le C ||\psi||_{C^1(Q)},$$

where $C < \infty$ does not depend on $(r, \Theta) \in Q_\eta$. Then the absolute value of the second integral in (8.43) does not exceed the quantity

$$C||\psi||_{C^1(Q)} \int_{\partial Q} d\Theta \int_{r_0}^{r_0 + \eta\Phi(\Theta)} |\zeta_\eta(r, \Theta)| dr \le C||\psi||_{C^1(Q)} \eta^{1/2} ||\zeta_\eta||_{L_2(Q_\eta)}.$$

This expression tends to zero as $\eta \to 0$ (it has been proved in the proof of Lemma 18, item B).

Now, we consider the first integral in (8.43). We write it as

$$-\int_{\partial Q} \frac{\psi(r_0, \Theta)}{\Phi(\Theta)} \left[B_{rr}(r_0 + \eta\Phi(\Theta), \Theta) - B_{rr}(r_0, \Theta) \right] T(\Theta) d\Theta - \quad (8.44)$$

$$-\int_{\partial Q} \frac{\psi(r_0, \Theta)}{\Phi(\Theta)} \left[B_{rr}(r_0, \Theta) T(\Theta) - B_{rr}(r_0, \Theta) \zeta_\eta(r_0, \Theta)) \right] d\Theta.$$

The first integral in (8.44) tends to zero as $\eta \to 0$ by virtue of the inequality

$$|B_{rr}(r_0 + \eta\Phi(\Theta), \Theta) - B_{rr}(r_0, \Theta)| \le C\eta,$$

($C < \infty$), which follows from the conditions for $b_{ij}(\mathbf{x})$.

By virtue of the uniform boundedness of the quantities $\dfrac{\partial B_{rr}}{\partial r}$ with respect to $(r, \Theta) \in Q_\eta$ (see conditions for $b_{ij}(\mathbf{x})$ and Remark 2, the second integral in (8.44) differs at most $C\eta||\zeta_\eta - \varphi^\eta||_{L_2(\partial Q)}$ from the integral

$$-\int_{\partial Q} \frac{\psi(r_0, \Theta)}{\Phi(\Theta)} B_{rr}(r_0, \Theta)(T(\Theta) - \varphi^\eta(r_0, \Theta)) d\Theta. \quad (8.45)$$

Note that from $\varphi^\eta(\mathbf{x}) \to \varphi^*(\mathbf{x})$ weakly in $H^1(Q)$ as $\eta \to 0$ and (8.39), it follows that [212]

$$\zeta_\eta(\mathbf{x}) \to \varphi^*(\mathbf{x}) \text{ in } L_2(\partial Q), \tag{8.46}$$

$$\varphi^\eta(\mathbf{x}) \to \varphi^*(\mathbf{x}) \text{ in } L_2(\partial Q)$$

as $\eta \to 0$.

Passing to limit as $\eta \to 0$ and taking into account the condition on the functions $b_{ij}(\mathbf{x})$, $\Phi(\mathbf{x})$ and (8.46), we arrive at (8.37). **Proof is completed**.

Theorem 9. *As $\delta \to 0$, the sequence $\{\varphi^\delta(\mathbf{x})\}$ of the solutions of the problem (8.2)– (8.4) converges weakly in $H^1(Q)$ (strongly in $L_2(Q)$) to the solution of the problem*

$$\int_Q a_{ij}(\mathbf{x})\frac{\partial \varphi^*}{\partial x_j}(\mathbf{x})\frac{\partial \psi}{\partial x_i}(\mathbf{x})d\mathbf{x} - \int_{\partial Q} K(\mathbf{x})(T(\mathbf{x}) - \varphi^*(\mathbf{x}))\psi(\mathbf{x})d\mathbf{x} + \tag{8.47}$$

$$+ \int_Q f(\mathbf{x})\psi(\mathbf{x})d\mathbf{x} = 0$$

for any $\psi(\mathbf{x}) \in H^1(Q)$, where

$$K(\mathbf{x}) = \frac{B_{rr}(\mathbf{x})}{\Phi(\mathbf{x})} = \frac{b_{ij}(\mathbf{x})n_i(\mathbf{x})n_j(\mathbf{x})}{\Phi(\mathbf{x})}, \tag{8.48}$$

or, in differential form,

$$\frac{\partial}{\partial x_i}\left(a_{ij}(\mathbf{x})\frac{\partial \varphi^*}{\partial x_j}\right) = f_i(\mathbf{x}) \text{ in } Q, \tag{8.49}$$

$$a_{ij}(\mathbf{x})\frac{\partial \varphi^*}{\partial x_j}(\mathbf{x})n_i = K(\mathbf{x})(T(\mathbf{x}) - \varphi^*(\mathbf{x})) \text{ on } \partial Q. \tag{8.50}$$

Proof. We pass to the limit in the equality (8.22) for the subsequence $\{\varphi^\eta(\mathbf{x})\}$ as $\eta \to 0$. Taking into account Lemmas 17–19, we conclude that for the limit function $\varphi^*(\mathbf{x})$ the equality (8.47) is satisfied for any $\psi(\mathbf{x}) \in C^\infty(Q)$. Since $C^\infty(Q)$ is dense in $H^1(Q)$ [212], then (8.47) is satisfied for any $\psi(\mathbf{x}) \in H^1(Q)$.

The problem (8.47) has a unique solution belonging to $H^1(Q)$ (it is verified in the standard manner taking into account that $K(\mathbf{x})$ defined by (8.48) is positive on ∂Q by virtue of the conditions on the functions $b_{ij}(\mathbf{x})$). Then the entire sequence $\{\varphi^\delta(\mathbf{x})\}$ converges to $\varphi^*(\mathbf{x})$ weakly in $H^1(Q)$ (thus, strongly in $L_2(Q)$ by virtue of the imbedding theorem [342]). **Proof is completed**.

The equalities (8.49) and (8.50) follow directly from (8.47).

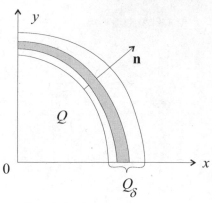

Figure 8.2. *A fragment of laminated cover.*

8.1.5. Transport property of thin laminated cover

Now, we consider a body with a multilayered cover, see Fig. 8.2, which are widely used in engineering.

We assume that the laminas are smooth (do not have microrelief like ribs, perforation, etc.). Then the local conductivity of the cover in longitudinal directions can be described by functions differentiating in the variables Θ. In the transversal direction, the local conductivity of the cover is a piecewise function. In order to include this case in our consideration, we introduce the functions b_{ij} of the form

$$b_{ij} = b_{ij}(r, \Theta) = b_{ij}^0 \left(\frac{r - r_0}{\delta}, \Theta \right), \tag{8.51}$$

where the function $b_{ij}^0(r, \Theta)$ satisfies the conditions:

$$\left| \frac{\partial b_{ij}^0(r, \Theta)}{\partial \theta_\alpha} \right| \leq C,$$

$C < \infty$ does not depend on the variables r, Θ and $b_{ij}^0(r, \Theta)$ is a piecewise continuous function with respect to the variable r, with the finite number of discontinuities of the first kind. The remaining conditions on b_{ij} are preserved without modification.

We can easily modify the method developed above for the problem under consideration. The modification has a technical character and consists in the following. The differentiating of b_{ij} with respect to the variable r has been used only in Lemma 19 at the integration by parts in (8.41). It is clear that the discontinuity of b_{ij} changes the limit value of the integral. In order to investigate the case, we introduce the trial function of the following form:

$$\phi(r, \Theta) = \begin{cases} \psi(r, \Theta) \in C^\infty(Q) \text{ in } Q, \\ \psi(r_0, \Theta) \left[1 - q(\Theta)\xi \left(\frac{r - r_0}{\delta}, \Theta \right) \right] \text{ in } Q_\delta. \end{cases} \tag{8.52}$$

We select the functions $\xi(r, \Theta)$ and $q(\Theta)$ in (8.52) from the condition that $\phi(r, \Theta)$ (8.52) belongs to $H_0^1(Q \bigcup Q_\delta)$. This condition is satisfied if

$$1 - q(\Theta)\xi(\Phi(\Theta), \Theta) = 0, \tag{8.53}$$

$$\xi(0, \Theta) = 0,$$

$$\xi(r, \Theta) \in H^1(Q_\delta), \ q(\Theta) \in C^1(\partial Q).$$

In addition, we want the equality

$$\frac{\partial \xi}{\partial r}(r, \Theta) = \frac{1}{B_{rr}\left(\dfrac{r - r_0}{\delta}, \Theta\right)} \tag{8.54}$$

to take place.

The equality (8.54) allows us to get rid of the discontinuity of the integrand in (8.41) and, as a consequence, to carry out the computations similar to the previous ones (it is demonstrated below). Thus, it is sufficient to verify that the functions $\xi(r, \Theta)$ and $q(\Theta)$ with the required properties exist. It can be done in the following way. On the basis of (8.53) and (8.54) we have

$$\xi(r, \Theta) = \int_{r_0}^{r} \frac{d\rho}{B_{rr}\left(\dfrac{\rho - r_0}{\delta}, \Theta\right)}, \tag{8.55}$$

where

$$q(\Theta) = \frac{1}{\displaystyle\int_{r_0}^{r_0 + \delta\Phi(\Theta)} \dfrac{d\rho}{B_{rr}\left(\dfrac{\rho - r_0}{\delta}, \Theta\right)}}. \tag{8.56}$$

Introducing the new variable

$$\eta = \frac{\rho - r_0}{\delta},$$

we can rewrite (8.56) as

$$q(\Theta) = \frac{\delta}{\displaystyle\int_{r_0}^{\Phi(\Theta)} \dfrac{d\eta}{B_{rr}(\eta, \Theta)}}. \tag{8.57}$$

From the equalities (8.55) and (8.56) it is seen that the selection of the trial function in the form (8.52) guarantees the differentiability of the expression

$$B_{rr}\left(\frac{r-r_0}{\delta},\Theta\right)\cdot\frac{\partial\phi}{\partial r}(r,\Theta)= \tag{8.58}$$

$$=\frac{1}{\delta}B_{rr}\left(\frac{r-r_0}{\delta},\Theta\right)\cdot\frac{q(\Theta)\psi(r_0,\Theta)}{B_{rr}\left(\dfrac{r-r_0}{\delta},\Theta\right)}=\frac{1}{\delta}q(\Theta)\psi(r_0,\Theta)$$

occurring in (8.52). After this, for the trial function (8.52) the proof of Lemma 19 goes though with the replacement of the quantity

$$K(\mathbf{x})=\frac{B_{rr}(\mathbf{x})}{\Phi(\mathbf{x})}$$

in (8.48) by

$$\mathcal{K}(\Theta)=\frac{1}{\displaystyle\int_0^{\Phi(\Theta)}\frac{d\eta}{B_{rr}(\eta)}}. \tag{8.59}$$

This can be seen from (8.58). As a result we arrive at the following statement.

Lemma 20. *For trial function $\phi(r,\Theta)$ (8.52) with a fixed $\psi(r,\Theta)$*

$$\delta\int_{Q_\delta}B_{rr}(r,\Theta)\frac{\partial\varphi^\delta}{\partial r}(r,\Theta)\frac{\partial\phi}{\partial r}(r,\Theta)drd\Theta\;\rightarrow \tag{8.60}$$

$$\rightarrow-\int_{\partial Q}\mathcal{K}(\Theta)\left(T(\Theta)-\varphi^*(r_0,\Theta)\right)\psi(r_0,\Theta)d\Theta$$

as $\delta\to 0$.

Theorem 10. *As $\delta\to 0$, the sequence $\{\varphi^\delta(\mathbf{x})\}$ of the solutions of the problem (8.2)–(8.4) with the functions b_{ij} determined by (8.51) converges weakly in $H^1(Q)$ (strongly in $L_2(Q)$) to the solution of the equation (8.49) with the boundary condition*

$$a_{ij}(\mathbf{x})\frac{\partial\varphi^*}{\partial x_j}n_i=\mathcal{K}(\mathbf{x})(T(\mathbf{x})-\varphi^*(\mathbf{x}))\;\;on\;\partial Q. \tag{8.61}$$

The proof is entirely similar to the proof of Theorem 9. We note that formula (8.59) is similar to formula for the homogenized thermal conductivity for laminated material in the direction perpendicular to laminas.

Transport characteristic of thin cover (membrane)

We consider a closed surface ∂Q of constant thickness δh, which separate domain Q from the outer space. Often, such surfaces are called membranes. We consider the diffusion problem and assume that concentration of the substance in ∂Q is equal to φ^+ and it is equal to φ^- outside G. The difference of the concentrations causes the substance infiltration through the membrane. The quantity of the substance infiltrating through the surface of unit area – flux, in the considered case is calculated as

$$flux = \frac{1}{|\partial Q|} \int_{\partial Q} \frac{\partial \varphi^*}{\partial \mathbf{n}} d\mathbf{x}.$$

The coefficient $K(\mathbf{x})$ from the equation (8.50) or the coefficient $\mathcal{K}(\mathbf{x})$ from the equation (8.61) describes flux through the cover when the potentials $T(\mathbf{x})$ and φ^* are applied to the opposite sides of the cover. Thus, they are the transport characteristics of the cover.

We consider the case where the cover is made of homogeneous isotropic material. In this case

$$a_{ij} = A\delta_{ij}, \ b_{ij} = B\delta_{ij},$$

where $\delta_{ii} = 1$ and $\delta_{ij} = 0$ if $i \neq j$.

Then (8.50) takes the form

$$A\frac{\partial \varphi^*}{\partial \mathbf{n}} = \frac{B}{h}(\varphi^- - \varphi^+) \tag{8.62}$$

and we obtain the following formula for computation of flux through a membrane

$$flux = \frac{B}{Ah}(\varphi^- - \varphi^+). \tag{8.63}$$

8.1.6. Numerical analysis of transport in a body with thin cover

We consider a body, whose geometry is generated by two circles. The first disk D_1 has the following characteristics: coordinate of center $\mathbf{x}_1 = (0, 0.1)$, radius $R_1 = 1.02$. The second disk D_2 has the following characteristics: coordinate of center $\mathbf{x}_1 = (0, 0)$, radius $R_1 = 1$.

The domain Q_1 lying between the disks is a thin ring of variable thickness. The thickness varies continuously from 0.01 on the top to 0.03 on the bottom. This is a model of cover. The domain Q lying in the disk D_2 is a model of body.

The boundary condition on $\partial Q_1 = D_2$ is the following:

$$T(\mathbf{x}) = 1 \text{ on the right semicircle,}$$

$$T(\mathbf{x}) = 0 \text{ on the left semicircle.}$$

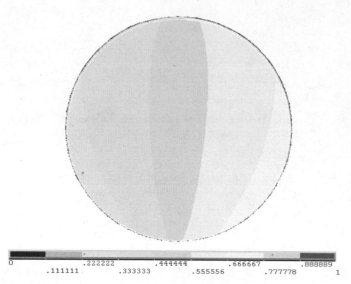

Figure 8.3. *Distribution of potential in a body with thin cover. Ratio = 100.*

The materials of the body and the cover are assumed to be homogeneous and isotropic. In this case

$$a_{ij} = A\delta_{ij}, \quad b_{ij} = B\delta_{ij},$$

where $\delta_{ii} = 1$ and $\delta_{ij} = 0$ if $i \neq j$.

In Fig. 8.3 numerically computed distribution of potential is presented for the ratio of the transport coefficients of the material of the body and material of the cover equal to

$$\frac{b}{a} = 100.$$

The nonsymmetry of the figure in the vertical direction is related to nonsymmetry of the cover in the vertical direction. The cover is more thin on the top. Then, in accordance with our previous computations, the transport characteristics

$$K(\mathbf{x}) = \frac{B_{rr}(\mathbf{x})}{\Phi(\mathbf{x})}$$

of the boundary are higher on the top. The distribution of potential displayed in Fig. 8.3 is in qualitative agreement with this prediction.

In Fig. 8.4 (left) numerically computed distribution of potential is presented for the ratio of the transport coefficients of the material of the body and material of the cover equal to

$$\frac{b}{a} = 1000.$$

In Fig. 8.4 (right) magnified fragments of ANSYS graphical output are presented for the left and the right fragments of the body with the cover.

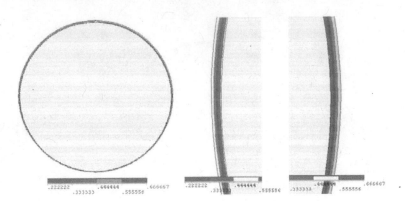

Figure 8.4. *Distribution of potential in a body with thin cover (left). Ratio = 100. The magnified left and right fragments of the body (right).*

In this case the cover has ten times more protecting characteristics than in the previous example. The distribution of potential is displayed in Fig. 8.4. It is seen that the change of potential occurs in the thin cover, see Fig. 8.4 (right).

8.2. Elastic bodies with thin underbodies layer (glued bodies)

An adhesive assembly is obtained by forming a thin adhesive layer between glued bodies. Usually, the adhesive substance is soft at least as compared with the glued bodies. Tables 8.1 and 8.2 present elastic moduli of some widely used structural and adhesive materials [217]. Since the problem involves large and small parameters, asymptotic method can be applied to the analysis of the problem.

Table 8.1. *Elastic Moduli of Structural Materials.*

Material	Young's modulus GPa
steel	200
glass	70
aluminum	60–70
oak wood	11

Table 8.2. *Elastic Moduli of Adhesive Materials.*

Material	Young's modulus GPa
nylon	2–4
epoxi	1–4
resins	0.01–0.1

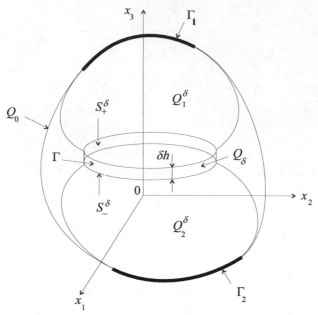

Figure 8.5. *Two glued bodies.*

8.2.1. Formulation of the problem

We consider two three-dimensional elastic bodies occupying domains Q_1^δ, Q_2^δ and thin domain Q_δ between these bodies, see Fig. 8.5. We assume that domain Q_δ is occupied by a soft material. We denote the characteristic thickness of the domain Q_δ by $\delta \ll 1$. This is our model of glued bodies. The domain Q_δ is a model of adhesive layer (glue layer).

We denote $a_{ijkl}(\mathbf{x})$ the tensor of elastic constants of the elastic bodies and $\delta b_{ijkl}(\mathbf{x})$ the tensor of elastic constants of the glue. It means that the ratio of elastic constants has the order of δ.

For simplicity (which does not lead to loss of generality of our consideration), we assume that the domain Q_δ (the adhesive layer) is a cylinder

$$Q_\delta = \{(\tilde{x}, x_3) : \tilde{x} \in S \subset R^2, -\delta h \le x_3 \le \delta h\}. \tag{8.64}$$

Here and afterward, $\tilde{x}$ means the first two coordinates of the vector $\mathbf{x}$:

$$\tilde{x} = (x_1, x_2).$$

As $\delta \to 0$, the domain Q_δ tends to the surface S called the adhesive surface or adhesive joint.

The following standard conditions are applied to the elastic constants (see, e.g., [118, 216]):

$$c_1|e_{ij}|^2 \le a_{ijkl}(\mathbf{x})e_{ij}e_{kl} \le c_2|e_{ij}|^2,$$

$$c_1|e_{ij}|^2 \le b_{ijkl}(\mathbf{x})e_{ij}e_{kl} \le c_2|e_{ij}|^2$$

for any $\mathbf{x} \in (Q_1^\delta \bigcup Q_2^\delta \bigcup Q_\delta)$ and any e_{ij} such that $e_{ij} = e_{ji}$;

$$a_{ijkl}(\mathbf{x}) = a_{klij}(\mathbf{x}), \quad a_{ijkl}(\mathbf{x}) = a_{ijlk}(\mathbf{x}) \in C^1(R^n),$$

$$b_{ijkl}(\mathbf{x}) = b_{klij}(\mathbf{x}), \quad b_{ijkl}(\mathbf{x}) = b_{ijlk}(\mathbf{x}) \in C^1(R^n),$$

$$\|a_{ijkl}\|_{C^1(R^3)}, \ \|b_{ijkl}\|_{C^1(R^3)} \leq c_3 < \infty.$$

Here constants $0 < c_1, c_2, c_3 < \infty$ do not depend on $\mathbf{x}$ and e_{ij}.

We assume that the boundaries of the domains Q_1^δ and Q_2^δ and the boundary of the surface S is as smooth as necessary (in particular, that they are sufficiently regular to apply the basic theorems from [113, 212]).

We denote

$$S_+^\delta = \{\mathbf{x} = (\tilde{x}, \delta h) : \tilde{x} \in S\}$$

and

$$S_-^\delta = \{\mathbf{x} = (\tilde{x}, -\delta h) : \tilde{x} \in S\}$$

the surfaces of contact (surfaces of adhesion) of the solid bodies Q_1^δ and Q_2^δ with the glue layer Q_δ.

On the surfaces $\Gamma_1 \subset \partial Q_1^\delta$ and $\Gamma_2 \subset \partial Q_2^\delta$, which do not cross the surfaces of adhesion, we state the following condition with respect to the displacements $\mathbf{u}^\delta$:

$$\mathbf{u}^\delta(\mathbf{x}) = 0$$

and we assume that the remaining surfaces are free (normal stresses on these surfaces are equal to zero):

$$a_{ijkl}(\mathbf{x})\mathrm{def}(\mathbf{u}^\delta)_{kl}(\mathbf{x})n_j = 0.$$

Here and afterward $\mathrm{def}(\mathbf{u})$ means the strain tensor [358], i.e.,

$$\mathrm{def}(\mathbf{u})_{ij} = \frac{1}{2}\left(\frac{\partial u_i}{\partial x_i} + \frac{\partial u_j}{\partial x_j}\right).$$

The domains Q_1^δ, Q_2^δ and Q_δ are not fixed. As $\delta \to 0$, the domain Q_δ tends to the surface S (the adhesive surface); the domains Q_1^δ and Q_2^δ tend to limit positions Q_1 and Q_2, respectively. We introduce domain Q_0, which contains the domains Q_1^δ, Q_2^δ and Q_δ for $\delta \to 0$ as shown in Fig. 8.5. The domain Q_0 does not depend on the parameter δ.

We introduce functional space V as closure in the norm

$$\|\mathbf{u}\|_{H^1(Q_0)} = \sqrt{\sum_{i,j=1}^{3} \int_{Q_0} \left|\frac{\partial u_i}{\partial u_j}(\mathbf{x})\right|^2 d\mathbf{x} + \sum_{i=1}^{3} \int_{Q_0} |u_i(\mathbf{x})|^2 d\mathbf{x}} \qquad (8.65)$$

of the set of infinitely differentiating functions equal to zero on Γ_1 and Γ_2.

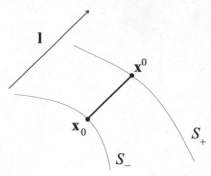

Figure 8.6. *On condition* l.

Another equivalent definition is

$$V = \{\mathbf{u}(\mathbf{x}) \in \{H^1(Q_0)\}^3 : \mathbf{u}(\mathbf{x}) = 0 \text{ on } \Gamma_1 \bigcup \Gamma_2\}.$$

If the system of the bodies under consideration is subjected to mass forces $\mathbf{f}(\mathbf{x})$, the displacements $\mathbf{u}^\delta(\mathbf{x})$ can be determined from the solution of the following minimization problem [118, 370]:

$$I_\delta(\mathbf{u}) \to \min, \ \mathbf{u}(\mathbf{x}) \in V, \tag{8.66}$$

where

$$I_\delta(\mathbf{u}) = \frac{1}{2} \int_Q a_{ijkl}(\mathbf{x}) \text{def}(\mathbf{u})_{ij}(\mathbf{x}) \text{def}(\mathbf{u})_{kl}(\mathbf{x}) d\mathbf{x} +$$

$$+ \frac{\delta}{2} \int_{Q_\delta} b_{ijkl}(\mathbf{x}) \text{def}(\mathbf{u})_{ij}(\mathbf{x}) \text{def}(\mathbf{u})_{kl}(\mathbf{x}) d\mathbf{x} + \int_{Q_1^\delta \bigcup Q_2^\delta \bigcup Q_\delta} \mathbf{f}(\mathbf{x}) \mathbf{u}(\mathbf{x}) d\mathbf{x}.$$

In the problem (8.66) both coefficient and domains of integration depend on the parameter δ. Our aim is deriving the limit problem for the problem (8.66) as $\delta \to 0$.

8.2.2. Estimates for solution of the problem (8.66)

We have analyzed a problem similar to the problem (8.66) in the previous section. In the previous section, we have considered the scalar problem. The problem under consideration is a vectorial problem. In addition, it involves the displacements $\mathbf{u}^\delta(\mathbf{x})$ in the form of strains $\text{def}(\mathbf{u}^\delta)$. This specific of the elasticity problem is well known [109, 118]. It also brings an additional problem to asymptotic analysis of the elasticity problem with thin layers. In particular, Lemma 11 cannot be expanded directly to the problem under consideration.

We formulate a condition on the geometry of the pair of the surfaces S_+ and S_-, see Fig. 8.6.

Definition 18. *We say that the pair of the surfaces S_+ and S_- satisfy condition l if the surface S_+ can be represented as a sum of measurable domains $\Gamma^1, \ldots, \Gamma^n$ such that*

a) for any Γ^i there exists a vector l such that the interval $[\mathbf{x}_0, \mathbf{x}^0]$, where $\mathbf{x}_0 \in S_-$, $\mathbf{x}^0 \in \Gamma^i$ and $[\mathbf{x}_0, \mathbf{x}^0]$ is parallel to the vector l, has the length not greater than CD, where constant $C < \infty$ does not depend on $\mathbf{x}^0$ and δ, and

$$D = \max_{\mathbf{x} \in S_+} \min_{\mathbf{y} \in S_-} |\mathbf{x} - \mathbf{y}|$$

is characteristic distance between the surfaces S_+ and S_-.

b) for any Γ^i there exists a set of linearly independent vectors $\{\mathbf{l}_\alpha\}$ $(\alpha = 1, 2, 3)$ such that every vector $\mathbf{l}_\alpha$ has the property described above for the vector l.

It is evident that the parallel planar surfaces S_+^δ and S_+^δ satisfy condition l.

We consider the interval $[\mathbf{x}_0, \mathbf{x}^0]$ described in the item *a*) of Definition 18. We call point $\mathbf{x}_0 \in S_+$ l-projection of the point $\mathbf{x}^0 \in S_-$ to the surface S_+ and define map

$$\mathbf{Pr}_{\mathbf{l}} : \mathbf{x}_0 \to \mathbf{x}^0 = \mathbf{Pr}_{\mathbf{l}}\mathbf{x}_0.$$

We call the set

$$\mathbf{Pr}_{\mathbf{l}}S = \{\mathbf{x}^0 = \mathbf{Pr}_{\mathbf{l}}\mathbf{x}_0 : \mathbf{x}_0 \in S \subset S_-\} \subset S_+$$

l-projection of the set $S \subset S_-$ to the surface S_+.

Lemma 21. *If surfaces S_+ and S_- satisfy condition l then for any $\mathbf{u}(\mathbf{x}) \in \{H^1(Q)\}$ the following inequality holds:*

$$\int_{S_+} |\mathbf{u}(\mathbf{x})|^2 d\mathbf{x} \leq \mathbf{C} \int_{S_-} |\mathbf{u}(\mathbf{x})|^2 d\mathbf{x} + CD \int_Q |\mathrm{def}(\mathbf{u})|^2(\mathbf{x})d\mathbf{x}, \qquad (8.67)$$

where $\mathbf{C} < \infty$ is determined by the constant C from Definition 18 and smoothness of the surfaces S_+ and S_- (the characteristic of the smoothness of the surfaces S_+ and S_- is the number γ introduced by formula (8.71) below).

Proof. In accordance with (8.65)

$$\frac{\partial u_i}{\partial u_j} + \frac{\partial u_j}{\partial u_i} = 2\mathrm{def}(\mathbf{u})_{ij}. \qquad (8.68)$$

We take a vector l described in Definition 18, item *a*) (we assume that l is not parallel to Ox_1x_2-plane) and denote by Γ_-^i the l-projection of the set $\Gamma^i \subset S_+$ to the surface S_-.

Forming contraction of tensor equation (8.68) with the tensor $l_i l_j$ (recall that l does not depend on $\mathbf{x}$), we obtain

$$\frac{d}{d\mathbf{l}}(\mathbf{l}\mathbf{u}) = \mathrm{def}(\mathbf{u})_{ij} l_i l_j, \qquad (8.69)$$

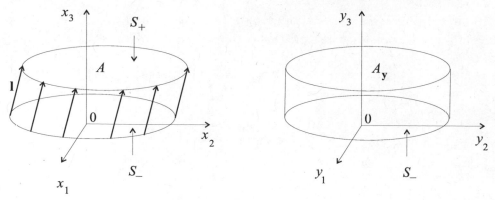

Figure 8.7. *The cylinders A (left) and $A_{\mathbf{y}}$ (right).*

where

$$\frac{d}{d\mathbf{l}} = l_i \frac{\partial}{\partial x_i}$$

is the operator of derivation in the direction of the vector $\mathbf{l}$.

We consider an interval $[\mathbf{x}_0, \mathbf{x}^0]$ described in Definition 18 and make parameterization of this interval of the form:

$$\mathbf{x} = \mathbf{x}_0 + (\mathbf{x}^0 - \mathbf{x}_0)\frac{x_3 + D}{2D}, \ \ x_3 \in [-D, D].$$

It is possible because $\mathbf{l}$ is not parallel to Ox_1x_2-plane.

We denote $F(\mathbf{x}) = \mathbf{u}(\mathbf{x})\mathbf{l}$. Integrating equality (8.69) over the interval $[\mathbf{x}_0, \mathbf{x}^0]$, we obtain

$$F(\mathbf{x}^0) = F(\mathbf{x}_0) + \int_{-D}^{D} \mathrm{def}(\mathbf{u})_{ij}(\mathbf{x})l_i l_j \Delta dx_3, \tag{8.70}$$

where

$$0 < \Delta = \left|\frac{\mathbf{x}_0 - \mathbf{x}^0}{2D}\right|^2 < \infty.$$

We consider cylinder A with the top base $S_+ \subset S_+^\delta$, the bottom base $S_- \subset S_-^\delta$ and the generatrix parallel to the vector $\mathbf{l}$, see Fig. 8.7. We make a change of variables $\mathbf{x} \to \mathbf{y}$, which transfer the inclined cylinder A to a cylinder $A_{\mathbf{y}}$ with the bottom base S_- and the generatrix parallel to Oz–axis. This change of variables is not degenerated, thus there exists $\gamma > 0$, such that

$$\left|\frac{\partial \mathbf{x}}{\partial \mathbf{y}}\right| \geq \gamma, \left|\frac{\partial \mathbf{y}}{\partial \mathbf{x}}\right| \geq \gamma, \left|\frac{\partial \mathbf{x}_\Gamma}{\partial \tilde{y}}\right| \geq \gamma, \left|\frac{\partial \tilde{y}}{\partial \mathbf{x}_\Gamma}\right| \geq \gamma. \tag{8.71}$$

In the new variables $\mathbf{y}$, equality (8.70) takes the form (as above, $\tilde{y} = (y_1, y_2)$)

$$F(\tilde{y}, D) = F(\tilde{y}, -D) + \int_{-D}^{D} \mathrm{def}(\mathbf{u})_{ij}(\mathbf{x})l_i l_j \Delta dy_3, \tag{8.72}$$

where $0 < \Delta < \infty$.

Applying the Cauchy-Bunyakovsky inequality to (8.72), we obtain the following inequality:

$$|F(\tilde{y}, D)|^2 \leq 2|F(\tilde{y}, -D)|^2 + CD \int_{-D}^{D} \text{def}(\mathbf{u})^2(\mathbf{y}) dy_3, \qquad (8.73)$$

where $C < \infty$. Here we use that $|\mathbf{l}| < \infty$ and $0 < \Delta < \infty$.

It is denoted

$$\text{def}(\mathbf{u})^2 = \sum_{i,j=1}^{3} \text{def}(\mathbf{u})_{ij}^2.$$

Integrating (8.73) over the base of the cylinder A (note that in the variables $\mathbf{y}$ the top and bottom bases of the cylinder $A_{\mathbf{y}}$ are congruent), we obtain

$$\int_{\Gamma_y^i} |F(\tilde{y}, D)|^2 d\tilde{y} \leq 2 \int_{\Gamma_{y-}^i} |F(\tilde{y}, -D)|^2 d\tilde{y} + c_6 D \int_{A_{\mathbf{y}}} \text{def}(\mathbf{u})^2(\mathbf{y}) d\mathbf{x}.$$

Returning to the original variables $\mathbf{x}$, we obtain

$$\int_{\Gamma^i} |F(\tilde{y}, D)|^2 \left|\frac{\partial \tilde{y}}{\partial \mathbf{x}_\Gamma}\right| (\mathbf{x}) d\mathbf{x} \leq$$

$$\leq \int_{\Gamma^i_-} |F(\tilde{y}, D)|^2 \left|\frac{\partial \tilde{y}}{\partial \mathbf{x}_\Gamma}\right| (\mathbf{x}) d\mathbf{x} + c_6 D \int_A \text{def}(\mathbf{u})^2(\mathbf{x}) \left|\frac{\partial \tilde{y}}{\partial \mathbf{x}}\right| (\mathbf{x}) d\mathbf{x} \leq$$

$$\leq \frac{1}{\gamma} \int_{S_-} |F(\mathbf{x})|^2 d\mathbf{x} + \frac{c_6}{\gamma} D \int_Q F^2(\mathbf{x}) \text{def}(\mathbf{u})^2(\mathbf{x}) d\mathbf{x}.$$

Here we use (8.71). In the force of (8.71), we also can write

$$\int_{\Gamma^i} |F|^2(\mathbf{x}) \left|\frac{\partial \tilde{y}}{\partial \mathbf{x}_\Gamma}\right| d\mathbf{x} \geq \gamma \int_{\Gamma^i} |F(\mathbf{x})|^2 d\mathbf{x}.$$

By virtue of this inequality we obtain

$$||F||_{L_2(\Gamma^i)}^2 \leq \tilde{c}||F||_{L_2(S_-)}^2 + \tilde{c}D||\text{def}(\mathbf{u})||_{L_2(Q)}^2, \qquad (8.74)$$

where $\tilde{c} < \infty$.

We use the notation

$$||\text{def}(\mathbf{u})||_{L_2(Q)}^2 = \sum_{i,j=1}^{3} \int_Q \text{def}(\mathbf{u})_{ij}^2(\mathbf{x}) d\mathbf{x}.$$

Since the number of the sets Γ_i $(i = 1, ..., n)$ is finite, we obtain from (8.74) the following inequality

$$||F||^2_{L_2(S_+)} \leq \tilde{c}_n ||F||^2_{L_2(S_-)} + \tilde{c}_n D ||\text{def}(\mathbf{u})||^2_{L_2(Q)}, \tag{8.75}$$

where $\tilde{c}_n < \infty$.

We have denoted $F(\mathbf{x}) = \mathbf{u}(\mathbf{x})\mathbf{l}$. For $\mathbf{l} = \mathbf{l}_\alpha$, where $\mathbf{l}_\alpha$ is a vector described in item b) of Definition 18, estimate (8.75) is valid for the function

$$F_\alpha(\mathbf{x}) = \mathbf{u}(\mathbf{x})\mathbf{l}_\alpha.$$

We take a system of linearly independent vectors $\{\mathbf{l}_\alpha, \alpha = 1, 2, 3\}$ described in item b) of Definition 18 and consider the corresponding functions $F_\alpha(\mathbf{x})$ $(\alpha = 1, 2, 3)$.

We write the equalities

$$F_\alpha(\mathbf{x}) = \mathbf{u}(\mathbf{x})\mathbf{l}_\alpha \ (\alpha = 1, 2, 3), \tag{8.76}$$

and denote $\mathbf{F}(\mathbf{x}) = (F_1(\mathbf{x}), F_2(\mathbf{x}), F_3(\mathbf{x}))$ - vector in R^3 and $\mathbf{L} = \{\mathbf{l}_1, \mathbf{l}_2, \mathbf{l}_3\}$ - 3×3 matrix formed of the vectors $\mathbf{l}_\alpha$ $(\alpha = 1, 2, 3)$. Using these notations, we can write the equalities (8.76) in the form

$$\mathbf{L}\mathbf{u}(\mathbf{x}) = \mathbf{F} \tag{8.77}$$

and consider (8.77) as a system of linear algebraic equations with respect to $\mathbf{u}(\mathbf{x})$.

The matrix $\mathbf{L}$ is invertible by virtue of Definition 18. The solution $\mathbf{u}(\mathbf{x})$ of the system (8.77) can be written as

$$\mathbf{u}(\mathbf{x}) = \mathbf{L}^{-1}\mathbf{F}(\mathbf{x}). \tag{8.78}$$

In accordance with the point b) of Definition 18

$$\det\mathbf{L} > 0, \tag{8.79}$$

$$\det\mathbf{L}^{-1} > 0.$$

From (8.79) it follows that

$$||\mathbf{u}||^2_{L_2(S_+)} \leq \tilde{C} \sum_{\alpha=1}^{3} ||\mathbf{F}||^2_{L_2(S_+)},$$

$$||\mathbf{F}||^2_{L_2(S_-)} \leq \tilde{C} \sum_{\alpha=1}^{3} ||\mathbf{u}||^2_{L_2(S_-)}, \tag{8.80}$$

where $\tilde{C} < \infty$.

Using (8.80), we obtain from (8.75) the following inequality

$$||\mathbf{u}||^2_{L_2(S_+)} \le C||\mathbf{u}||^2_{L_2(S_-)} + CD||\text{def}(\mathbf{u})||^2_{L_2(Q)}, \tag{8.81}$$

where $C < \infty$. **Proof is completed.**

Remark 3. If $S_+ = S^\delta_+$ and $S_- = S^\delta_-$ then in (8.67) $D = \delta$. If $S_+ = S^\delta_-$, $S_- = \Gamma_1$ or $S_+ = S^\delta_+$, $S_- = \Gamma_2$ then in (8.67) $D = 1$.

Lemma 22. *There exists $\delta_0 > 0$ such that for any function $\mathbf{u}(\mathbf{x}) \in H^1(Q^\delta_1 \bigcup Q^\delta_2 \bigcup Q_\delta)$ the following inequality holds:*

$$||\mathbf{u}||^2_{L_2(Q^\delta_1 \bigcup Q^\delta_2 \bigcup Q_\delta)} \le C \left(||\text{def}(\mathbf{u})||_{L_2(Q^\delta_1 \bigcup Q^\delta_2)} + \delta^{1/2}||\text{def}(\mathbf{u})||^2_{L_2(Q_\delta)} \right) \tag{8.82}$$

for any $\delta \in (0, \delta_0]$, where $C < \infty$ does not depend on δ.

Proof. In order to evaluate the integral over the cylinder $Q_\delta = S \times [-\delta h, \delta h]$, we apply Lemma 21 and Remark 3. With regard to Lemma 21 and Remark 3 we have

$$\int_{Q_\delta} |\mathbf{u}(\mathbf{x})|^2 d\mathbf{x} = \int_{-\delta h}^{\delta h} dx_3 \int_S |\mathbf{u}(\mathbf{x})|^2 d\tilde{x} \le \tag{8.83}$$

$$\le \int_{-\delta h}^{\delta h} dx_3 \left(C \int_S |\mathbf{u}(\mathbf{x})|^2 d\tilde{x} + C\delta \int_{-\delta h}^{\delta h} dx_3 \int_{Q_\delta} |\text{def}(\mathbf{u})(\mathbf{x})|^2 d\tilde{x} \right) \le$$

$$\le C\delta h \int_S |\mathbf{u}(\mathbf{x})|^2 d\mathbf{x} + C\delta^2 h \int_{Q_\delta} |\text{def}(\mathbf{u})|^2 d\mathbf{x}.$$

From (8.75) it follows that

$$||\mathbf{u}||_{L_2(Q_\delta)} \le c_7 \delta^{1/2} \left(||\mathbf{u}||_{L_2(S_-)} + \delta^{1/2}||\text{def}(\mathbf{u})||_{L_2(Q_\delta)} \right), \tag{8.84}$$

where $c_7 < \infty$. In (8.84) S_- means one of the surfaces S^δ_- or S^δ_+.

In similar manner, we obtain the following estimates for the domains Q^δ_1 and Q^δ_2

$$||\mathbf{u}||_{L_2(Q^\delta_i)} \le c_7 \delta^{1/2} \left(||\mathbf{u}||_{L_2(S_-)} + ||\text{def}(\mathbf{u})||_{L_2(Q_\delta)} \right) \ (i = 1, 2). \tag{8.85}$$

In (8.85) $S_- = \Gamma_i$ $(i = 1, 2)$.

Applying Lemma 21 in the case $S_- = \Gamma_1$ and $S_+ = S^\delta_+$ and taking into account that $\mathbf{u}(\mathbf{x}) = 0$ on Γ_1, we obtain the following estimate:

$$||\mathbf{u}||_{L_2(S^\delta_+)} \le C||\text{def}(\mathbf{u})||_{L_2(Q^\delta_1)} \tag{8.86}$$

and a similar estimate for the case when $S_- = \Gamma_1$ and $S_+ = S^\delta_+$:

$$||\mathbf{u}||_{L_2(S^\delta_-)} \le C||\text{def}(\mathbf{u})||_{L_2(Q^\delta_2)}. \tag{8.87}$$

Proof is completed.

The symmetric difference $A \bigtriangleup B$ of the domains A and B is introduced as [308]

$$A \bigtriangleup B = (A \setminus B) \bigcup (B \setminus A).$$

Lemma 23. *For the consequence $\{\mathbf{u}^\delta(\mathbf{x})\}$ of the solution of problem (8.66) the following estimates hold:*

$$\|\mathrm{def}(\mathbf{u}^\delta)\|_{L_2((Q_1 \bigcup Q_2) \bigtriangleup (Q_1^\delta \bigcup Q_2^\delta \bigcup Q^\delta))}, \ \delta^{1/2}\|\mathrm{def}(\mathbf{u}^\delta)\|_{L_2(Q^\delta)}, \qquad (8.88)$$

$$\|\mathbf{u}^\delta\|_{L_2(Q_1^\delta \bigcup Q_2^\delta \bigcup Q_\delta)}, \ \|\mathbf{u}^\delta\|_{H^1(Q_1^\delta \bigcup Q_2^\delta)}, \ \delta^{1/2}\|\mathbf{u}^\delta\|_{H^1(Q^\delta)} \leq c_8,$$

where $c_8 < \infty$ does not depend on δ.

Proof. Since $0 \in V$, we have the following inequality for the solution $\mathbf{u}^\delta(\mathbf{x})$ of problem (8.66):

$$I_\delta(\mathbf{u}^\delta) \leq I_\delta(0) = 0$$

for any $\delta > 0$.

By virtue of the condition on the coefficients $a_{ijkl}(\mathbf{x})$ and $b_{ijkl}(\mathbf{x})$ and estimate (8.82) we have

$$c_1\|\mathrm{def}(\mathbf{u}^\delta)\|^2_{L_2(Q_1^\delta \bigcup Q_2^\delta)} + c_1\delta\|\mathrm{def}(\mathbf{u}^\delta)\|^2_{L_2(Q^\delta)} \leq \qquad (8.89)$$

$$\leq \|\mathbf{f}\|_{L_2(R^3)}\|\mathbf{u}^\delta\|_{L_2(Q_1^\delta \bigcup Q_2^\delta \bigcup Q^\delta)} \leq$$

$$\leq \|\mathbf{f}\|_{L_2(R^3)} \left(\|\mathrm{def}(\mathbf{u}^\delta)\|_{L_2(Q_1^\delta \bigcup Q_2^\delta)} + \delta^{1/2}\|\mathrm{def}(\mathbf{u}^\delta)\|_{L_2(Q^\delta)} \right).$$

The set of solutions of the inequality

$$c_1(x^2 + y^2) \leq A(B + x + y)$$

is a bounded set in R^2 for $0 < c_1$, A, $B < \infty$, see proof of Lemma 14. By virtue of this note and estimate (8.89) we obtain the first two inequalities presented in (8.88). After that the third inequality presented in (8.88) follows from (8.82).

To justify the fourth inequalities presented in (8.88), we use the second Korn inequality [118, 199]. For the domains Q_1^δ and Q_2^δ, it has the form

$$\|\mathbf{u}\|_{H^1(Q_1^\delta \bigcup Q_2^\delta)} \leq c_9 \left(\|\mathrm{def}(\mathbf{u})\|_{L_2(Q_1^\delta \bigcup Q_2^\delta)} + \|\mathbf{u}\|_{L_2(Q_1^\delta \bigcup Q_2^\delta)} \right). \qquad (8.90)$$

The fourth inequalities presented in (8.88) follows from (8.90) and the first and third estimates from (8.88). The last estimate presented in (8.88) follows from the Korn inequality. **Proof is completed.**

The original problem has been formulated in variable domains. It is convenient to pass from the original problem to an equivalent problem in a fixed domain Q_0, see Fig. 8.5, which does not depend on δ. This new problem is called an extension (or continuation) of the original problem (8.88).

Using the results presented in [244], we can formulate the following proposition.

Lemma 24. *Let Q_δ, Q_0 be domains with C^1-smooth boundaries ∂Q_δ, ∂Q_0 such that*
1) $Q_\delta \subset Q_0$ for all δ;
2) there exist C^1 homeomorphism $F_\delta : Q_\delta \to Q_0$, such that

$$\|F_\delta\|_{C^1(Q_\delta)} \leq c, \quad \|F_\delta^{-1}\|_{C^1(Q_0)} \leq c,$$

where $c < \infty$ does not depend on δ.
Then there exists an extension operator

$$G_\delta : \{H^1(Q_\delta)\}^3 \to \{H^1(Q_0)\}^3,$$

such that for any function $\mathbf{u}(\mathbf{x}) \in \{H^1(Q_\delta)\}^3$
a) $G_\delta(\mathbf{u})(\mathbf{x}) = \mathbf{u}(\mathbf{x})$ for $\mathbf{x} \in Q_\delta$,
b) $\|G_\delta(\mathbf{u})\|_{H^1(Q_0)} \leq C\|\mathbf{u}\|_{H^1(Q_\delta)}$, where $C < \infty$ does not depend on δ.

We assume the domain Q_0 has C^1 boundary ∂Q_0. Then Lemma 24 can be applied to the pair of domains $Q_0^- = \{(\tilde{x}, x_3) \in Q_0 : x_3 \geq -\delta\}$ and Q_1^δ and to the pair of domains $Q_0^+ = \{(\tilde{x}, x_3) \in Q_0 : x_3 \leq \delta\}$ and Q_2^δ; $\delta > 0$. In accordance with Lemma 24, there exist the extension operators

$$. \ G_\delta : \{H^1(Q_1^\delta)\}^3 \to \{H^1(Q_0^+)\}^3,$$

and

$$G_\delta : \{H^1(Q_2^\delta)\}^3 \to \{H^1(Q_0^-)\}^3,$$

which have the properties described in the items *a)*, *b)* of Lemma 24.

By virtue of Lemma 24 and estimates (8.88) the following inequalities hold for the solution $\mathbf{u}^\delta(\mathbf{x})$ of problem (8.66):

$$\|G_\delta^+ \mathbf{u}^\delta\|_{H^1(Q_0^+)} \leq C, \quad \|G_\delta^- \mathbf{u}^\delta\|_{H^1(Q_0^-)} \leq C, \tag{8.91}$$

where $C < \infty$ does not depend on δ.

By virtue of (8.67) and (8.88),

$$\|\mathbf{u}^\delta\|_{L_2(\partial Q_1^\delta)} \leq c, \quad \|\mathbf{u}^\delta\|_{L_2(\partial Q_2^\delta)} \leq c, \tag{8.92}$$

where $c < \infty$ does not depend on δ.

From (8.83), (8.84), (8.91) and (8.92) we obtain that

$$\|G_\delta^+ \mathbf{u}^\delta\|_{L_2((Q_1 \cup Q_2) \triangle (Q_1^\delta \cup Q_2^\delta \cup Q^\delta))} \leq \delta^{1/2} c_{10}, \tag{8.93}$$

$$\|G_\delta^- \mathbf{u}^\delta\|_{L_2((Q_1 \cup Q_2) \triangle (Q_1^\delta \cup Q_2^\delta \cup Q^\delta))} \leq \delta^{1/2} c_{10},$$

where $c_{10} < \infty$ does not depend on δ.

Deriving (8.93), we takes into account that the domain $(Q_1 \bigcup Q_2) \triangle (Q_1^\delta \bigcup Q_2^\delta \bigcup Q^\delta)$ has the thickness of the order of δ. **Proof is completed**.

After the estimates (8.91) and (8.93) have been obtained, we can construct the limit problem corresponding to the original problem (8.66).

Lemma 25. *The sequence $G_\delta^+ \mathbf{u}^\delta(\mathbf{x})$ is weakly compact in $H^1(Q_0^+)$ and the sequence $G_\delta^- \mathbf{u}^\delta(\mathbf{x})$ is weakly compact in $H^1(Q_0^-)$. Then there exists a subsequence $\{\mathbf{u}^\eta(\mathbf{x})\} \subset \{\mathbf{u}^\delta(\mathbf{x})\}$ such that*

$$G_\eta^+ \mathbf{u}^\eta(\mathbf{x}) \to \mathbf{u}^*(\mathbf{x}) \ \ weakly \ in \ H^1(Q_0^+),$$

$$G_\eta^- \mathbf{u}^\eta(\mathbf{x}) \to \mathbf{u}^*(\mathbf{x}) \ \ weakly \ in \ H^1(Q_0^-)$$

as $\eta \to 0$.

For this subsequence

$$\|\mathbf{u}^\eta - \mathbf{u}^*\|_{L_2(|(Q_1 \bigcup Q_2) \backslash (Q_1^\eta \bigcup Q_2^\eta \bigcup Q^\eta|)} \to 0,$$

$$Q_1^\eta \bigcup Q_2^\eta \bigcup Q^\eta \to Q_1 \bigcup Q_2$$

as $\eta \to 0$.

Remark 4. The limit function $\mathbf{u}^*(\mathbf{x})$ belongs to $\{H^1(Q_0^+)\}^3$ and $\{H^1(Q_0^-)\}^3$, but $\mathbf{u}^*(\mathbf{x}) \notin \{H^1(Q_0^+ \bigcup Q_0^-)\}^3$, generally. In particular, the function $\mathbf{u}^*(\mathbf{x})$ can take different values on the opposite sides of the surface S.

Lemma 25 follows directly from estimate (8.91), compactness of the embedding $H^1(Q)$ to $L_2(Q)$ [341, 342, 392] and estimate (8.93).

In the following computations we use the variational form of the problem (8.66), which has the form

$$\int_{Q_1^\delta \bigcup Q_2^\delta} a_{ijkl}(\mathbf{x}) \frac{\partial u_k^\delta}{\partial x_l}(\mathbf{x}) \frac{\partial v_i}{\partial x_j}(\mathbf{x}) d\mathbf{x} + \delta \int_{Q^\delta} b_{ijkl}(\mathbf{x}) \frac{\partial u_k^\delta}{\partial x_l}(\mathbf{x}) \frac{\partial v_i}{\partial x_j}(\mathbf{x}) d\mathbf{x} + \quad (8.94)$$

$$+ \int_{Q_1^\delta \bigcup Q_2^\delta \bigcup Q^\delta} \mathbf{f}(\mathbf{x}) \mathbf{v}(\mathbf{x}) d\mathbf{x} = 0$$

for any $\mathbf{v}(\mathbf{x}) \in V$.

The variational equation (8.94) is derived from the minimization problem (8.66) in traditional way [94, 124] (it is necessary take into account symmetry of elastic constants tensor [216, 358] and definition of strains (8.68)).

8.2.3. Construction of special trial function

In order to compute the limit values of integrals in (8.94) as $\delta \to 0$, we consider the trial function of special form. We consider the following function:

$$
\mathbf{v}(\mathbf{x}) = \begin{cases} \mathbf{v}^{(1)}(\mathbf{x}) \in C^\infty(Q_0^+) \cap V \text{ in } Q_1^\delta, \\[2mm] \mathbf{v}^{(2)}(\mathbf{x}) \in C^\infty(Q_0^-) \cap V \text{ in } Q_2^\delta, \\[2mm] (\mathbf{v}^{(1)}(\tilde{x},\delta h) - \mathbf{v}^{(2)}(\tilde{x},-\delta h))\dfrac{x_3+\delta h}{2\delta h} + \mathbf{v}^{(2)}(\tilde{x},-\delta h) \text{ in } Q^\delta, \end{cases} \tag{8.95}
$$

where $\mathbf{v}^{(1)}(\tilde{x},x_3)$ and $\mathbf{v}^{(2)}(\tilde{x},x_3)$ are arbitrary functions belonging to the functional spaces indicated in (8.95).

It is evident that $\mathbf{v}(\mathbf{x}) \in V$ (we remind that $\tilde{x} = (x_1,x_2)$ and $(\tilde{x},x_3) = \mathbf{x}$).

Lemma 26. *The function* $\mathbf{v}(\tilde{x},x_3)$ *(8.95) with fixed* $\mathbf{v}_1(\tilde{x},x_3)$ *and* $\mathbf{v}_2(\tilde{x},x_3)$ *satisfies the following conditions:*

$$
|\mathbf{v}(\mathbf{x})|, \left|\frac{\partial \mathbf{v}(\mathbf{x})}{\partial x_\alpha}\right| \leq C \max(\|\mathbf{v}^{(1)}\|_{C^1(Q)}, \|\mathbf{v}^{(2)}\|_{C^1(Q)}) \text{ in } Q_1^\delta \bigcup Q_2^\delta \bigcup Q^\delta, \tag{8.96}
$$

$$
\frac{\partial \mathbf{v}}{\partial x_3}(\mathbf{x}) = \frac{\mathbf{v}^{(1)}(\tilde{x},\delta h) - \mathbf{v}^{(2)}(\tilde{x},-\delta h)}{2\delta h} \text{ in } Q_\delta, \tag{8.97}
$$

$$
\left|\frac{\partial \mathbf{v}(\mathbf{x})}{\partial x_3}\right| \leq C \max(\|\mathbf{v}^{(1)}\|_{C^1(Q)}, \|\mathbf{v}^{(2)}\|_{C^1(Q)}) \text{ in } Q_1^\delta \bigcup Q_2^\delta \tag{8.98}
$$

$(\alpha = 1, 2)$.

Proof. All the formulas (8.96)–(8.98) can be derived by differentiating (8.95) with regard to the inequality $|x_3| \leq \delta h$ in Q_δ.

8.2.4. The convergence theorem and the limit model

Lemma 27. *For the trial function* $\mathbf{v}(\tilde{x},x_3)$ *(8.95) with fixed* $\mathbf{v}^{(1)}(\tilde{x},x_3)$ *and* $\mathbf{v}^{(2)}(\tilde{x},x_3)$ *the following limits take place as* $\eta \to 0$:

$$
\int_{Q_1^\eta \bigcup Q_2^\eta} a_{ijkl}(\mathbf{x})\frac{\partial u_k^\eta}{\partial x_l}(\mathbf{x})\frac{\partial v_i}{\partial x_j}(\mathbf{x})d\mathbf{x} \to \int_{Q_1 \bigcup Q_2} a_{ijkl}(\mathbf{x})\frac{\partial u_k^*}{\partial x_l}(\mathbf{x})\frac{\partial v_i}{\partial x_j}(\mathbf{x})d\mathbf{x}, \tag{8.99}
$$

$$
\int_{Q_1^\eta \bigcup Q_2^\eta \bigcup Q^\eta} \mathbf{f}(\mathbf{x})\mathbf{v}(\mathbf{x})d\mathbf{x} \to \int_{Q_1 \bigcup Q_2} \mathbf{f}(\mathbf{x})\mathbf{v}(\mathbf{x})d\mathbf{x}. \tag{8.100}
$$

Proof. First, we demonstrate that

$$\int_{Q_1 \triangle Q_1^\eta} a_{ijkl}(\mathbf{x}) \frac{\partial (G_\eta^+ \mathbf{u}^\eta(\mathbf{x}))_k}{\partial x_l} \frac{\partial v_i}{\partial x_j}(\mathbf{x}) d\mathbf{x} \to 0 \qquad (8.101)$$

as $\delta \to 0$.

In accordance with (8.91),

$$\|G_\eta^+ \mathbf{u}^\eta\|_{H^1(Q_0^+)} \leq C,$$

where $C < \infty$ does not depend on η.

By virtue of the conditions on the coefficients $a_{ijkl}(\mathbf{x})$ and inequalities (8.96) and (8.98)

$$\left| a_{ijkl}(\mathbf{x}) \frac{\partial v_i}{\partial x_j}(\mathbf{x}) \right| \leq c_3 C \max(\|\mathbf{v}^{(1)}\|_{C^1(Q)}, \|\mathbf{v}^{(2)}\|_{C^1(Q)})$$

in the domain $Q_1^\eta \bigcup Q_2^\eta$. From these inequalities and the Cauchy-Bunyakovsky inequality, we obtain that the absolute value of the integral in (8.101) does not exceed

$$\|G_\eta^+ \mathbf{u}^\eta\|_{H^1(Q_0^+)} \left\| a_{ijkl}(\mathbf{x}) \frac{\partial v_i}{\partial x_j}(\mathbf{x}) \right\|_{L_2(Q_1 \triangle Q_1^\eta)} \leq$$

$$\leq C \left\| a_{ijkl}(\mathbf{x}) \frac{\partial v_i}{\partial x_j}(\mathbf{x}) \right\|_{L_\infty(Q_1 \triangle Q_1^\eta)} |Q_1 \triangle Q_1^\eta|^{1/2} C.$$

From this, taking into account that $Q_1 \triangle Q_1^\eta \to 0$ as $\eta \to 0$, we obtain (8.101). With regard to (8.101), the limit (8.99) is a consequence of Lemma 25.

The limit (8.100) follows from the following. The integral

$$\int_{(Q_1^\eta \bigcup Q_2^\eta) \triangle (Q_1 \bigcup Q_2)} \mathbf{f}(\mathbf{x}) \mathbf{v}(\mathbf{x}) d\mathbf{x} \to 0$$

because

$$|(Q_1^\eta \bigcup Q_2^\eta) \triangle (Q_1 \bigcup Q_2)| \to 0$$

as $\eta \to 0$.

For integral over Q^η, we have

$$\int_{Q^\eta} \mathbf{f}(\mathbf{x}) \mathbf{v}(\mathbf{x}) \leq \|\mathbf{f}\|_{L_2(Q^\eta)} \cdot \|\mathbf{v}\|_{L_2(Q^\eta)} \leq$$

$$\leq \|\mathbf{f}\|_{L_2(Q^\eta)} \max \left(|\mathbf{v}^{(1)}|_{C(Q_0^\downarrow)}, |\mathbf{v}^{(2)}|_{C(Q_0^-)} \right) |Q^\eta|^{1/2}.$$

Proof is completed.

Lemma 28. *For the trial function* $\mathbf{v}(\tilde{x}, x_3)$ *defined by (8.95) with fixed* $\mathbf{v}^{(1)}(\tilde{x}, x_3)$ *and* $\mathbf{v}^{(2)}(\tilde{x}, x_3)$ *the following limits hold*

$$\textbf{(A)} \quad \eta \int_{Q_\eta} b_{i\alpha k3}(\mathbf{x}) \frac{\partial u_i^\eta}{\partial x_\alpha}(\mathbf{x}) \frac{\partial v_k}{\partial x_3}(\mathbf{x}) d\mathbf{x} \to 0, \tag{8.102}$$

$$\textbf{(B)} \quad \eta \int_{Q_\eta} b_{i3k\alpha}(\mathbf{x}) \frac{\partial u_i^\eta}{\partial x_3}(\mathbf{x}) \frac{\partial v_k}{\partial x_\alpha} d\mathbf{x} \to 0,$$

$$\textbf{(C)} \quad \eta \int_{Q_\eta} b_{i\alpha k\beta}(\mathbf{x}) \frac{\partial u_i^\eta}{\partial x_\alpha}(\mathbf{x}) \frac{\partial v_k}{\partial x_\beta}(\mathbf{x}) d\mathbf{x} \to 0$$

as $\eta \to 0$.

Here Latin indices i, k *take values 1, 2, 3 and Greek indices* α, β *take values 1, 2.*

Proof. The expression in (8.94) consists of the sum of the terms indicated in the items A to C. We consider these cases separately.

Case A. With regard to (8.98) we have

$$\eta \int_{Q_\eta} b_{i\alpha k3}(\mathbf{x}) \frac{\partial u_i^\eta}{\partial x_\alpha}(\mathbf{x}) \frac{\partial v_k}{\partial x_3}(\mathbf{x}) d\mathbf{x} = \tag{8.103}$$

$$= \eta \int_{Q_\eta} b_{i\alpha k3}(\mathbf{x}) \frac{\partial u_i^\eta}{\partial x_\alpha}(\mathbf{x}) \frac{\mathbf{v}_1(\tilde{x}, \delta h) - \mathbf{v}_2(\tilde{x}, -\delta h)}{2h} d\mathbf{x}.$$

Integrate by part in (8.103) and taking into account that index α takes values 1, 2 and vector $\mathbf{n}$ normal to surfaces S_η^+ and S_η^- has the form $(0, 0, 1)$, we obtain that the integral in (8.103) is equal to

$$\eta \int_\Gamma b_{i\alpha k3}(\mathbf{x}) u_i^\eta(\mathbf{x}) \frac{\mathbf{v}_1(\tilde{x}, \delta h) - \mathbf{v}_2(\tilde{x}, -\delta h)}{2h} n_\alpha d\mathbf{x} + \tag{8.104}$$

$$+ \int_{Q_\eta} u_i^\eta(\mathbf{x}) \frac{\partial}{\partial x_\alpha} \left[b_{i\alpha k3}(\mathbf{x}) \frac{\mathbf{v}_1(\tilde{x}, \delta h) - \mathbf{v}_2(\tilde{x}, -\delta h)}{2h} \right] d\mathbf{x}.$$

Here $\Gamma = \partial Q_\eta \setminus (S_\eta^+ \bigcup S_\eta^-)$ is the lateral (not contacting with the glued bodies Q_1^η and Q_2^η) surface of the domain Q_η, see Fig. 8.5.

We estimate integrals in (8.104). By virtue of conditions formulated for the coefficients $b_{ijkl}(\mathbf{x})$, we have

$$\left| \frac{\partial}{\partial x_\alpha} \left(b_{i\alpha k3}(\mathbf{x}) \frac{\mathbf{v}_1(\tilde{x}, \delta h) - \mathbf{v}_2(\tilde{x}, -\delta h)}{2h} \right) \right| \le C,$$

where $C < \infty$ does not depend on $\mathbf{x} \in Q_\eta$.

Using (8.88) and the Cauchy-Bunyakovsky inequality, we obtain that the absolute value of the last integral in (8.104) does not exceed

$$C \int_{Q_\eta} |\mathbf{u}^\eta(\mathbf{x})| d\mathbf{x} \leq C |Q^\eta|^{1/2} ||\mathbf{u}^\eta||_{L_2(Q^\eta)}. \qquad (8.105)$$

The right-hand side of (8.105) tends to zero as $\eta \to 0$ by virtue of $|Q^\eta| \to 0$ as $\eta \to 0$.

By virtue of conditions for the coefficients $b_{ijkl}(\mathbf{x})$, estimate (8.81) and Remark 3, we obtain that the absolute value of the first integral in (8.104) does not exceed

$$C \int_\Gamma |\mathbf{u}^\eta(\mathbf{x})| d\mathbf{x} \leq C |\Gamma|^{1/2} ||\mathbf{u}^\eta||_{L_2(\Gamma)} \leq \qquad (8.106)$$

$$\leq |\Gamma|^{1/2} \left(||\mathrm{def}(\mathbf{u}^\eta)||_{L_2(Q_1^\eta \cap Q_2^\eta)} + \eta^{1/2} ||\mathrm{def}(\mathbf{u}^\eta)||_{L_2(Q_\eta)} \right).$$

The Cauchy-Bunyakovsky inequality is used to derive the last inequality in (8.106).

By virtue of (8.88) and $|\Gamma| \to 0$ as $\eta \to 0$, we have that the right-hand part of (8.106) tends to zero as $\eta \to 0$.

Case B. Using (8.96), the Cauchy-Bunyakovsky inequality and inequality $|Q_\eta| \leq C\eta$, we have

$$\left| \eta \int_{Q_\eta} b_{i3k\alpha}(\mathbf{x}) \frac{\partial u_i^\eta}{\partial x_3}(\mathbf{x}) \frac{\partial v_k}{\partial x_\alpha}(\mathbf{x}) d\mathbf{x} \right| \leq \qquad (8.107)$$

$$\leq C \max(||\mathbf{v}^{(1)}||_{C^1(Q)}, ||\mathbf{v}^{(2)}||_{C^1(Q)}) \eta \int_{Q_\eta} \left| \frac{\partial u_i^\eta}{\partial x_3}(\mathbf{x}) \right| d\mathbf{x} \leq$$

$$\leq C \max(||\mathbf{v}^{(1)}||_{C^1(Q)}, ||\mathbf{v}^{(2)}||_{C^1(Q)}) \eta^{3/2} ||\mathbf{u}^\eta||_{H^1(Q_\eta)}.$$

By virtue of (8.88) the right-hand side of (8.107) tends to zero as $\eta \to 0$.

Case C. Using (8.96), the Cauchy-Bunyakovsky inequality and inequality

$$|Q_\eta| \leq C\eta,$$

we have

$$\left| \eta \int_{Q_\eta} b_{i\alpha k\beta}(\mathbf{x}) \frac{\partial u_i^\eta}{\partial x_\alpha}(\mathbf{x}) \frac{\partial v_k}{\partial x_\beta}(\mathbf{x}) d\mathbf{x} \right| \leq \qquad (8.108)$$

$$\leq C ||\mathbf{v}||_{C^1(Q)} \eta \int_{Q_\eta} \left| \frac{\partial u_i^\eta}{\partial x_\alpha}(\mathbf{x}) \right| d\mathbf{x} \leq$$

$$\leq C ||\mathbf{v}||_{C^1(Q)} \eta^{3/2} ||\mathbf{u}^\eta||_{H^1(Q_\eta)}.$$

By virtue of (8.88) the right-hand side of (8.107) tends to zero as $\eta \to 0$.

Lemma 29. *For the trial function* $\mathbf{v}(\tilde{x}, x_3)$ *defined by (8.95) with fixed* $\mathbf{v}^{(1)}(\mathbf{x})$ *and* $\mathbf{v}^{(2)}(\mathbf{x})$ *the following limit holds:*

$$\eta \int_{Q_\eta} b_{i3k3}(\mathbf{x}) \frac{\partial u_i^\eta}{\partial x_3}(\mathbf{x}) \frac{\partial v_k}{\partial x_3}(\mathbf{x}) dx \to \int_S \frac{b_{i3k3}(\mathbf{x})}{2h} [u_i(\mathbf{x})]_S [v_k(\mathbf{x})]_S dx, \qquad (8.109)$$

as $\eta \to 0$.

Here $[f(\mathbf{x})]_S = f(\tilde{x}, +0) - f(\tilde{x}, -0)$ *is the difference of values of the function on the opposite sides of the surface* S.

Proof. By virtue of (8.98) we have

$$\eta \int_{Q_\eta} b_{i3k3}(\mathbf{x}) \frac{\partial u_i^\eta}{\partial x_3}(\mathbf{x}) \frac{\partial v_k}{\partial x_3}(\mathbf{x}) dx = \int_{Q_\eta} b_{i3k3}(\mathbf{x}) \frac{\partial u_i^\eta}{\partial x_3}(\mathbf{x}) \frac{v_k(\mathbf{x})}{2h} dx. \qquad (8.110)$$

Applying Fubini's theorem to domain

$$Q_\eta = \{(\tilde{x}, x_3) : \tilde{x} \in S, -\eta h \le x_3 \le \eta h\},$$

we obtain that the integral in (8.110) is equal to

$$\int_S d\tilde{x} \frac{v_k(\mathbf{x})}{2h} \int_{-\eta h}^{\eta h} b_{i3k3}(\mathbf{x}) \frac{\partial u_i^\eta}{\partial x_3}(\mathbf{x}) dx_3 = \qquad (8.111)$$

$$= \int_S d\tilde{x} \frac{v_k(\mathbf{x})}{2h} \left[b_{i3k3} u_i^\eta(\tilde{x}, x_3) \Big|_{x_3=-\eta h}^{x_3=\eta h} - \int_{-\eta h}^{\eta h} \frac{\partial b_{i3k3}(\mathbf{x})}{\partial x_3} u_i^\eta(\mathbf{x}) dx_3 \right].$$

We obtain the last expression in (8.111) by integrating by parts.

By virtue of the conditions formulated for the coefficients $b_{ijkl}(\mathbf{x})$, we have that

$$b_{ijkl}(\tilde{x}, x_3) \to b_{ijkl}(\tilde{x}, 0) \text{ in } C(Q_0)$$

as $x_3 \to 0$.

By virtue of the trace theorem for closely placed surfaces [199, 212] applied to the surfaces S_η^+ and S_η^- (which tend to the opposite sides of the surface S as $\eta \to 0$), we have

$$\mathbf{u}^\eta(\tilde{x}, -\eta h) \to \mathbf{u}^*(\tilde{x}, +0) \text{ in } L_2(S),$$

$$\mathbf{u}^\eta(\tilde{x}, \eta h) \to \mathbf{u}^*(\tilde{x}, -0) \text{ in } L_2(S)$$

as $\eta \to 0$.

Then, integrating the nonintegral term in (8.111) over S, we obtain the right-hand part of (8.109) as the limit value of the integral as $\eta \to 0$. We consider the remaining term in (8.111). By virtue of conditions formulated for the coefficients $b_{ijkl}(\mathbf{x})$, absolute value of this term does not exceed

$$C\int_S d\tilde{x}\int_{-\eta h}^{\eta h}|u_i^\eta(\mathbf{x})|dx_3 \leq C\int_{Q_\eta}|u_i^\eta(\mathbf{x})|dx \leq \qquad (8.112)$$

$$\leq C|Q_\eta|\cdot\|\mathbf{u}^\eta\|_{L_2(Q_1^\eta\bigcup Q_2^\eta\bigcup Q^\eta)}.$$

By virtue of (8.88) the right-hand part of (8.112) tends to zero as $\eta \to 0$. **Proof is completed**.

Theorem 11. *As $\delta \to 0$, the sequence $\{\mathbf{u}^\delta(\mathbf{x})\}$ of the solution of problem (8.66) tends (in the sense in Lemma 25) to solution of the following problem:*

$$\int_{Q_1\bigcup Q_2}a_{ijkl}(\mathbf{x})\frac{\partial u_k^*}{\partial x_l}(\mathbf{x})\frac{\partial v_i}{\partial x_j}(\mathbf{x})dx + \qquad (8.113)$$

$$+\int_S\frac{b_{i3k3}(\tilde{x},0)}{2h}[u_k^*(\mathbf{x})]_S[v_i(\mathbf{x})]_S dx + \int_{Q_1\bigcup Q_2}\mathbf{f}(\mathbf{x})\mathbf{v}(\mathbf{x})dx = 0.$$

for any $\mathbf{v} \in C^\infty(x_3 \neq 0)$.

Proof. We denote $C^\infty(x_3 \neq 0)$ the set of functions of infinitely differentiable for $x_3 \neq 0$ and vanish for $\mathbf{x} \in \Gamma_1\bigcup\Gamma_2$. A function from $C^\infty(x_3 \neq 0)$ can be discontinuous on the surface $x_3 = 0$.

We denote

$$Q_0^+ = \{\mathbf{x} \in Q_0^+ : x_1 \geq 0\},$$

$$Q_0^- = \{\mathbf{x} \in Q_0^+ : x_1 \leq 0\}.$$

and denote $\mathcal{V}^+$ the closure of the set $C(Q_0^+) \cap C^\infty(x_3 \neq 0)$ in the norm $H^1(Q_0^+)$, $\mathcal{V}^-$ the closure of the set $C(Q_0^-) \cap C^\infty(x_3 \neq 0)$ in the norm $H^1(Q_0^-)$. The problem (8.114) is solvable in $\mathcal{V} = \mathcal{V}^+ \cup \mathcal{V}^=$. Taking into account that $C^\infty(x_3 \neq 0)$ is dense in $\mathcal{V}$, we conclude that the equation (8.114) is satisfied for any $\mathbf{v}(\mathbf{x}) \in \mathcal{V}$. By virtue of the conditions formulated for the coefficients $a_{ijkl}(\mathbf{x})$ and $b_{ijkl}(\mathbf{x})$ (in particular, $b_{i3k3}(\mathbf{x})z_iz_k$ is a positively defined form), the bilinear form in the left-hand side of (8.114) for $\mathbf{v}(\mathbf{x}) = \mathbf{u}^*(\mathbf{x})$ becomes the positively defined quadratic form. Then, the problem (8.114) has a unique solution. Thus, Lemma 25 is valid for the sequence $\mathbf{u}^\delta(\mathbf{x})$. **Proof is completed**.

Integrating by parts in (8.114), we obtain the following equations of equilibrium and condition of interaction on the adhesive surface S:

$$\frac{\partial}{\partial x_j}\left(a_{ijkl}(\mathbf{x})\frac{\partial u_k^{*M}}{\partial x_l}\right) = f_i(\mathbf{x}) \text{ in } Q_M \ (M = 1,2), \tag{8.114}$$

$$\mathbf{u}^{*M}(\mathbf{x}) = 0 \text{ on } \Gamma_M,$$

$$a_{ijkl}(\mathbf{x})\frac{\partial u_k^{*M}}{\partial x_l}(\mathbf{x})n_j = 0 \text{ outside } \Gamma_M \text{ and } S, \tag{8.115}$$

$$a_{i3kl}(\mathbf{x})\frac{\partial u_k^{*M}}{\partial x_l}(\mathbf{x}) = \frac{b_{i3k3}(\mathbf{x})}{2h}\left(u_k^{*(1)}(\tilde{x}) - u_k^{*(2)}(\tilde{x})\right) \text{ on } S. \tag{8.116}$$

Here $\mathbf{u}^{*(1)} = \mathbf{u}^*(\tilde{x},+0)$ and $\mathbf{u}^{*(2)} = \mathbf{u}^*(\tilde{x},-0)$ are values of the function $\mathbf{u}^*$ on the opposite sides of the adhesive surface S.

It is seen from (8.116) that the elastic behavior of the adhesive joint is characterized by the stiffness matrix

$$K_{ij} = \frac{b_{i3k3}(\tilde{x},0)}{2h}, \tag{8.117}$$

which establishes the relationship between the normal stresses

$$\sigma_i = a_{i3kl}(\mathbf{x})\frac{\partial u_k^{*M}}{\partial x_l}$$

on the surfaces of the glued bodies Q_1 and Q_2 (in the case under consideration normal to the surface S is $\mathbf{n} = (0,0,1)$, the general case is discussed below) and vector of displacement of sides of the adhesive joint $\mathbf{u}^{*(1)} - \mathbf{u}^{*(2)}$.

8.2.5. Stiffness of adhesive joint in dependence on Poisson's ratio of glue

Now, we discuss some special problems related to the results obtained above. We consider the adhesive layer formed of isotropic homogeneous material (i.e., elastic constants $b_{ijkl} = const$ and have a special form indicated below). We consider properties of stiffness of the adhesive layer

Comparison of normal and longitudinal effective stiffnesses of adhesive joint

The normal and longitudinal stiffnesses of the adhesive joint K_{nn} and K_{ll} characterize the reactions, which appear when one displays the glued bodies in normal or longitudinal directions, see Fig. 8.8.

For an isotropic material [358]

$$b_{ijkl} = \frac{E\nu}{(1+\nu)(1-2\nu)}\delta_{ij}\delta_{kl} + \frac{E}{2(1+\nu)}(\delta_{ik}\delta_{jl} + \delta_{il}\delta_{jk}), \tag{8.118}$$

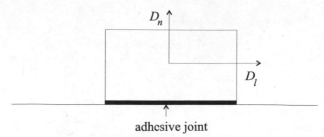

adhesive joint

Figure 8.8. *Reactions of glued bodies (n - normal direction, l - longitudinal direction).*

where E means Young's modulus and ν means Poisson's ratio.

In accordance to (8.117) and (8.118) we have that in the case under consideration the normal stiffness K_{nn} of the adhesive joint is

$$K_{nn} = \frac{b_{3333}}{2h} = \frac{\dfrac{E}{1-2\nu}}{2h} \qquad (8.119)$$

and the longitudinal stiffness K_{ll} of the adhesive joint is

$$K_{ll} = \frac{b_{i3i3}}{2h} = \frac{\dfrac{E}{1+\nu}}{2h} \; (i=1,2). \qquad (8.120)$$

As follows from (8.119) and (8.120), the normal stiffness and the longitudinal stiffness of the adhesive joint are not equal, generally. It means that the adhesive joint is not an isotropic object even if it is formed of isotropic material. We consider the ratio of normal and longitudinal stiffnesses of the adhesive joint:

$$\frac{K_{nn}}{K_{ll}} = \frac{1+\nu}{1-2\nu}. \qquad (8.121)$$

It is seen from (8.121) that for isotropic adhesive material, the ratio $\dfrac{K_{nn}}{K_{ll}}$ is determined by Poisson's ratio of the adhesive material. The plot of the function (8.121) is presented in Fig. 8.9. For positive Poisson's ratio,

$$\frac{K_{nn}}{K_{ll}} > 1.$$

Theoretically, Poisson's ratio can take values up to 0.5. Thus, the ratio (8.121) can take arbitrarily large values for the materials with $\nu \approx 0.5$. In the case $\nu \approx 0.5$ (it is the case of nearly uncompressible glue), the glued bodies demonstrate behavior similar to the behavior of two wetted sheets of glass. In this case the adhesive joint demonstrates strong normal reaction as compared with longitudinal reaction.

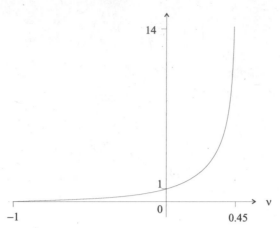

Figure 8.9. *Ratio of normal stiffness of adhesive joint to longitudinal stiffness as a function of Poisson's ratio of adhesive substance, function (8.121).*

It is known that there exist materials possessing negative Poisson's ratio. The existence of materials possessing negative Poisson's ratio was predicted theoretically [6, 178] in 1985. Two years later materials with negative Poisson's ratio were produced [201]. For adhesive material with negative Poisson's ratio,

$$\frac{K_{nn}}{K_{ll}} < 1.$$

The minimal possible value of $\frac{K_{nn}}{K_{ll}}$ is zero since the minimal possible value of Poisson's ratio is -1. Thus, for isotropic adhesive materials, the ratio $\frac{K_{nn}}{K_{ll}}$ can take arbitrary values from zero to infinity. This is the situation opposite the situation arising when $\nu \approx 0.5$. In this case the adhesive joint demonstrates weak normal reaction and strong longitudinal reaction (one as compared with another). Note again, that the value of Poisson's ratio equal to -1 is possible [6, 178, 201], but the authors have no information about glues with negative Poisson's ratio produced on regular basis.

It is seen from (8.121) that the normal and longitudinal stiffnesses of the adhesive joint are equal in the case $\nu = 0$, only. Most known material with Poisson's ratio equal to zero is cork.

The typical value of Poisson's ratio for usual adhesive materials lies in the interval 0.3 to 0.4 [217]. For such values of Poisson's ratio, the ratio $\frac{K_{nn}}{K_{ll}}$ takes values in the interval from 3 to 7.

Comparison of normal, longitudinal and torsional effective stiffnesses of adhesive joint

Glued bodies often work in the regime of torsion, see Fig. 8.10. We estimate the order of the torsional effective stiffnesses of adhesive joint. The relative displacements

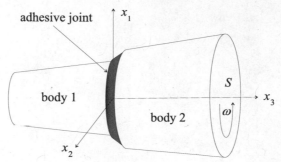

Figure 8.10. *Glued bodies working in the regime of torsion.*

are $\mathbf{u}^{*(1)}(\tilde{x}) - \mathbf{u}^{*(2)}(\tilde{x})$ in the case under consideration.

For an isotropic material [358]

$$u_1^{*(1)}(\tilde{x}) - u_1^{*(2)}(\tilde{x}) = -x_2\omega, \tag{8.122}$$

$$u_2^{*(1)}(\tilde{x}) - u_2^{*(2)}(\tilde{x}) = x_1\omega,$$

$$u_3^{*(1)}(\tilde{x}) - u_3^{*(2)}(\tilde{x}) = 0,$$

where ω is the angle of torsion

In the case under consideration the stresses σ_i^n on the adhesive surface S with normal $\mathbf{n} = (0, 0, 1)$ are

$$\sigma_i = \sigma_{ij}n_j = \sigma_{i3}^0\omega \quad (i = 1, 2), \tag{8.123}$$

$$\sigma_3 = 0,$$

where

$$\sigma_{ik}^0 = \frac{b_{i3k3}(\mathbf{x})}{2h}\left(u_k^{*(1)}(\tilde{x}) - u_k^{*(2)}(\tilde{x})\right) = -b_{i313}(\mathbf{x})x_2 + b_{i323}(\mathbf{x})x_1. \tag{8.124}$$

In the case under consideration, the torsional moment $\mathcal{M}$ is related to the angle of torsion ω by the formula

$$\mathcal{M} = M\omega,$$

where

$$M = \int_S (\sigma_{\mathbf{n}} \times \mathbf{x})d\mathbf{x}, \tag{8.125}$$

$$\sigma_{\mathbf{n}} = \sum_{i=1}^{3} \sigma_{i3}\mathbf{e}_i$$

and the cross "$\times$" in (8.125) means the vectorial product. In the case under consideration $\mathbf{x} = x_1\mathbf{e}_1 + x_2\mathbf{e}_2$.

Substituting (8.124) to (8.125) with regard to the symmetry of the elastic constants, we have

$$M = \int_S (b_{1313}x_2^2 + b_{2323}x_1^2)d\mathbf{x} = \frac{E}{1+\nu}\int_S(x_1^2 + x_2^2)d\mathbf{x}. \qquad (8.126)$$

In the case under consideration M is proportional to $\dfrac{E}{1+\nu}$ (i.e., the shearing elastic modulus). This means that torsional stiffness M of the adhesive joint is proportional to the longitudinal stiffness K_{ll}. Then, M behaves similarly to longitudinal stiffness K_{ll}. In particular, behavior of torsional stiffness as compared to normal stiffness is similar to behavior of longitudinal stiffness as compared to normal stiffness.

Soft adhesive joint

The elastic moduli δb_{ijkl} of adhesive material (glue) and elastic moduli a_{ijkl} of glued bodies have different orders. At the same time, the stiffnesses E_{ik} have the order of b_{ijkl} and they can be compared with elastic moduli a_{ijkl} of glued bodies as quantities of the same order.

Now, we consider the case when E_{ik} are small as compared with the elastic moduli a_{ijkl} of the glued bodies: $|E_{ik}| \ll |a_{ijkl}|$. In this case the condition (8.116) may be approximated by the condition

$$a_{ijkl}(\mathbf{x})\frac{\partial u_k^{*M}}{\partial x_l}(\mathbf{x})n_j = 0 \text{ on } S,$$

which corresponds to free surface S. This means that the glued bodies do not interact with one another.

Stiff adhesive joint

Now, we consider the case when stiffnesses E_{ik} are large as compared with the elastic moduli a_{ijkl} of the glued bodies: $|E_{ik}| \gg |a_{ijkl}|$. In this case the condition (8.116) may be approximated by the condition

$$a_{ijkl}(\mathbf{x})\frac{\partial u_k^{*1}}{\partial x_l}(\mathbf{x})n_j = a_{ijkl}(\mathbf{x})\frac{\partial u_k^{*2}}{\partial x_l}(\mathbf{x})n_j \text{ on } S,$$

$$\mathbf{u}^{*1}(\mathbf{x}) = \mathbf{u}^{*2}(\mathbf{x}) \text{ on } S,$$

which describes the perfect contact of the glued bodies on the surface S.

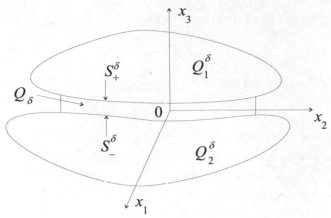

Figure 8.11. *Non-cylindrical adhesive layer between glued bodies.*

8.2.6. Adhesive joints of variable thickness or curvilinear joints

In the previous sections, we derived the limit model for the case of an adhesive joint of constant thickness with planar glued surfaces S_δ^+ and S_δ^+. It is possible to expand the results obtained to the case of adhesive joint of variable thickness and /or curved glued surfaces, see Fig. 8.11.

We assume that the surfaces S_δ^+ and S_δ^+ can be represented in the form of graphs of functions $\Phi_1(\tilde{x})$ and $\Phi_2(\tilde{x})$, see Fig. 8.11:

$$S_\delta^+ = \{(\tilde{x}, x_3) : \tilde{x} \in S, x_3 = \delta \Phi_1(\tilde{x})\},$$

$$S_\delta^- = \{(\tilde{x}, x_3) : \tilde{x} \in S, x_3 = -\delta \Phi_2(\tilde{x})\},$$

where $\Phi_1(\tilde{x})$ and $\Phi_2(\tilde{x})$ satisfy the following conditions:
  1. $0 < c_4 \le \Phi_\alpha(\tilde{x})$ $(\alpha = 1, 2)$ for any $(\tilde{x}) \in S$,
  2. $\|\Phi_\alpha\|_{C^1(S)} \le c_5 < \infty$ $(\alpha = 1, 2)$, where constants c_4 and c_5 do not depend on the parameter δ.

These conditions mean that we consider weakly wavy surfaces (with no fast oscillations like goffered, ribbed, etc.). It is possible to expand all the results obtained to this case if we change $2h$ for $\Phi_1(\mathbf{x}) + \Phi_2(\mathbf{x})$. The interaction condition (boundary condition on the glue surface S) has the form

$$a_{ijkl}(\mathbf{x}) \frac{\partial u_k^{*M}}{\partial x_l}(\mathbf{x}) n_j = \qquad (8.127)$$

$$= \frac{b_{ijkl}(\mathbf{x}) n_i(\mathbf{x}) n_j(\mathbf{x})}{\Phi_1(\mathbf{x}) + \Phi_2(\mathbf{x})} \left(u_k^{*1}(\tilde{x}, 0) - u_k^{*2}(\tilde{x}, 0) \right) \text{ on } S \ (M = 1, 2).$$

In accordance with (8.127), the elastic property of the adhesive layer is characterized by the stiffness matrix

$$\mathcal{K}_{ij}(\mathbf{x}) = \frac{b_{ijkl}(\mathbf{x})n_i(\mathbf{x})n_j(\mathbf{x})}{\Phi_1(\mathbf{x}) + \Phi_2(\mathbf{x})} = \frac{\delta b_{ijkl}(\mathbf{x})n_i(\mathbf{x})n_j(\mathbf{x})}{\delta(\Phi_1(\mathbf{x}) + \Phi_2(\mathbf{x}))}.$$

Appendix A

MATHEMATICAL NOTIONS USED IN THE ANALYSIS OF INHOMOGENEOUS MEDIA

For the convenience of the reader, we briefly summarize several basic notions used in the analysis of transport problems in composite materials and network approximation method which are used throughout the book.

Graphs

A graph (network) $\mathbf{G}$ is a set of items, which we call vertices or nodes, with connections between them, called edges. Two vertices $\mathbf{x}_i$ and $\mathbf{x}_j$ connected with an edge e_{ij} are called adjacent. The connections in a network can be represented by the collection of the edges $\{e_{ij}\}$ or by the adjacency matrix g_{ij} determined as

$$g_{ij} = \begin{cases} 1 \text{ if the } i\text{-th and } j\text{-th vertices are connected,} \\ 0 \text{ otherwise; } i, j = 1, ..., N; \end{cases} \tag{A.1}$$

where N is the number of the vertices of the network. The number N is called the size of the network.

A graph can be described by the set of its vertices $\mathbf{X} = \{\mathbf{x}_i, i = 1, ..., N\}$ and the set $\mathbf{E} = \{e_{ij}; i, j = 1, ..., N\}$ of its edges or the adjacency matrix $\mathbf{g} = \{g_{ij}; i, j = 1, ..., N\}$. Thus, graph can be represented in the form $\mathbf{G} = (\mathbf{X}, \mathbf{E})$ or $\mathbf{G} = (\mathbf{X}, \mathbf{g})$.

In many cases some numbers (called the weights of edges) are assigned in correspondence to the edges. A graph with weights is called a weighted graph. For a weighted graph, the definition (A.1) is modified as follows

$$c_{ij} = \begin{cases} \neq 0 \text{ if the } i\text{-th and } j\text{-th vertices are connected,} \\ 0 \text{ otherwise; } i, j = 1, ..., N. \end{cases} \tag{A.2}$$

The weights c_{ij} (A.2) include information about the adjacency matrix g_{ij} (A.1).

The edges of a graph can be directed. Graphs composed of directed edges are called directed (oriented) graphs. Directed weighted graphs are described by (A.2) with the weights taking positive or negative values. In specific problems the weights c_{ij} will represent fluxes in networks such as fluxes of fluid, electric current, heat, etc.

A weighted graph can be described by the set $\{\mathbf{x}_i, c_{ij}; i, j = 1, ..., N\}$.

A network (graph) is referred to as a connected one if each two vertices of the network are connected by a path consisting of edges in this network.

A loop in the network is a path, which begins and ends in the same vertex.

Functional spaces

Detailed exposition of the theory of functional spaces can be found in [164, 308, 392].

A normed complete linear space B is called Banach space. The linear space of continuous linear functionals on B forms dual space B^*, which is Banach space with the norm defined as (we omit arguments for functions belonging to abstract functional spaces and indicate arguments for functions belonging to specific functional spaces)

$$||\phi^*|| = \sup_{||\phi||_B=1} |\langle\phi^*, \phi\rangle|$$

where $\langle\phi^*, \phi\rangle$ means dual coupling (value of the functional $\phi^* \in B^*$ on function $\phi \in B$).

One can define space B^{**} as $B^{**} = (B^*)^*$. The space B is called reflexive if $B = B^{**}$.

Banach space B is called separable space if it contains a countable subset which is dense in B.

Banach space H with the norm defined as $||\varphi||_H = \sqrt{(\varphi, \varphi)_H}$, where $(\varphi, \phi)_H$ means the inner product in H, is called Hilbert space. In Hilbert space dual coupling can be associated with inner product.

A sequence of functions $\{\varphi_n\} \subset B$ is called converging to $\varphi \in B$ as $n \to \infty$ if $||\varphi_n - \varphi||_B \to 0$ as $n \to \infty$.

A sequence of functions $\{\varphi_n\} \subset B$ is called weakly converging to $\varphi \in B$ as $n \to \infty$ if $\langle\phi^*, \varphi_n\rangle \to \langle\phi^*, \varphi\rangle$ for any $\phi^* \in V^*$ as $n \to \infty$.

If space B is reflexive and a sequence $\{\varphi_n\} \subset B$ is bounded, then there exists a weakly converging subsequence $\{\varphi_k\} \subset \{\varphi_n\}$.

Convex sets and functionals

A set M of functions is called convex if for any $\varphi, \phi \in M$

$$\lambda\varphi + (1 - \lambda)\phi \in M$$

for any $0 \leq \lambda \leq 1$.

Functional $F : M \to R$ defined on convex set M is called convex if for any φ, $\phi \in M$, such that $\varphi \neq \phi$ and any $0 \leq \lambda \leq 1$

$$F(\lambda\varphi + (1 - \lambda)\phi) \leq \lambda F(\varphi) + (1 - \lambda)F(\phi).$$

If strict inequality holds, then the functional is called strictly convex.

Continuous convex functional $F(\varphi)$ defined on a convex closed set $M \subset B$ has the following property: if $\varphi_n \to \varphi$ weakly in B as $n \to \infty$ than $F(\varphi) \leq \liminf_{n \to \infty} F(\varphi_n)$, where "lim inf" denotes inferior limit [122, 164]. This property is called weak lower semicontinuity [122].

We consider the following minimization problem:

$$F(\varphi) \to \min, \ \varphi \in M. \tag{A.3}$$

If $F(\varphi)$ is a continuous convex functional and M is a convex closed subset of a reflexive functional space, then the problem (A.3) has at least one solution. If, in addition, functional $F(\varphi)$ is strictly convex, the problem (A.3) has a unique solution [122].

Distributions and distributional derivatives

Denote by $\mathcal{D}(Q)$ the set of infinitely differentiable functions with compact supports (a function has a compact support if it is equal to zero outside certain compact set $K \subset Q$). Topology (convergence) on $\mathcal{D}(Q)$ is introduced as follows: let $\{\phi_i(\mathbf{x}), \ i = 1, 2, ...\}$, $\phi(\mathbf{x}) \in \mathcal{D}(Q)$. Then the convergence $\phi_i(\mathbf{x}) \to \phi(\mathbf{x})$ in $\mathcal{D}(Q)$ as $i \to \infty$ means that supports of all functions belong to the same compact subset of Q, and $\phi_i(\mathbf{x})$ and all its derivatives converge uniformly to $\phi(\mathbf{x})$ and its corresponding derivatives.

Now, let T be a linear continuous functional determined on $\mathcal{D}(Q)$, i.e., a map (a rule) that assigns to every $\phi(\mathbf{x}) \in \mathcal{D}(Q)$ a number $\langle T, \phi \rangle$, which is the value of the functional T on this function $\phi(\mathbf{x}) \in \mathcal{D}(Q)$. This map is linear with respect to $\phi(\mathbf{x})$ and $\langle T, \phi_i \rangle \to \langle T, \phi \rangle$ if $\phi_i(\mathbf{x}) \to \phi(\mathbf{x})$ in $\mathcal{D}(Q)$ as $i \to \infty$.

Such functionals are called distributions on Q. The set of distributions is denoted by $\mathcal{D}'(Q)$.

Every locally integrable function $\varphi(\mathbf{x})$ on Q generates a distribution $\tilde{\varphi} \in \mathcal{D}'(Q)$ defined as follows

$$\langle \tilde{\varphi}, \phi \rangle = \int_Q \varphi(\mathbf{x})\phi(\mathbf{x})d\mathbf{x}.$$

If $T \in \mathcal{D}'(Q)$, its distributional derivative $\dfrac{\partial T}{\partial x_i} \in \mathcal{D}'(Q)$ is determined by the equality

$$\left\langle \frac{\partial T}{\partial x_i}, \phi \right\rangle = -\left\langle T, \frac{\partial \phi}{\partial x_i} \right\rangle \text{ for any } \phi(\mathbf{x}) \in \mathcal{D}(Q). \tag{A.4}$$

Formula (A.4) of distributional derivative calls to mind the interaction of derivation operation in the formula of integration by parts. It is not a matter of chance.
Historically, the notion of distribution derivative was inspired by the integration
formulas like the formula of integration by parts, see [323, 341, 342].

Sobolev functional spaces

We consider a function $\varphi(\mathbf{x})$ defined on the domain Q such that $|\varphi(\mathbf{x})|^p$ $(1 < p < \infty)$
is the Lebesque integrable [68, 308]. This set of such functions is denoted by $L_p(Q)$.
When equipped with the norm

$$||\varphi||_{L_p} = \left(\int_Q |\varphi(\mathbf{x})|^p d\mathbf{x} \right)^{1/p}$$

$L_p(Q)$ is a Banach functional space.

Widely used is the space $L_2(Q)$, which is a Hilbert space with the inner product

$$(\varphi, \phi)_{L_2} = \int_Q \varphi(\mathbf{x})\phi(\mathbf{x})d\mathbf{x}.$$

The functional space $L_\infty(Q)$ is introduced as the set of measurable functions
bounded almost everywhere [308]. The norm in $L_\infty(Q)$ is introduced as

$$||\varphi||_{L_\infty} = \operatorname{ess\ sup}_{\mathbf{x} \in Q} |\varphi(\mathbf{x})|.$$

For integrable function Hölder inequality takes place:

$$\int_Q |\varphi(\mathbf{x})\phi(\mathbf{x})d\mathbf{x}| \leq \left(\int_Q |\varphi(\mathbf{x})|^p d\mathbf{x} \right)^{1/p} \left(\int_Q |\phi(\mathbf{x})|^q d\mathbf{x} \right)^{1/q}, \qquad (A.5)$$

where $p > 1$ and

$$\frac{1}{p} + \frac{1}{q} = 1.$$

For $p = q = 2$, inequality (A.5) takes the form

$$\int_Q |\varphi(\mathbf{x})\phi(\mathbf{x})|d\mathbf{x} \leq \left(\int_Q |\varphi(\mathbf{x})|^2 d\mathbf{x} \right)^{1/2} \left(\int_Q |\phi(\mathbf{x})|^2 d\mathbf{x} \right)^{1/2} \qquad (A.6)$$

and it is referred to as the Cauchy-Bunyakovsky inequality.

The Sobolev functional space $W^{m,p}(Q)$ is the set of distributions which (with
all its distributional derivatives of order less than or equal to m) are generated by
functions from $L_p(Q)$. This is a Banach space with the norm

$$||\varphi||_{W^{m,p}} = \left(\int_Q \sum_{0 \leq m_1+...+m_N \leq m} \left| \frac{\partial^{m_1+...+m_N}\varphi(\mathbf{x})}{\partial x_1^{m_1}...\partial x_N^{m_N}} \right|^p d\mathbf{x} \right)^{1/p}. \qquad (A.7)$$

For $p = 2$, $W^{m,p}(Q)$ is a Hilbert space, which is denoted by $H^m(Q)$, with inner product

$$(f,g)_{H^m} = \int_Q \sum_{0 \le m_1 + \ldots + m_N \le m} \frac{\partial^{m_1 + \ldots + m_N} \varphi(\mathbf{x})}{\partial x_1^{m_1} \ldots \partial x_N^{m_N}} \cdot \frac{\partial^{m_1 + \ldots + m_N} \phi(\mathbf{x})}{\partial x_1^{m_1} \ldots \partial x_N^{m_N}} d\mathbf{x}.$$

For $m = 1$ the inner product in $H^1(Q)$ is

$$(\varphi, \phi)_{H^1} = \int_Q \left(\varphi(\mathbf{x})\phi(\mathbf{x}) + \nabla\varphi(\mathbf{x})\nabla\phi(\mathbf{x}) \right) d\mathbf{x}. \tag{A.8}$$

The norm in $H^1(Q)$ is

$$\|\varphi\|_{H^1} = \sqrt{\int_Q \left(\varphi^2(\mathbf{x}) + |\nabla\varphi(\mathbf{x})|^2 \right) d\mathbf{x}}.$$

The functional space $H^m(Q)$ may be also introduced as a closure of the functional space $C^\infty(Q)$ in the norm (A.7) [2]. The closure of the functional space $\mathcal{D}(Q)$ (the set of functions from $C^\infty(Q)$, which are equal to zero in a neighbor of the boundary ∂Q) in the norm (A.7) [2] is denoted by $H_0^m(Q)$. From these it follows that the sets $C^\infty(Q)$ and $\mathcal{D}(Q)$ are dense subsets of the functional spaces $H^m(Q)$ and $H_0^m(Q)$, correspondingly.

Hilbert space is reflexive [122, 309]. Functional spaces $L_p(Q)$ and $W^{m,p}(Q)$ are reflexive and separable for $0 < p < \infty$ [210]. The space $C(Q)$ is dense in $L_p(Q)$, the spaces $C^n(Q)$ and $C^\infty(Q)$ are dense in $L_p(Q)$ and $W^{m,p}(Q)$ ($n \ge m$).

It is possible to introduce the space $H^s(Q)$ for real (not necessary integer or even positive) values of s. In the special case where $Q = R^N$, $p = 2$, the space $H^s(R^N)$ can be introduced by using the Fourier transform. $H^s(R^N)$ is the set of all functions $\varphi(\mathbf{x}) \in L_2(R^N)$ such that their Fourier transforms

$$\hat{\varphi}(\zeta) = (2\pi)^{-\frac{N}{2}} \int_{R^N} \varphi(\mathbf{x}) e^{-i(\mathbf{x},\zeta)} d\mathbf{x}, \ \zeta = (\zeta_1, \ldots, \zeta_N) \in R^N$$

satisfy the condition $|(1 + |\zeta|^2)^{s/2} \hat{\varphi}(\zeta)| \in L_2(R^N)$.

The norm of a function $\varphi(\mathbf{x})$ in $H^s(R^N)$ is introduced as follows

$$\|\varphi\|_{H^s(R^N)} = \|(1 + |\zeta|^2)^{s/2} \hat{\varphi}(\zeta)\|_{L_2(R^N)}. \tag{A.9}$$

Traces of functions from $H^1(Q)$

The functions belonging to $H^1(Q)$ are not in general continuous and moreover they are defined almost everywhere (but not everywhere) in the domain Q. Thus assigning boundary values on ∂Q for such functions is not straightforward and requires special techniques.

The set $C^\infty(\bar{Q})$ of infinitely differentiable functions in $\bar{Q} = Q \bigcup \partial Q$ (we assume ∂Q is C^∞ smooth surface) is dense in $H^1(Q)$ [212].

Consider an $n-1$ dimensional smooth surface $\Gamma \subset \partial Q$. For a function $\varphi(\mathbf{x}) \in H^1(Q)$, the trace ("boundary values" of the function $\varphi(\mathbf{x})$ on Γ) $\varphi|_\Gamma(\mathbf{x})$ can be defined as follows.

For a given $\varphi(\mathbf{x}) \in H^1(Q)$ choose $\varphi_i(\mathbf{x}) \in C^\infty(Q)$ $(i = 1, 2, ...)$ such that

$$\varphi_i(\mathbf{x}) \to \varphi(\mathbf{x}) \text{ in } H^1(Q) \text{ as } i \to \infty. \qquad (A.10)$$

Then the sequence $\varphi_i(\mathbf{x})|_\Gamma(\mathbf{x})$ has a limit in $H^{1/2}(\Gamma)$ as $i \to \infty$ [212]. This limit is called the trace of the function $\varphi(\mathbf{x}) \in H^1(Q)$ on the surface Γ. Due to the density property the trace operator can be extended by continuity from $C^\infty(\bar{Q})$ to $H^m(Q)$ and thus it becomes a linear bounded (continuous) operator from $H^1(Q)$ to $H^{1/2}(\Gamma)$.

This definition of the trace operator allows integration by parts for functions from Sobolev functional spaces which follows immediately from the density of $C^\infty(\bar{Q})$ in $H^1(Q)$.

Note that from (A.10) it follows that

$$\varphi_i(\mathbf{x}) \to \varphi(\mathbf{x}) \text{ in } H^{1/2}(\Gamma)$$

and

$$\varphi_i(\mathbf{x}) \to \varphi(\mathbf{x}) \text{ in } L_2(\Gamma)$$

as $i \to \infty$.

The space of $H^1(Q)$ functions with zero trace is denoted by $H_0^1(Q)$.

In our studies we will use the following trace theorem [212, 355]. If domain Q has a C^2 smooth boundary and

$$\mathbf{v}(\mathbf{x}) \in L_2(Q) \text{ and } \frac{\partial \mathbf{v}}{\partial x_i}(\mathbf{x}) \in L_2(Q)$$

then the trace $\mathbf{v}(\mathbf{x})\mathbf{n}$ ($\mathbf{n}$ is the unit normal vector to ∂Q) is determined on $\Gamma \subseteq \partial Q$ as a function from $H^{-1/2}(\Gamma)$.

This trace theorem often will be used in the following form [212, 355]. If domain Q has C^2 smooth boundary and

$$\mathbf{v}(\mathbf{x}) \in L_2(Q) \text{ and } \operatorname{div}\mathbf{v}(\mathbf{x}) \in L_2(Q) \qquad (A.11)$$

then the trace $\mathbf{v}(\mathbf{x})\mathbf{n}$ is determined on $\Gamma \subseteq \partial Q$ as a function from $H^{-1/2}(\Gamma)$.

A particular case of (A.11) is divergence free functions from $L_2(Q)$:

$$\mathbf{v}(\mathbf{x}) \in L_2(Q) \text{ and } \operatorname{div}\mathbf{v}(\mathbf{x}) = 0. \qquad (A.12)$$

A detailed proof of the trace theorem for divergence free functions (A.12) can be found in [355].

Weak solutions of partial differential equations with discontinuous coefficients

An introduction to the theory of weak solutions of partial differential equations can be found in [212]. Here we present a weak formulation of an elliptic boundary-value problem.

Consider the following boundary-value problem

$$\frac{\partial}{\partial x_i}\left(a(\mathbf{x})\frac{\partial \varphi}{\partial x_i}\right) = f(\mathbf{x}) \text{ in } Q, \tag{A.13}$$

$$\frac{\partial \varphi}{\partial \mathbf{n}}(\mathbf{x}) = u_1(\mathbf{x}) \text{ on } \Gamma \subseteq \partial Q, \tag{A.14}$$

$$\varphi(\mathbf{x}) = u_2(\mathbf{x}) \text{ on } \partial Q \setminus \Gamma, \tag{A.15}$$

where

$$a(\mathbf{x}) \in L_\infty(Q) \tag{A.16}$$

and satisfy the coercivity condition

$$a(\mathbf{x}) \geq \gamma > 0 \text{ for all } \mathbf{x} \in Q. \tag{A.17}$$

The summation of repeated indices is assumed if not indicated otherwise.

If functions $a(\mathbf{x})$, $\varphi(\mathbf{x})$, $u_1(\mathbf{x})$, $u_2(\mathbf{x})$ and the surface ∂Q are sufficiently smooth, the problem (A.13) has a classical solution belonging to $C^2(Q)\bigcap C^1(\bar{C})$ [199]. Multiplying the differential equation (A.13) by arbitrary function $\phi(\mathbf{x}) \in C^\infty(Q)$ such that

$$\phi(\mathbf{x}) = 0 \text{ on } \partial Q \setminus \Gamma \tag{A.18}$$

and integrating by parts, we obtain

$$-\int_Q a(\mathbf{x})\frac{\partial \varphi}{\partial x_i}\frac{\partial \phi}{\partial x_i}d\mathbf{x} = \int_Q f(\mathbf{x})\phi(\mathbf{x})d\mathbf{x} - \int_\Gamma u_1(\mathbf{x})\phi(\mathbf{x})d\mathbf{x}. \tag{A.19}$$

for any $\phi(\mathbf{x}) \in C^\infty(Q)$ satisfying (A.18).

The equality (A.19) is defined for functions

$$a(\mathbf{x}) \in L_\infty(Q),$$

$$u_1(\mathbf{x}) \in H^{-1/2}(Q), u_2(\mathbf{x}) \in H^{1/2}(Q),$$

$$\varphi(\mathbf{x}), \phi(\mathbf{x}) \in H^1(Q). \tag{A.20}$$

Thus, it is possible to define the solution of the boundary-value problem (A.13) for non-differentiable (in particular, for piecewise continuous) coefficients $a(\mathbf{x})$. The solution of (A.19) is referred to as a generalized solution of the boundary-value problem (A.13)–(A.15) and it is understood in the sense of distributions.

Variational form of differential equations

Let H be a Hilbert functional space. Function $a(\varphi, \phi)$ defined on the product of the spaces $H \times H$ is called bilinear form on H if it is linear with respect to the first and the second variables. The bilinear form is called continuous if there exists a constant $C < \infty$ such that

$$|a(\varphi, \phi)| \leq C||\varphi||_H \cdot ||\phi||_H$$

for any $\varphi(\mathbf{x})$, $\phi(\mathbf{x}) \in H$.

Bilinear form is called symmetric form if for any $\varphi(\mathbf{x})$, $\phi(\mathbf{x}) \in H$

$$a(\varphi, \phi) = a(\phi, \varphi).$$

Bilinear form is called coercive if there exists a constant $c > 0$ such that

$$a(\varphi, \varphi) \geq c||\varphi||_H^2 \qquad (A.21)$$

for any $\varphi(\mathbf{x}) \in H$.

For any bilinear continuous, coercive symmetric form $a(\varphi, \phi)$ on $H \times H$ there exists a unique operator $L : H \to H^*$ such that

$$a(\varphi, \phi) = \langle L\varphi, \phi \rangle \qquad (A.22)$$

for any $\varphi(\mathbf{x}) \in H$.

If $u_1(\mathbf{x}) = 0$, the problem (A.13)–(A.15) is associated with the bilinear form

$$a(\varphi, \phi) = \int_Q a(\mathbf{x}) \nabla \varphi(\mathbf{x}) \nabla \phi(\mathbf{x}) d\mathbf{x} \qquad (A.23)$$

defined on the functional space

$$H = \{\varphi(\mathbf{x}) \in H^1(Q) : \varphi(\mathbf{x}) = 0 \text{ on } \partial Q \setminus \Gamma\}.$$

The form (A.23) satisfies the conditions above if $a(\mathbf{x})$ satisfies the conditions (A.16) and (A.17). Then the operator

$$L = -\frac{\partial}{\partial x_i}\left(a(\mathbf{x})\frac{\partial}{\partial x_i}\right)$$

can be treated as an operator acting from H to H^* [212].

The equation (A.19) takes the form

$$-\int_Q a(\mathbf{x})\frac{\partial \varphi}{\partial x_i}(\mathbf{x})\frac{\partial \phi}{\partial x_i}(\mathbf{x})d\mathbf{x} = \int_Q f(\mathbf{x})\phi(\mathbf{x})d\mathbf{x} \qquad (A.24)$$

for any $\phi(\mathbf{x}) \in H$.

The problem (A.19) is called the variational form of the problem (A.13)–(A.15). With regard to (A.22), the equation (A.24) can be written as

$$\langle -L\varphi, \phi \rangle = \langle -f, \phi \rangle \tag{A.25}$$

for any $\phi(\mathbf{x}) \in H$. Thus, we arrive at the equality

$$L\varphi = f(\mathbf{x}),$$

which must be treated as equality of the elements belonging to the space H^*.

The equation (A.24) is a necessary condition of minima of the quadratic functional

$$F(\varphi) = \frac{1}{2} \int_Q a(\mathbf{x}) |\nabla \varphi(\mathbf{x})|^2 d\mathbf{x} + \int_Q f(\mathbf{x}) \varphi(\mathbf{x}) d\mathbf{x} \tag{A.26}$$

on H.

The functional $F(\varphi)$ (A.26) is strictly convex under condition (A.17).

Under conditions (A.16) and (A.21) there exists unique $\varphi(\mathbf{x}) \in H$ such that (A.26) is satisfied (see, e.g., [113]).

If the solution of the problem (A.19) is sufficiently smooth, then it represents the classical solution of the problem (A.13)–(A.15).

Derivatives of functionals

The notion of derivatives can be introduced for functionals defined on functional spaces. It is a useful tool widely used both in pure mathematics and applications [101, 113, 148].

Let F be a functional from H into R. The limit (if it exists)

$$\lim_{\lambda \to 0} \frac{F(\varphi + \lambda\phi) - F(\varphi)}{\lambda} \tag{A.27}$$

is called the directional derivative of F at $\varphi(\mathbf{x})$ in the direction ϕ and it is denoted by $F'(\varphi, \phi)$.

If there exists $\varphi^* \in H^*$ such that:

$$F'(\varphi, \phi) = \langle \varphi^*, \phi \rangle \tag{A.28}$$

for any $\phi \in H$ we say that F is Gâteaux-differentiable at φ and call φ^* the Gâteaux differential at φ and denote it by $F'(\varphi)$.

The Gâteaux differential $F'(\varphi)$ is characterized by the equation

$$\lim_{\lambda \to 0} \frac{F(\varphi + \lambda\phi) - F(\varphi)}{\lambda} = \langle F'(\varphi), \phi \rangle \tag{A.29}$$

for any $\phi \in H$.

In this book we will deal with the integral functional of the form

$$F(\varphi) = \int_Q f(\varphi(\mathbf{x}))d\mathbf{x}. \tag{A.30}$$

If function $f(z)$ is differentiable with respect to the variable z, then

$$\langle F'(\varphi), \phi \rangle = \int_Q f'(\varphi(\mathbf{x}))\phi(\mathbf{x})d\mathbf{x}, \tag{A.31}$$

where $f'(z)$ means the derivative of the function $f(z)$ with respect to the variable z. Formula (A.31) is known from the calculus of variations as the first variation of the functional $F(\varphi)$ (A.30).

For quadratic functional

$$F(\varphi) = \int_Q c(\mathbf{x})|\nabla\varphi(\mathbf{x})|^2 d\mathbf{x}$$

one has

$$\langle F'(\varphi), \phi \rangle = \int_Q c(\mathbf{x})\nabla\varphi(\mathbf{x})\nabla\phi(\mathbf{x})d\mathbf{x}. \tag{A.32}$$

Integrating (A.32) by parts, one obtains the equality

$$F'(\varphi) = -\operatorname{div}(c(\mathbf{x})\nabla\varphi(\mathbf{x})) : H \to H^*$$

and the corresponding boundary conditions

$$\varphi(\mathbf{x}) = 0 \text{ on } \partial Q \setminus \Gamma,$$

$$\frac{\partial\varphi}{\partial\mathbf{n}}(\mathbf{x}) = 0 \text{ on } \Gamma.$$

In the partial case, when $H = H_0^1(Q)$ (i.e. $\Gamma = \emptyset$), one obtains the equality

$$F'(\varphi) = -\operatorname{div}(c(\mathbf{x})\nabla\varphi(\mathbf{x})) : H_0^1(Q) \to H^{-1}(Q)$$

with the boundary condition

$$\varphi(\mathbf{x}) = 0 \text{ on } \partial Q.$$

Appendix B

DESIGN OF LAMINATED MATERIALS AND CONVEX COMBINATIONS PROBLEM

Here, we briefly summarize results concerning convex combinations problem and its application to the design of laminated composite materials possessing required homogenized characteristics.

Design of laminated composite materials

The problem of design of composite material possessing required homogenized characteristics is an inverse problem with respect to the problem of computation of homogenized characteristics. The statement of the problem is the following. We assume the composite has periodic structure with the periodicity cell δY and the coefficients $c_{ij}(\mathbf{y})$, which describe the structure of composite, may take values from the set

$$M = \{f(\mathbf{y}) \in L_\infty(Y)\}$$

The design problem is formulated as follows: one has to find the distribution of materials (that is described by the function $c_{ij}(\mathbf{y})$, $\mathbf{y} \in Y$) which gives composite material the necessary homogenized characteristics C_{ij}. In other words (see Section 1.1.1), it is necessary to find $c_{ij}(\mathbf{y}) \in M$, such that

$$\int_Y (c_{ij}(\mathbf{y}) + c_{ik}(\mathbf{y})N^j_{,ky}(\mathbf{y}))d\mathbf{y} = C_{ij}|Y|, \tag{B.1}$$

where $N^j(\mathbf{y})$ represents a solution of the problem

$$\begin{cases} \left(c_{ij}(\mathbf{y})N^k_{,jy} + c_{ik}(\mathbf{y})\right)_{,iy} = 0 \text{ in } Y, \\ N^k(\mathbf{y}) \text{ is periodic in } \mathbf{y} \text{ with periodicity cell } Y. \end{cases} \tag{B.2}$$

In the general case, solution of the problem (B.1) and (B.2) is the subject matter of the theory known as topology design (or topology optimization). This theory was

283

started in [28] and then numerous papers and some books were written on the theme. The fundamental contributions to the topology design are the monographs [27, 29]. Although great progress was made in the field, the topology design problem is not solved completely. Algorithms constructing partial solutions of the problem (B.1), (B.2) [27, 28, 29] and similar problems [3, 82, 148, 173] were developed. But, to the best knowledge of the authors, for now there exists no algorithm constructing the general solution (i.e., the set of all solutions) of the problem (B.1), (B.2).

The unique case when the topology design problem is solved completely is the case of laminated materials [159]. In this case function $N^k(\mathbf{y})$ depends on one spatial variable (say $z = y_3$) and the cellular problem (B.2) takes the form

$$
\begin{cases}
\dfrac{d}{dz}\left(c_{i3}(z)\dfrac{dN^k}{dz} + c_{ik}(z)\right) = 0 \text{ in } [0,1], \\[2mm]
N^k(0) = N^k(1), \\[2mm]
c_{i3}(0)\dfrac{dN^k}{dz}(0) = c_{i3}(1)\dfrac{dN^k}{dz}(1).
\end{cases}
\tag{B.3}
$$

The formula (1.23) for computation of the homogenized transport characteristics of composite material takes the form

$$
C_{ij} = \left\langle c_{ij}(z) + c_{i3}(z)\frac{dN^j}{dz}(z)\right\rangle,
\tag{B.4}
$$

where

$$
\langle \bullet \rangle = \int_0^1 \bullet\, dz
$$

is the average value over the periodicity cell of laminated composite.

The problem (B.3) is written for the periodicity cell $[0,1]$. It does not lead to a loss of generality of our consideration because it is always possible to transform the arbitrary periodicity cell $[0, L]$ to the interval $[0,1]$. The problem (B.3) can be solved in explicit form (see for details [159]). Determining function $N^k(\mathbf{y})$ from (B.2), we then substitute it to (B.4). As a result, we obtain formulas for computation of the homogenized characteristics of laminated composite (formulas for various homogenized characteristics of laminated composite materials can be found in [159]). The formulas have the following structure:

$$
C_{ij} = F_{ij}(\langle f_{ij}(c_{kl}(z))\rangle),
\tag{B.5}
$$

where F_{ij} and f_{ij} are known algebraic functions.

Solving (B.5) with respect to $\langle f_{ij}(c_{kl}(z))\rangle$, we reduce the original problem to solution of a system of integral equations

$$
\int_0^1 f_{ij}(c_{kl}(z))dz = A_{ij}
\tag{B.6}
$$

with respect to the functions $c_{kl}(y) \in L_\infty([0,1])$. Usually, the functions $c_{kl}(y)$ are subjected to the condition $a_{kl} \le c_{kl}(y) \le b_{kl}$ or similar restrictions.

We note that although the cellular problem (B.3) is linear, the functions f_{ij} are usually nonlinear, see examples in [159].

The convex combinations problem

We consider a map $\mathcal{P}$ from the set

$$U = \{\mathbf{u}(z) \in L_\infty([0,1]) : a_i \le u_i(z) \le b_i, i = 1, ..., n\} \qquad (B.7)$$

into R^m, defined as

$$\mathcal{P} : \mathbf{u}(z) \in U \rightarrow \int_0^1 \mathbf{f}(\mathbf{u}(z))dz, \qquad (B.8)$$

with function

$$\mathbf{f} : R^n \rightarrow R^m.$$

It is known [307] (see also [159]) that the image of the set U in the mapping $\mathcal{P}$ (B.8) is a convex hull of the set

$$\Sigma = \{\mathbf{y} = \mathbf{f}(\mathbf{u}) \in R^m : \mathbf{u} \in U\} \qquad (B.9)$$

and any point belonging to the image of the set U under the mapping $\mathcal{P}$ can be obtained as value of the functional $\mathcal{P}$ on a piecewise-constant function not taking more than $m + 1$ different values.

If a finite number of materials is available, the set U is changed for the set

$$U_d = \{\mathbf{u}(z) \in L_\infty([0,1]) : u_i(z) \in \{a_j, j = 1, ..., n\}\}.$$

The image of the set U_d in the mapping $\mathcal{P}$ is the polyhedron $\text{conv}\Sigma_d$, where

$$\Sigma_d = \{\mathbf{f}(\mathbf{u}_i) \in R^m, i = 1, ..., n\}, \qquad (B.10)$$

is a finite set of points.

Any point belonging to the image of the set U_d under the mapping $\mathcal{P}$ can be obtained as value of the functional $\mathcal{P}$ on a piecewise-constant function not taking more than $m + 1$ different values.

Thus, the necessary and sufficient condition of solvability of the problem

$$\int_0^1 \mathbf{f}(\mathbf{u}(z))dz = \mathbf{A} \qquad (B.11)$$

with respect to the function $\mathbf{u}(z)$ is the following:

The problem (B.11) is solvable in U if and only if $\mathbf{A} \in \text{conv}\Sigma$ ("conv" means the convex hull [307]).

The problem (B.11) is solvable in U_d if and only if **A** belongs to the polyhedron convΣ_d.

Problem (B.11) on the set U_d can be written in the form of the convex combinations problem, formulated as follows: one has to find the general solution (i.e. the set $\Lambda(A)$ of all solutions) of the problem

$$\sum_{i=1}^{n} \mathbf{y}_i \lambda_i = \mathbf{A}, \tag{B.12}$$

$$\sum_{i=1}^{n} \lambda_i = 1, \ \lambda_1, ..., \lambda_n \geq 0. \tag{B.13}$$

Note that the vector equation (B.12) contains m scalar equations.

One can see an analogy between the convex combination problem and the problem of enumeration of vertexes of a polyhedron [120]. Some methods for the solution of the enumeration problem were developed (see, e.g., survey [18]). The double description method is mentioned [67, 262]. The problem is related to determining positive solutions of systems of linear equations [83, 84, 85, 240].

Application of the convex combinations problem technique to design of laminated composites can be found in [13, 159, 175].

The general solution $\Lambda(\mathbf{A})$ of the convex combinations problem was constructed in [181]. An algorithm presented in [181] (see also [159]) is closest to the Chernikov–Chernikova convolution algorithm [83, 84, 85]. We briefly present the idea of the convolution algorithm.

A simplex in R^m spanned by $m \leq k+1$ points with nonzero volume in R^m is called a non-degenerate one.

The convex hull $\Pi = \text{conv}\{\mathbf{y}_i, i = 1, ..., n\}$ is a polyhedron, which contains a finite number of non-degenerate simplexes. Denote by $\{\Pi_\eta, \eta = 1, ..., N\}$ those non-degenerate simplexes of the polyhedron Π that contain **A** (if $\mathbf{A} \notin \Pi$, then $\Lambda(\mathbf{A}) = \emptyset$).

A vector $\mathbf{s}_\eta(\mathbf{A}) = (s_{\eta 1}, ..., s_{\eta n}) \in R^n$ whose nonzero coordinates are the barycentric coordinates of the point **A** relative to a non-degenerate simplex Π_η is called a simplex solution of the problem (B.12), (B.13) corresponding of the simplex Π_η.

It is known (see, e.g., [181]) that $\Lambda(\mathbf{A}) = \text{conv}\{\mathbf{s}_\eta(\mathbf{A}), \eta = 1, ..., N\}$, in other words every solution $\lambda = (\lambda_1, ..., \lambda_n)$ of the convex combinations problem (B.12), (B.13) is given by the formula

$$\lambda = \sum_{i=1}^{N} \mathbf{s}_\eta \mu_\eta, \tag{B.14}$$

$$\sum_{i=1}^{N} \mu_\eta = 1, \ \mu_1, ..., \mu_N \geq 0 \tag{B.15}$$

with suitable values of $\mu_1, ..., \mu_N$.

For one-dimensional polytope (an interval), the set of all non-degenerate simplex can be constructed in a simple way and all simplex solutions can be computed in explicit form, see [159, 181].

We consider the first equation from (B.12) with the conditions (B.13) as a one-dimensional convex combinations problem. Its solution is given by (B.14), (B.15). We substitute (B.14) in the equalities (B.12). The first equality in (B.12) will be satisfied. The remaining $m - 1$ equalities with condition (B.15) will form a convex combinations problem with respect to the variables $\mu_1, ..., \mu_N$. Repeating this procedure, we reduce the number of equations at every step and finally obtain the solution of the convex combinations problem (if it exists). The detailed computational formulas can be found in [159, 181].

The new procedure added to convolution algorithm in [181] (see also [159]) is truncation (reducing) of solutions at every step of the convolution algorithm. Namely, at the p-th step of the convolution algorithm, we remove all solutions, which have more than $p + 1$ non-zero coordinates. This modification gives the algorithm the optimal computation property [181]. Note that the criterion of optimality in [159, 181] differs from the criterion based on the number of operations [77]. The optimality of algorithm presented in [159, 181] means that at every step of the algorithm, the set of so called simplicial solutions is minimal with probability one.

It is evident that the problems (B.6) can be written in the form (B.11).

Two-dimension convex combinations problem

The dimension of the convex combinations problem is determined by the number m. In two-dimensional ($m = 2$) case Σ (B.9) is a planar set and Σ_d (B.10) is a finite set of points on a plane. It is easy to draw the sets conv Σ and conv Σ_d. Solution of the convex combination problem (B.12), (B.13) also can be carried out by using a graphical method (for details see [159]).

REFERENCES

[1] A. Acrivos and E. Chang. A model for estimating transport quantities in two-phase materials. *Phys. Fluids*, 29(3):3–4, 1986.

[2] R.A. Adams. *Sobolev Spaces*. Academic Press, New York, 1975.

[3] G. Alaire. *Shape Optimization by Homogenization Method*. Springer-Verlag, Berlin, Heidelberg, New York, 2002.

[4] H.-D. Alber. Evolving microstructure and homogenization. *Continuum Mech. Thermodyn.*, 12(4):235–287, 2000.

[5] G. Allaire. *Shape Optimization by the Homogenization Method*. Springer-Verlag, Berlin, 2002.

[6] R.F. Almgren. An isotropic three-dimensional structure with Poisson's ratio $=-1$. *J. Elasticity*, 15:427–430, 1985.

[7] V. Ambegaokar, B.I. Halperin, and J.S. Langer. Hopping conductivity in disordered systems. *Phys. Rev. B*, 4(8):2612–2620, 1971.

[8] A. Aminian and Y. Rahmat-Samii. On the determination of the effective permittivity of metamaterials: a spectral FDTD approach. *Antennas and Propagation Society International Symposium, 2005 IEEE, 3–8 July 2005*, 2B:324–327, 2005.

[9] I.V. Andrianov, V.V. Danishevs'kyy, and A.L. Kalamkarov. Asymptotic analysis of effective conductivity of composite materials with large rhombic fibers. *Composite Struct.*, 56(33):229–234, 2002.

[10] I.V. Andrianov, V.V. Danishevs'kyy, and S. Tokarzewski. Two-point quasi-fractional approximants for effective conductivity of a simple cubic lattice of spheres. *Int. J. Heat Mass Transfer*, 39(11):2349–2352, 1996.

[11] I.V. Andrianov, G.A. Starushenko, V.V. Danishevskiy, and S. Tokarzewski. Homogenization procedure and Pade approximants for effective heat conductivity of composite material with cylindrical inclusions having square cross-section. *Proc. R. Soc. London*, A, 455:3401–3413, 1999.

[12] Ch. Ang, Z. Yu, R. Guo, and A.S. Bhalla. Calculation of dielectric constant and loss of two-phase composites. *J. Appl. Phys*, 93(6):3475–3480, 2003.

[13] B.D. Annin, A.L. Kalamkarov, A.G. Kolpakov, and V.Z. Parton. *Analysis and Design of Composite Materials and Structural Elements (in Russian)*. Nauka, Novosibirsk, 1993.

[14] N. Antonic, C.J. van Duijn, W. Jäger, and A. Mikelic, eds. *Multiscale Problems in Science and Technology*. Springer-Verlag, Berlin, Heidelberg, New York, 2002.

[15] F. Aurenhammer and R. Klein. Voronoi diagrams. *In: Handbook of Computational Geometry*, Elsevier Science Publ., Amsterdam, 2000.

[16] A. Averbakh, A. Shauly, A. Nir, and R. Semiat. Slow viscous flows of highly concentrated suspensions – part I. Laser-Doppler velocitometry in rectangular ducts. *Int. J. Multiphase Flow*, 23:409–424, 1997.

[17] A. Averbakh, A. Shauly, A. Nir, and R. Semiat. Slow viscous flows of highly concentrated suspensions – part II. Particle migration, velocity and concentration profiles in rectangular ducts. *Int. J. Multiphase Flow*, 23:613–629, 1997.

[18] D. Avis, D. Bremner, and R. Seidel. How good are convex hull algorithms? *Comp. Geometry*, 7:265–301, 1997.

[19] Sh. Axler, P. Bourdon, and R. Wade. *Harmonic Function Theory*. Springer-Verlag, Berlin, 2001.

[20] I. Babŭshka, B. Anderson, P. Smith, and K. Levin. Damage analysis of fiber composites. Part I. Statistical analysis on fiber scale. *Comput. Methods Appl. Mech. Engrg.*, 172:27–77, 1999.

[21] N.S. Bakhvalov and G.P. Panasenko. *Homogenization: Averaging Processes in Periodic Media*. Kluwer Academic Publ., Dordrecht, 1989.

[22] E.J. Barbero. *Finite Element Analysis of Composite Materials*. CRC Press, Boca Raton, 2007.

[23] G.K. Batchelor and R.W. O'Brien. Thermal or electrical conduction through a granular material. *Proc. R. Soc. London*, A, 335:313–333, 1977.

[24] Ch.E. Baukal. *Heat Transfer in Industrial Combustion*. CRC Press, Boca Raton, 2000.

[25] M. Bellieud and G. Bouchitt'e. Homogenization of a soft elastic material reinforced by fibers. *Asymptotic Anal.*, 32(2):153–183, 2002.

[26] L. Benabou, M. Naıt Abdelaziz, and N. Benseddiq. Effective properties of a composite with imperfectly bonded interface. *Theoretical and Applied Fracture Mechanics*, 41:15–20, 2004.

[27] M.P. Bendsøe. *Optimization of Structural Topology, Shape and Material.* Springer-Verlag, Berlin, Heidelberg, New York, 1995.

[28] M.P. Bendsøe and N. Kikuchi. Generating optimal topologies in structural design using a homogenization method. *Comp. Meth. Appl. Mech. Engng*, 71:95–112, 1988.

[29] M.P. Bendsøe and O. Sigmund. *Topology Optimization.* Springer-Verlag, Berlin, Heidelberg, New York, 2004.

[30] A. Bensoussan, J.-L. Lions, and G. Papanicolaou. *Asymptotic Analysis for Periodic Structures.* North-Holland Publ., Amsterdam, 1978.

[31] M.J. Beran. *Statistical Continuum Theories.* John Wiley & Sons, New York, 1968.

[32] V.L. Berdichevskij. *Variational Principles of Continuum Mechanics (in Russian).* Nauka, Moscow, 1983.

[33] D.J. Bergman, E. Duering, and M. Murat. Discrete network models for the low-field Hall effect near a percolation threshold: Theory and simulation. *J. Stat. Phys.*, 1(58):1–43, 1990.

[34] D.J. Bergman and K.J. Dunn. Bulk effective dielectric constant of a composite with periodic micro-geometry. *Phys. Rev. B*, 45:13262–13271, 1992.

[35] L. Berlyand, L. Borcea, and A. Panchenko. Network approximation for effective viscosity of concentrated suspensions with complex geometries. *SIAM J. Math. Anal.*, 36(5):1580–1628, 2005.

[36] L. Berlyand, G. Cardone, Y. Gorb, and G. Panasenko. Asymptotic analysis of an array of closely spaced absolutely conductive inclusions. *Networks and Heterogen. Media*, 3(1):353–377, 2006.

[37] L. Berlyand, D. Golovaty, A. Movchan, and J. Phillips. Transport properties of densely packed composites. Effect of shapes and spacings of inclusions. *Quart. J. Mech. Appl. Math.*, 57(4):1–34, 2004.

[38] L. Berlyand, Y. Gorb, and A. Novikov. Discrete network approximation for highly-packed composites with irregular geometry in three dimensions. *In: Multiscale Methods in Science and Engineering*, 44. B. Engquist, P. Lotstedt and O. Runborg, eds. Springer-Verlag, Berlin, Heidelberg, New York:21–58, 2005.

[39] L. Berlyand and A. Kolpakov. Justification of the net model for a high-contrast structure and its application to randomly filled composite. *First SIAM–EMS Conf. Applied mathematics in our changing world*, Berlin, Germany:51, 2001.

[40] L. Berlyand and A. Kolpakov. Net model for a system of the dense packing particles. *EUROPHYSICS – Conf. Comput. Phys. Aachen, Germany. Publ. Series J. von Neumann Inst. for Comput.*, 8:A50, 2001.

[41] L. Berlyand and A. Kolpakov. Network approximation in the limit of small interparticle distance of the effective properties of a high-contrast random dispersed composite. *Arch. Rational Mech. Anal.*, 159(3):179–227, 2001.

[42] A.-L. Bessoud, F. Krasucki, and M. Serpilli. Inclusions élastiques de grande rigidité de type plaque ou coque. *C. R. Mathematique*, 346(11/12):697–702, 2008.

[43] D. M. Bigg. Thermally conductive polymer composition. *Polymer Composites*, 7:125–139, 1986.

[44] D. Blachandr, A. Gaudiello, and G. Griso. Junction of a periodic family of elastic rods with a 3d plate. Part I. *J. Math. Pures Appl.*, 88:1–33, 2007.

[45] L. Boccardo and P. Marcellini. Sulla convergenza della soluzioni di disequazioni variazionali. *Ann. Mat. Pura Appl.*, 110:137–159, 1976.

[46] R.T. Bonnecaze and J.F. Brady. The effective conductivity of random suspensions of spherical particles. *Proc. R. Soc. London*, A 432:445–465, 1991.

[47] L. Borcea. Asymptotic analysis of quasi-static transport in high contrast conductive media. *SIAM J. Appl. Math.*, 2(59):597–635, 1998.

[48] L. Borcea, J.G. Berryman, and G. Papanicolaou. Matching pursuit for imaging high-contrast conductivity. *Inverse Problems*, 15:811–849, 1999.

[49] L. Borcea and G. Papanicolaou. Network approximation for transport properties of high contrast conductivity. *Inverse Problems*, 4(15):501–539, 1998.

[50] M. Born and K. Huang. *Dynamical Theory of Crystal Lattices*. Oxford University Press, Oxford, 1954.

[51] M. Born and E. Wolf. *Principles of Optics. 5th Ed.* Pergamon, New York, 1975.

[52] G. Bouchitte, G. Buttazzo, and P. Suquet. *Calculus of Variations, Homogenizations and Continuum Mechanics*. World Scientific, Singapore, 1994.

[53] N. Bourbaki. *Integration*. Springer-Verlag, Berlin, 2003.

[54] A. Bourgeat, A. Mikelic, and S. Wright. Stochastic two-scale convergence in the mean and applications. *J. Reine Angew. Math.*, 456:19–51, 1994.

[55] A. Bourgeat and A. Piatnitski. Approximations of effective coefficients in stochastic homogenization. *Ann. Inst. H. Poincare*, 40:153–165, 2004.

[56] C. Boutin. Study of permeability by periodic and self-consistent homogenization. *Eur. J. Mech. A / Solids*, 19:603–632, 2000.

[57] A. Braides. *Γ-convergence for Beginners*. Oxford Univer. Press, Oxford, 2002.

[58] A. Braides and A. Garroni. Homogenization of periodic nonlinear media with stiff and soft inclusions. *Math. Models Methods Appl. Sci.*, 5:543–564, 1995.

[59] M. Brelot. *Elements de la Theorie Classique de Potencial*. Centre de Doc. Univ. Paris, 1961.

[60] M. Briane and D. Manceau. Duality results in the homogenization of two-dimensional high-contrast conductivities. *Networks and Heterogen. Media*, 3:509–522, 2008.

[61] L.J. Broutman and R.H. Krock, eds. *Composite Materials. Vol. 1-8*. Academic Press, New York, London, 1974.

[62] D.A.G. Bruggeman. Berechnung verscheidenerphysikalischer Konstanten von heterogenen Substanzen. I. Dielekrizitätkonstanten und Leitfähigkeiten der Mischkörper aus isotropen Substanzen. *Ann. Phys.*, 416(8):665–679, 1935.

[63] D.A.G. Bruggeman. Berechnung verscheidenerphysikalischer Konstanten von heterogenen Substanzen. II. Dielekrizitätkonstanten und leitfähigkeiten von vielkristallen der nichtregularen Systeme. *Ann. Phys.*, 417(7):645–672, 1936.

[64] D.A.G. Bruggeman. Berechnung verscheidenerphysikalischer Konstanten von heterogenen Substanzen. III. Die elastischen Konstanten der quasiisotropen Mischkorper aus isotropen Substanzen. *Ann. Phys.*, 421(2):160–176, 1937.

[65] O. Bruno. The effective conductivity of strongly heterogeneous composites. *Proc. R. Soc. London*, A 433:353–381, 1991.

[66] P. Bujard, G. Kuhnlein, S. Ino, and T. Shiobara. Thermal conductivity of molding compounds for plastic packaging. *IEEE Trans. Comp. Pack. Manu. Tech.*, A17:527–532, 1994.

[67] E. Burger. Uber homogene lineare Ungleichungssysteme. *Z. Angew. Math. und Mech.*, 36:135–139, 1956.

[68] J.C. Burkill. *The Lebesque Integral*. Cambridge University Press, Cambridge, 2004.

[69] V. Buryachenko. *Micromechanics of Heterogeneous Materials.* Springer-Verlag, Berlin, Heidelberg, New York, 2007.

[70] D. Caillerie. Sur la comportement limite d'une inclusion mince de grande rigidite dans un corps elestique. *C. R. Acad. Sci. Paris*, Ser. A., 287:675–678, 1978.

[71] G. Caloz, M. Costabel, M. Dauge, and G. Vial. Asymptotic expansion of the solution of an interface problem in a polygonal domain with thin layer. *Asymptotic Analysis*, 50(1/2):121–173, 2006.

[72] L. Carbone and R. De Arcangelis. *Unbounded Functionals in the Calculus of Variations: Representation, Relaxation, and Homogenization.* CRC Press, Boca Raton, 2001.

[73] L. Carbone, R. De Arcangelis, and U. De Maio. Homogenization of media with periodically distributed conductors. *Asymptot. Anal.*, 2:157–194, 2000.

[74] L.A. Carlsson, D.F. Adamsm, and R.B. Pipes. *Experimental Characterization of Advanced Composite Materials. 3rd Ed.* CRC Press, Boca Raton, 2002.

[75] P. Cartraud and T. Messager. Computational homogenization of periodic plate-like structures. *Int. J. Solids Struct.*, 43(3/4), 2006.

[76] Ch. Chang and R.L. Powell. Effect of particle size distribution on the rheology of concentrated bimodal suspension. *J. Rheol.*, 38:85–98, 1994.

[77] B. Chazelle. An optimal convex hull algorithm in any fixed dimension. *Discrete Comput. Geometry*, 10:377–409, 1993.

[78] G.A. Chechkin, A.L. Piatnitski, and A.S. Shamaev. *Homogenization Method and Applications.* American Math. Soc., Providence, 2007.

[79] Y. Chen, X. Dong, J. Li, and Y. Wang. Dielectric properties of $Ba_{0.6}Sr_{0.4}TiO_3$/ Mg_2SiO_4/MgO composite ceramics. *J. Appl.Phys.*, 98:064–107, 2005.

[80] H. Cheng and L. Greengard. On the numerical evaluation of electrostatic fields in dense random dispersions of cylinders. *J. Comput. Phys.*, 136:626–639, 1997.

[81] H. Cheng and L. Greengard. A method of images for the evaluation of electrostatic fields in system of closely spaced conducting cylinders. *SIAM J. Appl. Math.*, 50:122–141, 1998.

[82] A. V. Cherkaev. *Variational Methods for Structural Optimization.* Springer-Verlag, Berlin, Heidelberg, New York, 2000.

[83] S.N. Chernikov. *Linear Inequalities (in Russian).* Nauka, Moscow, 1968.

[84] S.N. Chernikov. *Lineare Ungleichungen*. Deutscher Verlag der Wissenschaft, Berlin, 1971.

[85] N.V. Chernikova. Algorithm for finding a general formula for the non negative solutions of a system of linear equations. *Comput. Math. Mathem. Phys.*, 4:151–156, 1964.

[86] Ph.D. Chinh. Overall properties of planar quasisymmetric randomly inhomogeneous media: Estimates and cell models. *Phys. Rev. E*, 56:652–660, 1997.

[87] L. Chiter. Singular perturbation problems. A study by A-convergence. *Res. J. Appl. Sci.*, 2 (5):579–583, 2007.

[88] T.-W. Chou and F.K. Ko, eds. *Textile Structural Composites*. Elsevier Science Publ., Amsterdam, 1989.

[89] R.M. Christensen. *Mechanics of Composite Materials*. John Wiley & Sons, New York, 1979.

[90] P. G. Ciarlet. *Plates and Junctions in Elastic Multi-Structures. An Asymptotic Analysis*. Macon, Paris, 1990.

[91] D. Cioranescu and P. Donato. *An Introduction to Homogenization*. Oxford University Press, Oxford, 2000.

[92] R.J.E. Clausius. *Die mechanische Behandlung der Electricität*. F. Vieweg, Braunschweig, 1879.

[93] E. Cosserat and F. Cosserat. *Théorie des Corps Déformables*. Hermann, Paris, 1909.

[94] R.S. Courant and D. Hilbert. *Methods of Mathematical Physics*. John Wiley & Sons, New York, 1953.

[95] W.A. Curtin and H. Scher. Mechanical modeling using a spring networks. *J. Mater. Res.*, 5:554–562, 1990.

[96] A. Damlamian, D. Lukkassen, A. Meidell, and A. Piatnitski. *Multi Scale Problems and Asymptotic Analysis*. Gakuto Intern. Series, Tokyo, 2006.

[97] E.M. Daya, B. Braikat, N. Damil, and M. Potier-Ferry. Modelisation of modulated vibration modes of repetitive structures. *J. Comput. Appl. Math.*, 168(1/2):117–124, 2004.

[98] R. De Arcangelis, A. Gaudiello, and G. Paderni. Some cases of homogenization of linearly coercive gradient constrained variational problems. *Math. Models Methods Appl. Sci.*, 7:901–940, 1996.

[99] E. De Giorgi and S. Spagnolo. Sulla converzenca delli integrali dell energia per operatori ellittico del secondo ordine. *Boll. Unione Mat. Ital.*, 8:391–411, 1973.

[100] G. Del Maso. *An introduction to Γ-convergence*. Birkhäuser, Boston, Basel, Stuttgart, 1993.

[101] J.A. Dieudonne. *Treatise on Analysis*. Academic Press, New York, 1969.

[102] J. Ding, H.E. Warriner, and J.A. Zasadzinski. Viscosity of two-dimensional suspensions. *Phys. Rev. Lett.*, 88(16):168102.1–168102.4, 2002.

[103] P. Donato, L. Faella, and S. Monsurró. Homogenization of the wave equation in composites with imperfect interface: a memory effect. *J. Math. Pures Appl.*, 2:119–143, 2007.

[104] L. Dormieux, D. Kondo, and F.-J. Ulm. *Microporomechanics*. John Wiley & Sons, New York, 2006.

[105] A. Douglis and L. Nirenberg. Interior estimates for elliptic systems of partial differential equations. *Comm. Pure Appl. Math.*, 8:503–538, 1955.

[106] W.T. Doyle. The Clasius–Mossotti problem for cubic arrays of spheres. *J. Appl. Phys.*, 49:795–797, 1978.

[107] J.E. Drummon and M.I. Tahir. Laminar viscous flow through regular arrays of parallel solid cylinders. *Int. J. Multiphase Flow*, 10:515–540, 1984.

[108] G. Duvaut. Analyse fonctionelle et mechanique des mulieux continus. *In: Proc. 14th IUTAM Congress*, North-Holland, Amsterdam:119–132, 1977.

[109] G. Duvaut and J.-L. Lions. *Les Inequations en Mechanique et Physique*. Dunod, Paris, 1972.

[110] A.M. Dykhne. Conductivity of a two-dimensional two-phase system. *Sov. Physics (J. Exper. Teor. Phys.)*, 32(63):63–65, 1971.

[111] A. Einstein. Eine neue Bestimmung der Molekuldimensionen. *Ann. Phys.*, 324:289–306, 1906.

[112] A. Einstein. Berichtigung zu meiner Arbeit: Eine neue Bestimmung der Molekuldimensionen. *Ann. Phys.*, 339:591–592, 1911.

[113] I. Ekeland and R. Temam. *Convex Analysis and Variational Problems*. North-Holland, Amsterdam, 1976.

[114] Yu. P. Emetz. Electrical characteristics of three-component dielectric composites with close-packed inclusions. *J. Appl. Mech. Thech. Phys.*, 42(4):133–213, 2001.

[115] R. Ewing, O. Iliev, R. Lazarov, I. Rybak, and J. Willems. *An Efficient Approach for Upscaling Properties of Composite Materials with High Contrast of Coefficients.* Berichte des Fraunhofer ITWM, Nr. 132, Kaiserslautern, 2008.

[116] Z. Fan. A microstructural approach to the effective transport properties of multiphase composites. *Philos. Mag.*, 73(6):1663–1684, 1996.

[117] M. Faraday. *Experimental Research in Electricity.* Richard & John Edward Taylor, London, 1839.

[118] G. Fichera. *Existence Theorems in Elasticity.* Springer-Verlag, Berlin, New York, 1972.

[119] J.E. Flaherty and J.B. Keller. Elastic behavior of composite media. *Comm. Pure Appl. Math.*, 26:565–580, 1973.

[120] J.B.J. Fourier. Solution d'une question particuiere du calcul des inegalities. *In: Oeuvres II*, Gauthier-Villars, Paris:317–328, 1890.

[121] N.A. Frenkel and A. Acrivos. On the viscosity of concentrated suspension of solid spheres. *Chem. Engng Sci.*, 22:847–853, 1967.

[122] S. Fučik and A. Kufner. *Nonlinear Differential Equations.* Elsevier Science Publ., Amsterdam, 1980.

[123] Y. C. Fung and Pin Tong. *Classical and Computational Solid Mechanics.* World Scientific, Singapore, 2001.

[124] H. Gajewski, K. Gröger, and K. Zacharias. *Nichtlineare Operatorleichngen und Operatordifferential-gleichungen.* Academie-Verlag, Berlin, 1974.

[125] R.H. Gallagher. *Finite Element Analysis: Fundamentals.* Prentice Hall, Englewood Cliffs, NJ, 1975.

[126] L. Gao and Z. Li. Effective medium approximation for two-component nonlinear composites with shape distribution. *J. Phys.: Condensed Matter*, 15:4397–4409, 2003.

[127] E.J. Garboczi and J.F. Douglas. Intrinsic conductivity of objects having arbitrary shape and conductivity. *Phys. Rev. E*, 53(6):6169–6180, 1996.

[128] J.C.M. Garnett. Colours in metal glasses and metal films. *Trans. Royal Soc.*, CCIII:385–420, 1904.

[129] A. Gaudiello. Homogenization of an elliptic transmission problem. *Adv. Math. Sci. Appl.*, 5 (2):639–657, 1995.

[130] A. Gaudiello, R. Monneau, J. Mossino, F. Murat, and A. Sili. On the junction of elastic plates and beams. *ESAIM Control Optim. Calc. Var.*, 13:419–457, 2007.

[131] R.F. Gibson. *Principles of Composite Material Mechanics. 2nd Ed.* CRC Press, Boca Raton, 2007.

[132] R.P. Gilbert and A. Panchenko. *Homogenization: Applications to the Biological and Physical Sciences.* Chapman & Hall/CRC, Boca Raton (in preparation).

[133] R. Glowinski, T.W. Pan, T.I. Hesla, D.D. Joseph, and J. Périaux. A fictitious domain approach to the direct numerical simulation of incompressible viscous flow past moving rigid bodies: Application to particulate flow. *J. Comput. Phys.*, 169(2):363 – 426, 2001.

[134] K.M. Golden. Electrical transport properties of high contrast composite materials. *In: Proc. Fourth Intern. Conf. Composites Engng (D. Hui, ed.)*, Int. Community Composites Engng:363–364, 1997.

[135] H. Goto and H. Kuno. Flow of suspensions containing particles of two different sizes through a capillarity tube. II. Effect of the particles size ratio. *J. Rheol.*, 28:197–205, 1984.

[136] A.L. Graham. On the viscosity of suspension of solid particles. *Appl. Sci. Res.*, 37:275–286, 1981.

[137] L. Greengard and M. Moura. On the numerical evaluation of electrostatic fields in composite materials. *Acta Numerica*, 3:379–410, 1994.

[138] G. Grimet. *Percolation.* Springer-Verlag, Berlin, Heidelberg, New York, 1992.

[139] G.Q. Gu. Conductivity of composites containing anisotropic graded fibers. *Journal of Composite Materials*, 39(2):127–145, 2005.

[140] F. Guérin. Microwave chiral materials: A review of experimental studies and some results on composites with ferroelectric ceramic inclusions. *Progr. Electromagnet. Res.*, 9:219263, 1994.

[141] I. Guiassu and H. Raszillier. Optimal approximation of the electrostatic capacity matrix of two conducting spheres by a short-distance asymptotic expansion. *IMA J. Appl. Math.*, 43(2):185–193, 1989.

[142] R. Guinovart-Díaz, J. Rodríguez-Ramos, R. Bravo-Castillero, and F.J. Sabina. Modeling of three-phase fibrous composite using the asymptotic homogenization method. *Mech. Adv. Mater. Struct.*, 10(4):319–333, 2003.

[143] P.K. Gupta and A.R. Cooper. Topologically disordered networks of rigid polytopes. *J. Non-Crystal. Solids*, 123(14):14–21, 1990.

[144] N.A. Hakeem, H.I. Abdelkader, N.A. El-Sheshtawi, and I.S. Eleshmawi. Spectroscopic, thermal and electrical investigations of PVDF films filled with $BiCl_3$. *J. Appl. Pol. Sci.*, 102:2125–2131, 2006.

[145] J. Happel. Viscous flow relatively to arrays of cylinders. *AIChE J*, 5:174–177, 1959.

[146] H. Hasimoto. On the periodic fundamental solution of the Stokes equations and their application to viscous flow past a cubic array of spheres. *J. Fluid Mech.*, 5:317–328, 1959.

[147] D.P.H. Hasselman, K.Y. Donaldson, and J.R. Thomas Jr. Effective thermal conductivity of uniaxial composite with cylindrically orthogonal carbon fibers and interfacial thermal barrier. *J. Comp. Mater.*, 27:637–544, 1993.

[148] E.J. Haug, K.K. Choi, and V. Komkov. *Design Sensitivity Analysis of Structural Systems*. Academic Press, Orlando, 1986.

[149] B. Heron, J. Mossino, and C. Picard. Homogenization of some quasilinear problems for stratified media with low and high conductivities. *Differ. Integral Equ.*, 1:157–178, 1994.

[150] R. Hill. Elastic properties of reinforced solids: Some theoretical principles. *J. Mech. Phys. Solids*, 11:357–372, 1963.

[151] R. Hill. Characterization of thermally conductive epoxy composite fillers. *Emerging Packing Tech.*, Surface Mount Tech. Symp.:125–131, 1996.

[152] R.F. Hill and P.H. Supancic. Thermal conductivity of platelet-filled polymer composite. *J. Am. Cer. Soc.*, 85:851–857, 2002.

[153] E.J. Hinch. *Perturbation Methods*. Cambridge University Press, Cambridge, 1991.

[154] T.J.R. Hughes. *The Finite Element Method*. Prentice Hall, Englewood Cliffs, NJ, 1987.

[155] K. Ichikawa, ed. *Functionally Graded Materials in the 21st Century: A Workshop on Trends and Forecasts*. Springer-Verlag, Berlin, Heidelberg, New York, 2000.

[156] D.J. Jeffrey and A. Acrivos. The rheological properties of suspensions of rigid particles. *AIChE J*, 22:417–432, 1976.

[157] V.V. Jikov, S.M. Kozlov, and O.A. Oleinik. *Homogenization of Differential Operators and Integral Functionals.* Springer-Verlag, Berlin, Heidelberg, New York, 1994.

[158] L. Jylhä and A.H. Sihvola. Tunability of granular ferroelectric–dielectric composites. *Progr. Electromagnet. Res.*, 78:189–207, 2008.

[159] A.L. Kalamkarov and A.G. Kolpakov. *Analysis, Design and Optimization of Composite Structures.* John Wiley & Sons, Chichester, 1997.

[160] L.V. Kantorovich and G.P. Akilov. *Functional Analysis in Normed Spaces.* Pergamon Press, Oxford, 1964.

[161] S. Kanzaki, M. Shimada, K. Komeya, and A. Tsuge. Recent progress in the synergy ceramics project. *Key Eng. Mater.*, 161–163:437–442, 1999.

[162] F.C. Karal Jr. and J.B. Keller. Effective dielectric constant, permeability, and conductivity of a random medium and the velocity and attenuation coefficient of coherent waves. *J. Math. Phys.*, 7:661–670, 1966.

[163] B.L. Karihaloo and J. Wang. On the solution of doubly periodic array of cracks. *Mech. Mater.*, 26(4):209–212, 1987.

[164] T. Kato. *Perturbation Theory for Linear Operators.* Springer-Verlag, New York, 1976.

[165] H. Keller and D. Sachs. Calculations of the conductivity of a medium containing cylindrical inclusions. *J. Appl. Phys.*, 35:537–538, 1964.

[166] J.B. Keller. Conductivity of a medium containing a dense array of perfectly conducting spheres or cylinders or nonconducting cylinders. *J. Appl. Phys.*, 4(34):991–993, 1963.

[167] J.B. Keller. A theorem on the conductivity of a composite medium. *J. Math. Phys.*, 5:548–549, 1964.

[168] M. Kellomaki, J. Astrom, and J. Timonen. Rigidity and dynamics of random spring networks. *Phys. Rev. Lett.*, 77:2730–2733, 1996.

[169] A. Kelly and Yu.N. Rabotnov, eds. *Handbook of Composites.* North-Holland, Amsterdam, 1988.

[170] H. Kesten. *Percolation Theory for Mathematicians.* Birkhäuser, Boston, Basel, Stuttgart, 1992.

[171] G.S. Khizha, I.B. Vendik, and E.A. Serebryakova. *Microwave Phase Shifters Based on p-i-n Diodes (in Russian).* Radio i Svyas, Moscow, 1984.

[172] A.S. Kobayashi, ed. *Handbook on Experimental Mechanics.* Prentice Hall, Englewood Cliffs, NJ, 1987.

[173] A.A. Kolpakov. Design of laminated plate possessing the required stiffnesses using the minimal number of materials and layers. *J. Elasticity*, 86:245–261, 2007.

[174] A.A. Kolpakov. Numerical verification of the existence of the energy-concentration effect in a high-contrast heavy-charged composite material. *J. Engng Phys. Thermophysics*, 4:812–819, 2007.

[175] A.A. Kolpakov and A.G. Kolpakov. Solution of the laminated plate design problem: New problems and algorithms. *Computers & Structures*, 83(12-13):964–975, 2005.

[176] A.A. Kolpakov and A.G. Kolpakov. Asymptotic of capacity of a system of closely placed bodies. Tamm's shielding effect and network models. *Doklady Physics*, 415(2):188–192, 2007.

[177] A.A. Kolpakov, A.G. Kolpakov, and S.I. Rakin. Effective permittivity, tunability, loss and commutation quality of high contrast composites. *In: Recent Advances in Dielectric Materials*, Ai Huang Ed. Nova Publ., New York, 2009.

[178] A.G. Kolpakov. On determination of the average characteristics of elastic frameworks. *J. Appl. Math. Mech.*, 49:739–745, 1985.

[179] A.G. Kolpakov. Asymptotic behavior of the first boundary-value problem for elliptic equations in domains with a thin coating. *Siberian Math. J.*, 6:931–940, 1988.

[180] A.G. Kolpakov. Glued bodies. *Differential Equations*, 8:1131–1139, 1992.

[181] A.G. Kolpakov. The solution of the convex combinations problem. *Comput. Math. Mathem. Phys.*, 8:1183–1188, 1992.

[182] A.G. Kolpakov. *Stressed Composite Structures: Homogenized Models for Thin-Walled Nonhomogeneous Structures with Initial Stresses.* Springer-Verlag, Berlin, Heidelberg, New York, 2004.

[183] A.G. Kolpakov. Asymptotic behavior of the conducting properties of high-contrast media. *J. Appl. Mech. Thech. Phys.*, 3:412–422, 2005.

[184] A.G. Kolpakov. Effective conductance of a high-contrast, random-structure composite. Numerical simulation. *J. Engng Phys. Thermophysics*, 6:170–177, 2005.

[185] A.G. Kolpakov. The asymptotic screening and network models. *J. Engng Phys. Thermophysics*, 2:39–47, 2006.

[186] A.G. Kolpakov. Convergence of solutions for a network approximation of the two-dimensional Laplace equation in a domain with a system of absolutely conducting disks. *Comput. Math. Mathem. Phys.*, 46(9):1601–1610., 2006.

[187] A.G. Kolpakov. Averaged characteristics of a nonlinear composite. *J. Engng Phys. Thermophysics*, 81(4):801–812, 2008.

[188] A.G. Kolpakov. Enhancing the controllability of a composite dielectric. *J. Appl. Mech. Thech. Phys.*, 5:823–831, 2008.

[189] A.G. Kolpakov. Tuneability amplification factor and loss of high-contrast composites. *Philos. Mag.*, 89(3):263–283, 2009.

[190] A.G. Kolpakov and S.I. Rakin. Effective tunability of composite in strong fields. *J. Engng Phys. Thermophysics (in print)*.

[191] A.G. Kolpakov, A.K. Tagantsev, L. Berlyand, and A. Kanareykin. Nonlinear dielectric response of periodic composite materials. *J. Electroceramics*, 18:129–137, 2007.

[192] J. Koplik. Creeping flow in two-dimensional networks. *J. Fluid Mech.*, 119:219–247, 1982.

[193] V.A. Kovtunenko. *Variational Methods in Theory of Cracks with Constraints (in Russian)*. Doctor of phys.–math. sciences dissertation. M.A. Lavrent'ev Institute of hydrodynamics of Siberian branch of Russian Academy of Sci., Novosibirsk, 2007.

[194] S.M. Kozlov. Averaging of random structures. *Soviet Doklady Math.*, 241 (5):1016–11019, 1978.

[195] S.M. Kozlov. Averaging of random operators. *Math. USSR Sbornik*, 37:167–180, 1980.

[196] S.M. Kozlov. Geometric aspects of averaging. *Russian Math. Surveys*, 2(44):91–144, 1989.

[197] H. Kuchling. *Physics*. VEB Fachbuchverlag, Leipzig, 1980.

[198] Y.W. Kwon, D.H. Allen, and R. Talreja, eds. *Multiscale Modeling and Simulation of Composite Materials and Structures*. Springer-Verlag, Berlin, Heidelberg, New York, 2007.

[199] O.A. Ladyzhenskaya. *The Boundary Value Problems of Mathematical Physics*. Springer-Verlag, New York, 1985.

[200] O.A. Ladyzhenskaya and N.N. Ural'tseva. *Linear and Quasilinear Elliptic Equations*. Academic Press, New York, London, 1968.

[201] R. Lakes. Foam structures with negative Poisson's ratio. *Science*, 235:1038, 1987.

[202] H. Lamb. *Hydrodynamics*. Dover Publ., New York, 1991.

[203] L.D. Landau and E.M. Lifshitz. *Electrostatics of Continuum Media*. Pergamon Press, Oxford, 1984.

[204] F. Lebon, R. Rizzoni, and S. Ronel-Idrissi. Asymptotic analysis of some non-linear soft thin layers. *Computers & Structures*, 82(23-26):1929–1938, 2004.

[205] M. Lenczner. Homogenization of linear spatially periodic electronic circuits. *Networks and Heterogen. Media*, 1(3):407–494, 2000.

[206] F. Lenè and D. Leguillon. Homogenized constitutive low for a partially cohesive composite material. *Int. J. Solids Structures*, 18:413–458, 1982.

[207] T. Levinski. Dynamical tests of accuracy of Cosserat models for honeycomb gridworks. *Z. Angew. Math. Mech.*, 210(68):210–212, 1988.

[208] Y.Y. Li and M. Vogelius. Gradient estimates for solutions to divergence form elliptic equations with discontinuous coefficients. *Arch. Rational Mech. Anal.*, 153(2):91–151, 2000.

[209] L. Lia, X.M. Chen, and X.Ch. Fan. Microwave dielectric properties of $MgTiO_3SrTiO_3$ layered ceramics. *J. Europ. Ceramic Soc.*, 26(13):2817–2821, 2006.

[210] E.H. Lieb and M. Loss. *Analysis. 2nd ed.* American Math. Soc., Providence, 2001.

[211] J.-L. Lions. A remark on certain computational aspects of the homogenization method in composite materials. *In: Vychislitel'nye metody v matematike, geofizike i optimal'nom upravlenii*, Nauka, Novosibirsk:5–19, 1978.

[212] J.-L. Lions and E. Magenes. *Non-Homogeneous Boundary Value Problems and Applications. Vol. 1, 2.* Springer-Verlag, Berlin, Heidelberg, New York, 1973.

[213] W.K. Liu, E.G. Karpov, and H.S. Park. *Nano Mechanics and Materials: Theory, Multiscale Methods and Applications.* John Wiley & Sons, New York, 2006.

[214] H.A. Lorentz. *Theory of Electrons*. Teubner, Leipzig, 1916.

[215] L. Lorenz. Uber die refractionconstante. *Ann. Phys. Chem.*, 11:70–103, 1880.

[216] A.E.H. Love. *A Treatise on the Mathematical Theory of Elasticity*. Oxford University Press, Oxford, 1929.

[217] G. Lubin, ed. *Handbook of Composites*. Van Nostrand, New York, 1982.

[218] M. Mabrouk and A. Boughammoura. Homogenization of a strongly heterogeneous periodic elastic medium. *Int. J. Engng Sci.*, 41:817–843, 2002.

[219] M. Mabrouk and H. Samadi. Homogenization of a heat transfer problem in a highly heterogeneous periodic medium. *Int. J. Engng Sci.*, 40:1233–1250, 2002.

[220] S.F. Makaruk, V.V. Mityushev, and S.V. Rogosin. An optimal design problem for two-dimensional composite materials. A constructive approach. *In: Analytic Methods of Analysis and Differential Equations*, A.A. Kilbas and S.V. Rogosin, eds. Cambridge Sci. Publ. Cottenham, Cambridge:153–167, 2006.

[221] P. Marcellini. Un teorema di passagio de limite per la somma di convesse. *Boll. Unione Mat. Ital.*, 4:107–124, 1975.

[222] P. Marcellini. Periodic solution and homogenization of nonlinear variational problems. *Ann. Mat. Pura App*, 117 (1978):139–152, 1978.

[223] P. Marcellini and C. Sbordone. Sur quelque de *G*-convergence et d'homogenisation non-lineare. *C. R. Acad. Sci. Paris*, 284:535–537, 1977.

[224] A. Marino and S. Spagnolo. Un tipo di appromazioni dell'operatore $\sum_{ij} D_i(a_{ij}(\bar{x})D_j)$ con operatori $\sum_i D_i(\beta(\bar{x})D_i)$. *Ann. Scuola Norm. Sup. Pisa*, 23(3):657–673, 1969.

[225] K.Z. Markov. Elementary micromechanics of heterogeneous media. *In: Heterpgeneous Media: Micromechanics Modeling Methods and Simulation*, K. Markov, L. Preziosi, eds. Birkhäuser, Basel:1–162, 2000.

[226] C.F. Matt and M.E. Cruz. Effective thermal conductivity of composite materials with 3-d microstructures and interfacial thermal resistance. *Numer. Heat Transfer, Part A*, 53(6):577–604, 2008.

[227] J.C. Maxwell. *Treatise on Electricity and Magnetism*. Clarendon Press, Oxford, 1873.

[228] V. Maz'ya, S. Nazarov, and B. Plamenevskij. *Asymptotic Theory of Elliptic Boundary Value Problems in Singularly Perturbed Domains. Vol. 1, 2.* Birkhäuser, Boston, Basel, Stuttgart, 2000.

[229] V.G. Maz'ya, S.A. Nazarov, and B.A. Plamenevskiĭ. The Dirichlet problem in domains with thin cross connections. *Siberian Math. J.*, 25(2):161–179, 1984.

[230] L. E. McAllister and W.L. Lachman. Multidirectional carbon–carbon composites. *In: Handbook of Composites. V.4. Fabrication of Composites*, A. Kelly and S.T. Mileiko, eds. North-Holland, Amsterdam:109–176, 1983.

[231] D.R. McKenzie and R.C. McPhedran. Exact modeling of cubic lattice permittivity and conductivity. *Nature*, 265:128–129, 1977.

[232] D.R. McKenzie, R.C. McPhedran, and G.H. Derrik. The conductivity of lattice of spheres II. The body centered and face centered lattices. *Proc. R. Soc. London*, A, 362:211–232, 1978.

[233] R. McPhedran. Transport property of cylinder pairs of the square array of cylinders. *Proc. R. Soc. London*, A, 408:31–43, 1986.

[234] R. McPhedran, L. Poladian, and G.W. Milton. Asymptotic studies of closely spaced, highly conducting cylinders. *Proc. R. Soc. London*, A, 415:195–196, 1988.

[235] R.C. McPhedran. Transport properties of cylinder pairs and of the square array of cylinders. *Proc. R. Soc. London*, A, 408:31–43, 1986.

[236] R.C. McPhedran and D.R. McKenzie. The conductivity of lattice of spheres I. The simple cubic lattice. *Proc. R. Soc. London*, A, 359:45–63, 1978.

[237] R.C. McPhedran and N.A. Nicorovici. Effective dielectric constant of arrays of elliptical cylinders. *Physica A*, 241:173–178, 1997.

[238] D.B. Melrose and R.C. McPhedran. *Electromagnetic Processes in Dispersive Media*. Cambridge University Press, Cambridge, 1991.

[239] R.E. Meredith and C.W. Tobias. Resistance to potential flow through a cubical array of spheres. *J. Appl. Physics*, 31(7):1270–1273, 1960.

[240] Yu. I. Merzlyakov. On the existence of positive solutions of systems of linear equations. *Russian Math. Surveys*, 18(3):179–186, 1963.

[241] J.C. Michel, H. Moulinec, and P. Suquet. A computational method based on augmented Lagrangians and fast Fourier transforms for composites with high contrast. *Comp. Modeling Engng Sci.*, 1(2):79–88, 2000.

[242] J.C. Michel, H. Moulinec, and P. Suquet. A computational scheme for linear and non-linear composites with arbitrary phase contrast. *Int. J. Numer. Meth. Engng.*, 52:139–160, 2002.

[243] A. Mielke, ed. *Analysis, Modeling and Simulation of Multiscale Problems*. Springer-Verlag, Berlin, Heidelberg, New York, 2006.

[244] V.P. Mikhailov. *Partial Differential Equations*. Mir Publ., Moscow, 1978.

[245] M.J. Miksis. Effective dielectric constant of nonlinear composite material. *SIAM J. Appl. Math.*, 43:1140–1155, 1983.

[246] G.W. Milton. *The Theory of Composites*. Cambridge University Press, Cambridge, 2002.

[247] L.L. Mishnaevsky Jr. and S. Schmauder. Continuum mesomechanical finite element modeling in materials development: A state-of-the-art review. *Appl. Mech. Rev.*, 54(1):49–74, 2001.

[248] V. Mityushev. Transport properties of doubly periodic arrays of circular cylinders and optimal design problems. *Appl. Math. Optim.*, 44:17–31, 2001.

[249] V. Mityushev and P.M. Adler. Longitudinal permeability of a doubly periodic rectangular array of cylinders. I. *Z. Angew. Math. Mech.*, 82:335–345, 2002.

[250] V. Mityushev and P.M. Adler. Longitudinal permeability of a doubly periodic rectangular array of cylinders. II. An arbitrary distribution of cylinders inside the unit cell. *Z. Angew. Math. Phys.*, 53:486–517, 2002.

[251] V. Mityushev and S.V. Rogozin. *Constructive Methods for Linear and Nonlinear Boundary Value Problems of the Analytic Functions Theory*. Chapman & Hall/CRC, Boca Raton, 2000.

[252] V.V. Mityushev. *Private communication*.

[253] V.V. Mityushev. Transport properties of doubly-periodic arrays of circular cylinders. *Z. Angew. Math. Mech.*, 77:115–120, 1997.

[254] V.V. Mityushev, E. Pesetskaya, and S.V. Rogosin. Analytical methods for heat conduction in composites and porous media. *In: Cellular and Porous Materials: Thermal Properties Simulation and Prediction*, Wiley-VCH, Weinheim:17–31, 2008.

[255] Y. Miyamoto, W.A. Kaysser, B.H. Rabin, A. Kawasaki, and R.G. Ford, eds. *Functionally Graded Materials: Design, Processing and Applications*. Kluwer Academic Publ., Boston, 1999.

[256] S. Mizohata. *The Theory of Partial Differential Equations*. Cambridge University Press, Cambridge, 1973.

[257] S. Molchanov. Ideas in the theory of random media. *Acta Applicandae Mathematicae*, 22(2/3):139–282, 1991.

[258] S. Molchanov. Lectures on random media. *In: Lectures on probability theory. Lect. Notes in Math. 1581*, Springer-Verlag. Berlin, Heidelberg, New York:242–411, 1994.

[259] J. Molyneux. Effective permittivity of a polycrystalline dielectric. *J. Math. Phys.*, 11(4):1172–1184, 1970.

[260] C.B. Morrey. Second order elliptic equations in several variables and holder continuity. *Math. Z.*, 72:146–164, 1959.

[261] O.F. Mossotti. Discussione analitica sull'influenza che l'azione di un mezzo dielettrico ha sulla distribuzione dell'elettricità alla superficie di più corpi elettrici disseminati in esso. *Memorie di Matematica e di Fisica della Societa Italiana delle Scienze*, XXIV, Parte seconda (Modena):49–74, 1850.

[262] T.S. Motzkin, H. Raiffa, G.L. Thompson, and R.M. Thrall. The double description method. *In: Contributions to the theory of games II*, Princeton Univ. Press. 8:51–73, 1953.

[263] D. Moulton and J. Pelesko. Thermal boundary conditions: an asymptotic analysis. *Heat and Mass Transfer*, 44(7):795–803, 1983.

[264] A.B. Movchan, N.V. Movchan, and C.G. Poulton. *Asymptotic Models of Fields in Dilute and Densely Packed Composites*. Imperial College Press, London, 2002.

[265] M. Mukhopadhyay. *Mechanics of Composite Materials and Structures*. Universities Press (India), 2005.

[266] J. Nash. Continuity of solutions of parabolic and elliptic equations. *Amer. J. Math.*, 80(4):931–954., 1958.

[267] P.S. Neelakanta. *Handbook of Electromagnetic Materials: Monolithic and Composite Versions and their Applications*. CRC Press, Boca Raton, 1995.

[268] M.E.J. Newman. The structure and functions of complex networks. *SIAM Rev.*, 45(2):167–256, 2003.

[269] N.A. Nicorovici and R.C. McPhedran. Transport properties of arrays of elliptical cylinders. *Phys. Rev. E*, 54:1945–1957, 1996.

[270] N.A. Nicorovici, R.C. McPhedran, and G.W. Milton. Transport properties of a three-phase composite material: The square array of coated cylinders. *Proc. R. Soc. London*, A 442:599–620, 1993.

[271] L.F. Nielsen. *Composite Materials: Properties as Influenced by Phase Geometry*. Springer-Verlag, Berlin, Heidelberg, New York, 2005.

[272] A.K. Noor. Continuum modeling for repetitive structures. *Appl. Mech. Rev.*, 41(7):285–296, 1988.

[273] A.K. Noor, N.S. Anderson, and W.H. Greene. Continuum models for beam-like and plate-like lattice structures. *AIAA Journal*, 16(12):1219–1228, 1978.

[274] V.V. Novikov and Ch. Friedrich. Viscoelastic properties of composite materials with random structure. *Phys. Rev. E*, 72:021506-1–021506-9, 2005.

[275] V.V. Novozilov. On relationship between average values of stress tensor and strain tensor in statistically isotropic elastic bodies. *Appl. Math. Mech.*, 34(1), 1970.

[276] K.C. Nunan and J.B. Keller. Effective elasticity tensor for a periodic composite. *J. Mech. Phys. Solids*, 32:259–280, 1984.

[277] K.C. Nunan and J.B. Keller. Effective velocity of periodic suspension. *J. Fluid Mech.*, 142:269–287, 1984.

[278] O.A. Oleinik, G.A. Iosifian, and A.S. Shamaev. *Mathematical Problems in Elasticity and Homogenization.* North-Holland, Amsterdam, 1992.

[279] G.P. Panasenko. Homogenization of fields in composite materials with high modulus reinforcement (in Russian). *Vestnik Moskovskogo Universiteta. Comput. Math. Cybern.*, 2:20–27, 1983.

[280] G.P. Panasenko. Homogenization of processes in strongly non-homogeneous media. *Soviet Doklady Physics*, 298(1):76–79, 1988.

[281] G.P. Panasenko. Asymptotic solutions of the elasticity theory system of equations for lattice and skeletal structures. *AMS Sbornik Math.*, 75(1):85–110, 1993.

[282] G.P. Panasenko. Homogenization of lattice-like domains: *L*-convergence. *In: Nonlinear Partial Differential Equations and Their Applications: College de France Seminar*, D. Cioranescu, J.-L. Lions, eds. CRC Press, Boca Raton:259–280, 1998.

[283] G.P. Panasenko. *Multi-Scale Modeling for Structures and Composites.* Springer-Verlag, Berlin, Heidelberg, New York, 2005.

[284] G.P. Panasenko and G. Virnovsky. Homogenization of two-phase flow: high contrast of phase permeability. *C.R. Mecanique*, 331:9–15, 2003.

[285] M. Panfilov. *Macroscale Models of Flow Through Highly Heterogeneous Porous Media.* Springer-Verlag, Berlin, Heidelberg, New York, 2000.

[286] G.C. Papanicolaou and S.R.S. Varadhan. Boundary value problems with rapidly oscillating random coefficients. *Seria Coll. Janos Bolyai*, North-Holland, Amsterdam, 27:835–873, 1981.

[287] V.Z. Parton and B.A. Kudryavtsev. *Engineering Mechanics of Composite Structures.* CRC Press, Boca Raton, 1993.

[288] G.A. Pavliotis and A.M. Stuart. *Multiscale Methods: Averaging and Homogenization.* Springer-Verlag, Berlin, Heidelberg, New York, 2008.

[289] J.N. Pernin and E. Jacquet. Elasticity and viscoelasticity in highly heterogeneous composite medium: Threshold phenomenon and homogenization. *Int. J. Engng Sci.*, 39:1655–1689, 2001.

[290] J.N. Pernin and E. Jacquet. Elasticity in highly heterogeneous composite medium: Threshold phenomenon and homogenization. *Int. J. Engng Sci.*, 39:755–798, 2001.

[291] W.T. Perrins, R.C. McPhedran, and D.R. McKenzie. Transport properties of regular arrays of cylinders. *Proc. R. Soc. London*, A, 369:207–225, 1979.

[292] E.V. Pesetskaya. Effective conductivity of composite materials with random positions of cylindrical inclusions: finite number inclusions in the cell. *Applicable Anal.*, 84(8):843–865, 2005.

[293] I.V. Petrov, I.I. Stonev, and F.V. Babalievskii. Correlated two-component percolation. *J. Phys. A*, 24(18):4421–4426, 1991.

[294] H. Pham Huy and E. Sanchez-Palencia. Phenomenes de transmission a travers des couches minces de conductivite eleve. *J. Math. Anal. Appl.*, 47:284–309, 1974.

[295] S.D. Poisson. Memoire sur la theorie du magnetisme. *Memoires de l'Academie Royale des Sciences de l'Institute de France*, 5:247–338, 1826.

[296] L. Poladian. Asymptotic behavior of the effective dielectric constants of composite materials. *Proc. R. Soc. London*, A, 426:343–360, 1989.

[297] G. Pólya and G. Szëgo. Inequalities for the capacity of a condenser. *Amer. J. Math.*, 67:1–32, 1945.

[298] G. Pólya and G. Szëgo. *Isoperimetric Inequalities in Mathematical Physics.* Princeton Univ. Press, Princeton, 1951.

[299] P. Ponte Castañeda, G. deBotton, and G. Li. Effective properties of nonlinear inhomogeneous dielectrics. *Phys. Rev. B*, 46:4387–4394, 1992.

[300] P. Ponte Castañeda, J.J. Telega, and B. Gambin, eds. *Nonlinear Homogenization and its Applications to Composites, Polycrystals and Smart Materials: Proc. NATO Adv. Res. Workshop Math., Phys. and Chem.* Springer-Verlag, Berlin, Heidelberg, New York, 2004.

[301] A.J. Poslinski, M.E. Ryan, R.K. Gupta, S.G. Seshadri, and F.J. Frechette. Rheological behavior of filled polymeric systems II. The effect of bimodal size distribution of particulates. *J. Rheol.*, 32:751–771, 1988.

[302] S. Prager. Diffusion and viscous flow in concentrated suspension. *Physica*, 50:129–139, 1963.

[303] G.I. Pshenichnov. *A Theory of Latticed Plates and Shells.* World Scientific, Singapore, 1993.

[304] J. Purczynski. Capacity estimation by means of Ritz and Trefftz's methods. *Archiv fur Elektrotechnik*, 59:269–274, 1977.

[305] A. Reuss. Berchung der fiessgrenze von mischkristallen auf grund der plastiziatsbedingung fur einkristalle. *Z. Angew. Math. Mech.*, 9:49–58, 1929.

[306] D.A. Robinson and S.F Friedman. Effect of particle size distribution on the effective dielectric permittivity of saturated granular media. *Water Resour. Res.*, 37(1):33–40, 2001.

[307] R.T. Rockafellar. *Convex Analysis.* Princeton University Press, 1970.

[308] W. Rudin. *Principles of Mathematical Analysis.* McGraw-Hill, New York, 1964.

[309] W. Rudin. *Functional Analysis.* McGraw-Hill, New York, 1992.

[310] I. Runge. Zur elektrischer leitfähigkeit metallischer aggregate. *Z. tech. Phys.*, 6(2):61–68, 1925.

[311] N. Rylko. Transport properties of the regular array of highly conducting cylinders. *J. Engrg. Math.*, 38:1–12, 2000.

[312] N. Rylko. Structure of the scalar field around unidirectional circular cylinders. *Proc. R. Soc. London*, A 464:391–407, 2008.

[313] K. Sab. On the homogenization and the simulation of random materials. *Eur. J. Mech., A/Solids*, 11 (5):585–607, 1992.

[314] F. J. Sabina and C. E. Garza-Hume. Universal relations for three-dimensional thermal, electric and magnetic properties. *Rev. Mexic. Fís.*, 48 (4):335–338, 2002.

[315] M. Sahimi. *Heterogeneous Materials, Vol. 1, 2.* Springer-Verlag, New York, Berlin, Heidelberg, 2003.

[316] Sh. Salon and M.V.K. Chari. *Numerical Methods in Electromagnetism.* Academic Press, New York, London, 1999.

[317] E. Sanchez-Palencia. Problems de perturbations lies aux phenomenes de conduction a travers des couches minces de grande resistivity. *J. Math. Pure Appl.*, 53:251–270, 1974.

[318] E. Sanchez-Palencia. *Non-Homogeneous Media and Vibration Theory.* Springer-Verlag, Berlin, 1980.

[319] A.S. Sangani and A. Acrivos. On the effective thermal conductivity and permeability of regular arrays of spheres. *In: Macroscopic Properties of Disordered Media*, Springer-Verlag, Berlin:216–225, 1982.

[320] A.S. Sangani and A. Acrivos. The effective conductivity of a periodic array of spheres. *Proc. R. Soc. London*, A, 386:263–275, 1983.

[321] J. Schauder. Uber lineare elliptiche Differentialgleichungen zveiter Ordnung. *Math. Z.*, 38:257–282, 1934.

[322] W.H.A. Schilders and E.J.W. Ter Maten. *Numerical Methods in Electromagnetics.* North-Holland, Amsterdam, 2005.

[323] L. Schwartz. *Theorie des Destribution.* Hermann, Paris, 1966.

[324] L.M. Schwartz, D.L. Johnson, and S. Feng. Vibration modes in granular materials. *Phys. Rev. Lett.*, 52(831):831–834, 1984.

[325] P.K. Senatorov. The coefficient stability of a solution of the Dirichlet problem for the equation $div(k(x)gradu) = -f(x)$. *Differential Equations*, 7:1414–1418, 1971.

[326] T. Senba and T. Suzuki. *Applied Analysis: Mathematical Methods in Natural Science.* Imperial College Press, London, 2004.

[327] L.C. Sengupta and S. Sengupta. Breakthrough advantages in low loss, tunable dielectric materials. *Mat. Res. Innovat.*, 2:278–282, 1999.

[328] L.C. Sengupta, S. Stowell, E. Ngo, and S. Sengupta. Thick film fabrication of ferroelectric phase shifter materials. *Integr. Ferroelectrics*, 13:203–214, 1996.

[329] V.O. Sherman, A.K. Tagantsev, and N. Setter. Model of a low-permittivity and high-tunability ferroelectric based composite. *Appl. Phys. Lett.*, 90(16):162901–1–162901–3, 2007.

[330] V.O. Sherman, A.K. Tagantsev, N. Setter, D. Iddles, and T. Price. Permittivity, tunability and loss in ferroelectrics for reconfigurable high frequency electronics. *In: Electroceramic-Based MEMS*, N. Setter, ed. Springer-Verlag, Berlin:217–234, 2005.

[331] V.O. Sherman, A.K. Tagantsev, N. Setter, D. Iddles, and T. Price. Ferro-electric–dielectric tunable composites. *J. Appl. Phys.*, 99:074104, 2006.

[332] T. Shikata and D.S. Pearson. Viscoelastic behavior of concentrated spherical suspensions. *J. Rheol.*, 38:601–616, 1994.

[333] I. Shiota and Y. Miyamoto, eds. *Functionally Graded Materials*. Elsevier Science Publ., Amsterdam, 1996.

[334] B. Shivamoggi. *Perturbation Methods for Differential Equations*. Birkhäuser, Boston, Basel, Stuttgart, 2002.

[335] A. Sihvola. *Electromagnetic Mixing Formulas and Applications*. IEEE Publishing, London, 1999.

[336] A. Sili. Homogenization of the linearized system of elasticity in anisotropic heterogeneous thin cylinders. *Math. Meth. Appl. Sci.*, 25:263–288, 2002.

[337] I.B. Simonenko. Electrostatics problems for inhomogeneous medium: A case of thin dielectric with high dielectric constant: I. *Differential Equations*, 10:301–309, 1974.

[338] I.B. Simonenko. Electrostatics problems for inhomogeneous medium: A case of thin dielectric with high dielectric constant: II. *Differential Equations*, 11:1870–1878, 1975.

[339] J.C. Slater and N.H. Frank. *Electromagnetism*. Dover, New York, 1969.

[340] W.R. Smythe. *Static and Dynamical Electricity. 2nd Ed.* McGraw-Hill, New York, Toronto, London, 1950.

[341] S.L. Sobolev. On the boundary value problem for polyharmonic functions *(in Russian)*. *Matem. Zbornik*, 2(3):465–499, 1937.

[342] S.L. Sobolev. *Partial Differential Equations of Mathematical Physics*. Pergamon Press, Oxford, 1964.

[343] S. Spagnolo. Sul limite delle soluzioni di problemi di Cauchy relativi all'equazioni del calore. *Ann. Scuola Norm. Sup. Pisa*, 21(3):657–699, 1967.

[344] S. Spagnolo. Sul limite delle soluzioni di equazioni paraboliche et elittiche. *Ann. Scuola Norm. Sup. Pisa*, 22(3):577–597, 1968.

[345] D. Stauffer and A. Aharony. *Introduction to Percolation Theory*. Taylor & Francis, London, 1992.

[346] D. Stroud and P.M. Hui. Nonlinear susceptibility of granular materials. *Phys. Rev. B*, 37(15):8719–8724, 1988.

[347] D. Stround and V.E. Wood. Decoupling approximation for the nonlinear-optical response of composite media. *J. Opt. Soc. Am. B*, 6:778–786, 1989.

[348] J.W. Strutt. On the influence of obstacles arranged in rectangular order upon the properties of the medium. *Philos. Mag.*, 34 (241):481–491, 1892.

[349] W.M. Suen, S.P. Wong, and K. Young. The lattice model of heat conduction in a composite material. *J. Phys. D: Appl. Phys.*, 12:1325–1338, 1979.

[350] A.K. Tagantsev, V.O. Sherman, K.F. Astafiev, J. Venkatesh, and N. Setter. Ferroelectric materials for microwave tunable applications. *J. Electroceramics*, 11:5–66, 2003.

[351] D.R.S. Talbot and J.R. Willis. Upper and lower bounds for the overall properties of a nonlinear composite dielectric I. Random microgeometry. *Proc. R. Soc. London*, A, 447:365–384, 1994.

[352] D.R.S. Talbot and J.R. Willis. Upper and lower bounds for the overall properties of a nonlinear composite dielectric II. Periodic microgeometry. *Proc. R. Soc. London*, A, 447:385–396, 1994.

[353] I.E. Tamm. *Fundamentals of the Theory of Electricity. 1st Ed. (in Russian)*. Nauka, Moscow, 1927.

[354] I.E. Tamm. *Fundamentals of the Theory of Electricity*. Mir Publ., Moscow, 1979.

[355] R. Temam. *Navier–Stokes Equations*. North-Holland Publ., Amsterdam, New York, Oxford, 1979.

[356] J.F. Thovert and A. Acrivos. The effective thermal conductivity of a random polydispersed suspension of spheres to order c^2. *Chem. Eng. Comm.*, 82:177–191, 1989.

[357] J.F. Thovert, I. C. Kim, S. Torquato, and A. Acrivos. Bounds on the effective properties of polydispersed suspensions of spheres: An evaluation of two relevant morphological parameters. *J. Appl. Phys.*, 67:6088–6098, 1990.

[358] S. Timoshemko and J.N. Goodier. *Theory of Elasticity*. McGraw–Hill, New York, 1951.

[359] S. Torquato. *Random Heterogeneous Materials*. Springer-Verlag, Berlin, Heidelberg, New York, 2002.

[360] L. Trabucho and J.M. Viaño. *Mathematical Modeling of Rods*. Elsevier, Amsterdam, 1996.

[361] T.D. Tsiboukis. Estimation of electromagnetic field parameters by dual energy methods. *Archiv fur Electotechnik*, 68:183–189, 1985.

[362] J.C. Van der Werff, C.G. de Kruif, C. Blom, and J. Mellema. Linear viscoelastic behavior of dense hard-sphere dispersions. *Phys. Rev. A*, 39:795–807, 2005.

[363] U. van Rienen, M. Günther, and D. Hecht, eds. *Numerical Methods in Computational Electrodynamics*. Springer, Berlin, 2001.

[364] A.H. Van Tuyl. Asymptotic expansions with error bounds for the coefficients of capacity and induction of two spheres. *SIAM J. Math. Anal.*, 27(3):782–804, 1996.

[365] J.R. Vinson and R.L. Sierakowski. *The Behavior of Structures Composed of Composite Materials*. Springer-Verlag, Berlin, 2002.

[366] V.S. Vladimirov. *Equations of Mathematical Physics*. Mir Publ., Moscow, 1984.

[367] W. Voigt. *Lehrbuch der Krystallphysik*. Teubner, Stuttgart, Leipzig, 1910.

[368] J.L. Volakis, B.C. Usner, and K. Sertel. Hybrid volume-surface integral equation method for high contrast composite media and metamaterials. *Antennas and Propagation Soc. Interna. Symp.*, IEEE:87–90, 2006.

[369] B.L. Wang and Y.-W. Mai. A periodic array of cracks in functional graded materials subjected to thermo-mechanical loading. *Intern. J. Engng Sci.*, 43(5/6):432–446, 2005.

[370] K. Washizu. *Variational Methods in Elasticity and Plasticity*. Pergamon Press, Oxford, New York, 1992.

[371] E. Weber. *Electromagnetic Fields*. John Wiley & Sons, New York, 1950.

[372] J. Wermer. *Potential Theory*. Lect. Notes in Math. Springer-Verlag, Berlin, 1974.

[373] K.W. Whites. Static permittivity of a multiphase system of spheres. *Antennas and Propagation Society International Symposium, 1999. IEEE*, 3:1934 – 1937, 1999.

[374] K.W. Whites. Permittivity of a multiphase and isotropic lattice of spheres at low frequency. *J. Appl. Phys.*, 88(4):1962–1970, 2000.

[375] O. Wiener. Die theorie des mischkörpers für das feld der stationären strömung. *Abh. Math. Phys. K1. Saechs. Akad. Wiss.*, 32:505–604, 1912.

[376] J.R. Willis. Bounds and self-consistent estimates for the overall properties of anisotropic composites. *J. Mech. Phys. Solids*, 25:185–202, 1977.

[377] J.R. Willis. Variational estimates for the overall response of an inhomogeneous nonlinear dielectric. *In: Homogenization and Effective Moduli of Materials and Media*, J.-L. Eriksen, D. Kinderlehrer, R. Kohn, and J.-L. Lions, eds. Springer-Verlag, New–York:247–263, 1986.

[378] J.R. Willis. Lectures on mechanics of random media. *In: Mechanics of Random and Multiscale Structures, CISM Lecture Notes*, D. Jeulin and M. Ostoja-Starzewski, eds. Springer-Verlag, Wein, New York:221–267, 2002.

[379] J. Wilson. *Experimental Solid Mechanics*. McGraw-Hill, New York, 1993.

[380] S.A. Wilson and R.W. Whatmore. Electric field structuring of piezoelectric composite materials. *J. Korean Phys. Soc.*, 32:S1204–S1206, 1998.

[381] K.W. Wojciechowski, K.V. Tretiakov, A.C. Brańka, and M. Kowalik. Elastic properties of two-dimensional hard disks in the close-packing limit. *J. Chem. Phys.*, 119(2):939–946, 2003.

[382] R. Wojnar, R. Bytner, and A. Galka. Effective properties of elastic composites subject to thermal fields. *In: Thermal Stresses*, R. B. Hetnarski, ed. Lastran Corp. Publ. Div. Rochester:257–465, 1999.

[383] F. Wu and K.W. Whites. Computation of static effective permittivity for a multiphase lattice of cylinders. *Electromagn.*, 21(2):97–114, 2001.

[384] F. Wu and K.W. Whites. Quasi-static effective permittivity of periodic composites containing complex shaped dielectric particles. *IEEE Trans. Antennas Propag.*, 49(8):1174–1182, 2001.

[385] www.ansys.com. *ANSYS 6.0 User Manual*. ANSYS Inc.

[386] H. Xu, J. Zhong, X. Liu, J. Chen, and D. Shen. Ferroelectric and switching behavior of poly(vinyliden fluoride-trifluoroethylene) copolymer ultrathin films with polypyrrole interface. *Appl. Phys. Lett.*, 90:092903, 2007.

[387] Y. Yan, J. Li, and L.M. Sander. Fracture growth in 2-D elastic networks with Born model. *Europhys. Lett.*, 10:7–13, 1989.

[388] C.S. Yang and P.M. Hui. Effective nonlinear response in random nonlinear resistor networks: numerical studies. *Phys. Rev. B*, 44:12559–12561, 1991.

[389] J.G. Yardley, A.J. Reuben, and R.C. McPhedran. The transport properties of layers of elliptical cylinders. *Proc. R. Soc. London*, A 457:395–423, 2001.

[390] R.H.T. Yeh. Variational principles of elastic moduli of composite materials. *J. Appl. Phys.*, 41(8):3353–3356, 1970.

[391] R.H.T. Yeh. Variational principles of transport properties of composite materials. *J. Appl. Phys.*, 41(1):224–226, 1970.

[392] K. Yosida. *Functional Analysis*. Springer-Verlag, Berlin, 1971.

[393] V.V. Yurinski. Average of an elliptic boundary problem with random coefficients. *Siberian Math. J.*, 21:470–482, 1980.

[394] V.V. Zagoskin, V.M. Nesterov, E.A. Zamotrinskaya, and T.G. Mikhailova. Dielectric properties of moist disperse materials in the microwave range. *Russian Phys. J.*, 24(7):644–647, 1981.

[395] P.M. Zeng, X.S. Hui, D.J. Bergmanm, and D. Stroud. Mean field theory for weakly nonlinear composites. *Physica A*, 157 (192), 1989.

[396] X.S. Zeng, D.J. Bergmanm, P.M. Hui, and D. Stroud. Effective-medium theory for weakly nonlinear composites. *Phys. Rev. B*, 38:10970–10973, 1988.

[397] V.V. Zhikov, S.M. Kozlov, O.A. Oleinik, and K. T'en Ngoan. Averaging and *G*-convergence of differential operators. *Russian Math. Surveys*, 34(5):69–147, 1979.

[398] R.W. Zimmerman. Effective conductivity of a two-dimensional medium containing elliptical inclusions. *Proc. R. Soc. London*, A 452:1713–1727, 1996.

[399] M. Zuzovsky and H. Brenner. Effective conductivities of composite materials composed of cubic arrangement of spherical particles embedded in an isotropic matrix. *Z. Angew. Math. Phys.*, 28:979–992, 1977.

SUBJECT INDEX

AUTHOR INDEX

Andrianov I.V.   xiii
Berlyand L.   xiii, 33
Berryman J.G.   26, 27, 29, 30
Borcea L.   26, 27, 29, 30
Chernikov   286
Chernikova   286
Clausius R.J.E.   2
Delaunay B.   69
Dykhne A.M.   106
Faraday M.   2
Gaudiello A.   xiii
Hill R.   2
Kirchhoff G.R.   50
Keller J.B.   18, 18, 21, 113
Kozlov S.M.   26, 27
Ladizenskay O.A.   60, 92
Lorenz L.   2
Marino A.   2
Maxwell J.C.   14, 18,
McPhedran R.   15
Meredith R.E.   18
Mityushev V.V.   xiii, 15, 44
Mossotti O.F.   2
Papanicolaou G.   26, 27, 29, 30
Poisson S.D.   2
Rakin S.I.   xiii
Reuss A.   2, 65
Schauder J.   60
Spagnolo, S.   2
Strutt J.W. (Lord Rayleigh)   14, 18
Tamm I.E.   xii, 21, 58, 107
Tobias C.W.   18
Voight W.   2, 65
Voronoi G.   69

Ural'tseva N.N.   60, 92
Zikov V.V.   xiii

321